NUREG-1800, Rev. 2

United States Nuclear Regulatory Commission

Protecting People and the Environment

Standard Review Plan for Review of License Renewal Applications for Nuclear Power Plants

Final Report

Manuscript Completed: December 2010
Date Published: December 2010

Office of Nuclear Reactor Regulation

ABSTRACT

The Standard Review Plan for Review of License Renewal Applications for Nuclear Power Plants (SRP-LR) provides guidance to U.S. Nuclear Regulatory Commission (NRC) staff reviewers in the Office of Nuclear Reactor Regulation. These reviewers perform safety reviews of applications to renew nuclear power plant licenses in accordance with Title 10 of the Code of Federal Regulations Part 54. The principal purposes of the SRP-LR are to ensure the quality and uniformity of staff reviews and to present a well-defined base from which to evaluate applicant programs and activities for the period of extended operation. The SRP-LR also is intended to make regulatory information widely available to enhance communication with interested members of the public and the nuclear power industry and to improve public and industry understanding of the staff review process. The safety review is based primarily on the information provided by the applicant in a license renewal application. Each of the individual SRP-LR sections addresses (a) who performs the review, (b) the matters that are reviewed, (c) the basis for review, (d) the way the review is accomplished, and (e) the conclusions that are drawn.

TABLE OF CONTENTS

LIST OF TABLES

LIST OF CONTRIBUTORS

Division of License Renewal, Office of Nuclear Reactor Regulation

B. Holian	Division Director
M. Galloway	Deputy Division Director
S. Lee	Deputy Division Director
L. Lund	Deputy Division Director
R. Auluck	Branch Chief
J. Dozier	Branch Chief
D. Pelton	Branch Chief
A. Hiser	Senior Level
R. Gramm	Team Leader
H. Ashar	Structural Engineering
M. Banic	Mechanical Engineering
A. Buford	Structural Engineer
C. Cho	Administrative Assistant
J. Davis	Materials Engineering
C. Doutt	Electrical Engineering
B. Elliot	Materials Engineering
A. Erickson	General Engineer
S. Figueroa	Licensing Assistant
B. Fu	Materials Engineering
J. Gavula	Mechanical Engineering
W. Holston	Mechanical Engineering
E. Keegan	Project Manager
I. King	Licensing Assistant
R. Li	Electrical Engineering
J. Medoff	Mechanical Engineering
S. Min	Materials Engineering
D. Nguyen	Electrical Engineering
V. Perin	Mechanical Engineering
A. Prinaris	Mechanical Engineering
L. Regner	Project Manager
B. Rogers	Reactor Engineer
S. Sakai	Project Manager
A. Sheikh	Structural Engineering
W. Smith	Mechanical Engineering
R. Sun	Mechanical Engineering
R. Vaucher	Mechanical Engineering
A. Wong	Mechanical Engineering

C.Y. Yang	Materials Engineering
L. Yee	Administrative Assistant
O. Yee	Mechanical Engineering

Office of Nuclear Reactor Regulation

G. Casto	Branch Chief
T. Chan	Branch Chief
M. Khanna	Branch Chief
A. Klein	Branch Chief
T. Lupold	Branch Chief
M. Mitchell	Branch Chief
R. Taylor	Branch Chief
G. Wilson	Branch Chief

R. Hardies	Senior Level – Materials Engineering
K. Karwoski	Senior Level – Steam Generators
K. Manoly	Senior Level – Structural Engineering

D. Alley	Materials Engineering
J. Bettle	Mechanical Engineering
T. Cheng	Structural Engineering
G. Cheruvenki	Materials Engineering
J. Collins	Mechanical Engineering
R. Davis	Materials Engineering
S. Gardocki	Mechanical Engineering
M. Hartzman	Materials Engineering
K. Hoffman	Mechanical Engineering
N. Iqbal	Fire Protection Engineering
A. Johnson	Mechanical Engineering
S. Jones	Mechanical Engineering
B. Lee	Mechanical Engineering
R. Mathew	Electrical Engineering
P. Patniak	Mechanical Engineering
G. Perciavello	Mechanical Engineering
A. Tsirigotis	Mechanical Engineering
M. Yoder	Chemical Engineering
E. Wong	Chemical Engineering

Region I

G. Meyer   Mechanical Engineering
M. Modes   Mechanical Engineering

Office of Nuclear Regulatory Research

A. Csontos   Branch Chief
M. Gavrilas   Branch Chief
R. Hogan   Branch Chief
T. Koshy   Branch Chief
M. Salley   Branch Chief
R. Tregoning   Senior Level – Materials Engineering

S. Aggarwal   Electrical Engineering
J. Burke   Mechanical Engineering
G. Carpenter   Materials Engineering
H. Graves   Structural Engineering
A. Hull   Materials Engineering
B. Lin   Structural Engineering
L. Ramadan   Electrical Engineering
G. Stevens   Materials Engineering
D. Stroup   Fire Protection Engineering
G. Wang   Mechanical Engineering

Advanced Technologies and Laboratories International, Inc. (ATL)

K. Makeig   Project Manager

K. Chang   Mechanical Engineering
O. Chopra   Materials Engineering
W. Jackson   Mechanical Engineering
D. Jones   Programming (Project Enhancement Corp.)
M. May   Mechanical Engineering
A. Ouaou   Structural Engineering
E. Patel   Mechanical Engineering
J. Davis   Materials Engineering
R. Royal   Electrical Engineering
T. Brake   Technical Editing

ABBREVIATIONS

ACAR	aluminum conductor aluminum alloy reinforced
ACSR	aluminum conductor steel reinforced
AFW	auxiliary feedwater
AMP	aging management program
AMR	aging management review
ANL	Argonne National Laboratory
ANSI	American National Standards Institute
ASME	American Society of Mechanical Engineers
ASTM	American Society for Testing and Materials
ATWS	anticipated transients without scram
B&W	Babcock & Wilcox
BWR	boiling water reactor
BWRVIP	Boiling Water Reactor Vessel and Internals Project
CASS	cast austenitic stainless steel
CDF	core damage frequency
CE	Combustion Engineering
CFR	Code of Federal Regulations
CLB	current licensing basis
CRD	control rod drive
CUF	cumulative usage factor
DBA	design basis accident
DBE	design basis event
DG	Draft Regulatory Guide
DLR	Division of License Renewal
DOR	Division of Operating Reactors
ECCS	emergency core cooling system
ECT	eddy current testing
EDG	emergency diesel generator
EFPY	effective full power year
EMA	equivalent margins analysis
EOL	end-of-life
EPDM	ethylene propylene diene monosomer
EPR	ethylene-propylene rubber
EPRI	Electric Power Research Institute
EPU	extended power uprate
FAC	flow-accelerated corrosion
FR	Federal Register
FSAR	Final Safety Analysis Report
FSER	Final Safety Evaluation Report
GALL	Generic Aging Lessons Learned
GE	General Electric
GL	generic letter
GSI	generic safety issue

HAZ	heat-affected zone
HDPE	high-density polyethylene
HELB	high-energy line break
HPCI	high-pressure coolant injection
HPSI	high-pressure safety injection
HVAC	heating, ventilation, and air conditioning
I&C	instrumentation and control
IASCC	Irradiation-assisted stress corrosion cracking
IEEE	Institute of Electrical and Electronics Engineers
IGA	intergranular attack
IGSCC	intergranular stress corrosion cracking
IN	information notice
INPO	Institute of Nuclear Power Operations
IPA	integrated plant assessment
IPE	individual plant examination
IPEEE	individual plant examination of external events
IR	insulation resistance
ISI	inservice inspection
LCD	liquid crystal display
LCO	limiting conditions of operation
LED	light-emitting diode
LER	licensee event report
LOCA	loss of coolant accident
LR	license renewal
LRA	license renewal application
LTOP	low-temperature overpressure protection
MEB	metal enclosed bus
MIC	microbiologically-influenced corrosion
MEAP	material/environment/aging effect/program as summarized on AMR line-items
MRV	minimum required value
NDE	nondestructive examination
NDT	nil-ductility temperature
NEI	Nuclear Energy Institute
NFPA	National Fire Protection Association
NPS	nominal pipe size
NRC	Nuclear Regulatory Commission
NRR	NRC Office of Nuclear Reactor Regulation
NSAC	Nuclear Safety Analysis Center
NSR	nonsafety-related
NSSS	nuclear steam supply system
ODSCC	outside diameter stress corrosion cracking
OE	operating experience
OM	operation and maintenance
OMB	Office of Management and Budget

P&ID	piping and instrument diagrams
PLL	predicted lower limit
PM	Project Manager
PRA	probabilistic risk analysis
PT	penetrant testing
P-T	pressure-temperature
PTLR	pressure-temperature limit reports
PTS	pressurized thermal shock
PWR	pressurized water reactor
PWSCC	primary water stress corrosion cracking
QA	quality assurance
RCIC	reactor core isolation cooling
RCPB	reactor coolant pressure boundary
RCS	reactor coolant system
RG	Regulatory Guide
RPV	reactor pressure vessel
RT	reference temperature
RV	reactor vessel
SBO	station blackout
SC	structures and components
SCC	stress corrosion cracking
SER	safety evaluation report
SG	steam generator
S/G	standards and guides
SOC	statements of consideration
SOER	significant operating experience report
SR	safety-related
SR	silicon rubber
SRM	staff requirements memorandum
SRP	standard review plan
SRP-LR	standard review plan for license renewal
SS	stainless steel
SSC	systems, structures, and components
SSE	safe shutdown earthquake
TC	thermocouples (nozzles)
TLAA	time-limited aging analysis
UFSAR	updated final safety analysis report
USE	upper-shelf energy
USI	unresolved safety issue
UT	ultrasonic testing
UUSE	unirradiated upper-shelf energy
UV	ultraviolet
VHP	vessel head penetration (nozzles)
WSLR	within scope of license renewal

XPLE   cross-linked polyethylene

INTRODUCTION

The "Standard Review Plan for Review of License Renewal Applications for Nuclear Power Plants" (SRP-LR) provides guidance to Nuclear Regulatory Commission (NRC) staff reviewers in the Office of Nuclear Reactor Regulation (NRR). These reviewers perform safety reviews of applications to renew nuclear power plant licenses in accordance with Title 10 of the Code of Federal Regulations (CFR) Part 54. The principal purposes of the SRP-LR are to ensure the quality and uniformity of staff reviews and to present a well-defined base from which to evaluate applicant programs and activities for the period of extended operation. The SRP-LR also is intended to make regulatory information widely available to enhance communication with interested members of the public and the nuclear power industry and to improve their understanding of the staff review process.

The safety review is based primarily on the information provided by the applicant in a license renewal application (LRA). The NRC regulation in 10 CFR 54.4 defines what is within the scope of the license renewal rule. The NRC regulation in 10 CFR 54.21 requires that each license renewal application include an integrated plant assessment (IPA), current licensing basis (CLB) changes during review of the application by the NRC, an evaluation of time-limited aging analyses (TLAAs), and a final safety analysis report (FSAR) supplement.

In addition to the technical information required by 10 CFR 54.21, an LRA must contain general information (10 CFR 54.19), necessary technical specification changes (10 CFR 54.22), and environmental information (10 CFR 54.23). The application must be sufficiently detailed to permit the reviewers to determine (a) whether there is reasonable assurance that the activities authorized by the renewed license will continue to be conducted in accordance with the CLB and (b) whether any changes made to the plant's CLB to comply with 10 CFR Part 54 are in accordance with the Atomic Energy Act of 1954 and NRC regulations.

Before submitting an LRA, an applicant should have analyzed the plant to ensure that actions have been or will be taken to (a) manage the effects of aging during the period of extended operation (this determination should be based on the functionality of structures and components that are within the scope of license renewal and that require an aging management review) and (b) evaluate TLAAs. The LRA is the principal document in which the applicant provides the information needed to understand the basis upon which this assurance can be made.

10 CFR 54.21 specifies, in general terms, the technical information to be supplied in the license renewal application. NRC Regulatory Guide (RG) 1.188, "Standard Format and Content for Applications to Renew Nuclear Power Plant Operating Licenses," endorses the Nuclear Energy Institute (NEI) guidance in NEI 95-10, "Industry Guidelines for Implementing the Requirements of 10 CFR Part 54 - The License Renewal Rule." NEI 95-10 provides guidance on the format and content of an LRA. SRP-LR sections are keyed to and numbered according to the section numbers in NRC RG 1.188.

During the review of the initial LRAs, NRC staff and the applicants have found that most of the programs to manage aging that are credited for license renewal are programs already in use by the applicants. In a staff paper (SECY 99-148), "Credit for Existing Programs for License Renewal," dated June 3, 1999, the staff described options and provided a recommendation for crediting existing programs to improve the efficiency of the license renewal process. In a staff requirements memorandum (SRM) dated August 27, 1999, the NRC approved the staff recommendation and directed the staff to focus the review guidance in the SRP-LR on areas where existing programs should be augmented for license renewal. Under the terms of the

SRM, the SRP-LR references a "Generic Aging Lessons Learned (GALL) Report," which evaluates existing programs generically, to document (a) the conditions under which existing programs are considered adequate to manage identified aging effects without change and (b) the conditions under which existing programs should be augmented for this purpose.

The GALL Report (NUREG-1801) should be treated as an approved topical report. The NRC reviewers should not re-review a matter described in the GALL Report, but should find an application acceptable with respect to such a matter when the application references the GALL Report and when the evaluation of the matter in the GALL Report applies to the plant. However, reviewers should ensure that the material presented in the GALL Report is applicable to the specific plant involved and that the applicant has identified specific programs, as described and evaluated in the GALL Report, if they rely on the report for license renewal.

The SRP-LR is divided into four major chapters: (a) Administrative Information; (b) Scoping and Screening Methodology for Identifying Structures and Components Subject to Aging Management Review and Implementation Results; (c) Aging Management Review Results; and (d) Time-Limited Aging Analyses. The appendices to the SRP-LR list branch technical positions. The SRP-LR addresses various site conditions and plant designs and provides complete procedures for all of the areas of review pertinent to each of the SRP-LR sections. For any specific application, NRC reviewers may select and emphasize particular aspects of each SRP-LR section, as appropriate for the application. In some cases, the major portion of the review of a plant program or activity may be done on a generic basis (with the owners' group of that plant type) rather than in the context of reviews of particular applications from utilities. In other cases, a plant program or activity may be sufficiently similar to that of a previous plant that a complete review of the program or activity is not needed. For these and similar reasons, reviewers need not carry out in detail all of the review steps listed in each SRP-LR section in the review of every application.

The individual SRP-LR sections address (a) who performs the review, (b) the matters that are reviewed, (c) the basis for review, (d) the way the review is accomplished, and (e) the conclusions that are sought. One of the objectives of the SRP-LR is to assign review responsibilities to the appropriate NRR branches. Each SRP-LR section identifies the branch that has the primary review responsibility for that section. In some review areas, the primary branch may require support; the branches that are assigned these secondary review responsibilities also are identified for each SRP-LR section.

Each SRP-LR section is organized into the following six subsections, generally consistent with NUREG-0800, "Standard Review Plan for the Review of Safety Analysis Reports for Nuclear Power Plants" (March 2007, with individual sections subsequently revised as needed).

1. Areas of Review

This subsection describes the scope of review, that is, what is being reviewed by the branch that has primary review responsibility. It contains a description of the systems, structures, components, analyses, data, or other information that is reviewed as part of the license renewal application. It also contains a discussion of the information needed or the review expected from other branches to permit the primary review branch to complete its review.

2. Acceptance Criteria

This subsection contains a statement of the purpose of the review, an identification of applicable NRC requirements, and the technical basis for determining the acceptability of programs and activities within the area of review of the SRP-LR section. The technical bases consist of specific criteria, such as NRC regulatory guides, codes and standards, and branch technical positions.

Consistent with the approach described in NUREG-0800, the technical bases for some sections of the SRP-LR can be provided in branch technical positions or appendices as they are developed and can be included in the SRP-LR.

3. Review Procedures

This subsection discusses the way the review is accomplished. It is generally a step-by-step procedure that the reviewer follows to provide reasonable verification that the applicable acceptance criteria have been met.

4. Evaluation Findings

This subsection presents the type of conclusion that is sought for the particular review area (e.g., the reviewers' determination as to whether the applicant has adequately identified the aging effects and the aging management programs credited with managing the aging effects). For each section, a conclusion of this type is included in the safety evaluation report (SER), in which the reviewers publish the results of their review. The SER also contains a description of the review, including which aspects of the review were selected or emphasized; which matters were modified by the applicant, required additional information, will be resolved in the future, or remain unresolved; where the applicant's program deviates from the criteria provided in the SRP-LR; and the bases for any deviations from the SRP-LR or exemptions from the regulations.

5. Implementation

This subsection discusses the NRC staff's plans for using the SRP-LR section.

6. References

This subsection lists the references used in the review process.

The SRP-LR incorporates the staff experience in the review of license renewal applications. It may be considered a part of a continuing regulatory framework development activity that documents current methods of review and provides a basis for orderly modifications of the review process in the future. The SRP-LR is revised and updated periodically, as needed, to incorporate experience gained during recent reviews, to clarify the content or correct errors, to reflect changes in relevant regulations, and to incorporate modifications approved by the NRR Director. A revision number and publication date are printed in a lower corner of each page of each SRP-LR section. Because individual sections will be revised as needed, the revision numbers and dates may not be the same for all sections.

CHAPTER 1

ADMINISTRATIVE INFORMATION

1.1 DOCKETING OF TIMELY AND SUFFICIENT RENEWAL APPLICATION

Review Responsibilities

Primary - Program responsible for license renewal projects

Secondary - Branches responsible for technical review, as appropriate

1.1.1 Areas of Review

This section addresses (a) the review of the acceptability of a license renewal application for docketing in accordance with 10 CFR 2.101 and the requirements of 10 CFR Part 54 and (b) whether a license renewal application is timely and sufficient, which allows the provisions of 10 CFR 2.109(b) to apply. Application of this regulation, written to comply with the Administrative Procedures Act, means that the current license will not expire until the U.S. Nuclear Regulatory Commission (NRC) makes a final determination on the license renewal application.

The review described in this section is not a detailed, in-depth review of the technical aspects of the application. The docketing and subsequent finding of a timely and sufficient renewal application does not preclude the NRC reviewers from requesting additional information as the review proceeds, nor does it predict the NRC's final determination regarding the approval or denial of the renewal application. A plant's current license will not expire upon the passing of the license's expiration date if the renewal application was found to be timely and sufficient. During this time, and, until the renewal application has been finally determined by the NRC, the licensee must continue to perform its activities in accordance with the facility's current licensing basis (CLB), including all applicable license conditions, orders, rules, and regulations.

To determine whether an application is acceptable for docketing, the following areas of the license renewal application are reviewed.

1.1.1.1 Docketing and Sufficiency of Application

The license renewal application is reviewed for acceptability for docketing as a sufficient application in accordance with 10 CFR 2.101, 10 CFR Part 51, and 10 CFR Part 54.

1.1.1.2 Timeliness of Application

The timeliness of a license renewal application is reviewed in accordance with 10 CFR 2.109(b).

1.1.2 Acceptance Criteria

1.1.2.1 Docketing/Sufficiency of Application

The NRC staff determines acceptance for docketing and sufficiency on the basis of the required contents of an application, established in 10 CFR 2.101, 10 CFR 51.53(c), 54.17, 54.19, 54.21, 54.22, 54.23, and 54.4. A license renewal application is sufficient if it contains the reports, analyses, and other documents required in such an application.

1.1.2.2 Timeliness of Application

In accordance with 10 CFR 2.109(b), a license renewal application is timely if it is submitted at least 5 years before the expiration of the current operating license (unless an exemption is granted) and if it is determined to be sufficient.

1.1.3 Review Procedures

A licensee may choose to submit plant-specific reports addressing portions of the license renewal rule requirements for NRC review and approval prior to submitting a renewal application. An applicant may incorporate (by reference) these reports or other information contained in previous applications for licenses or license amendments, statements, or correspondence filed with the NRC, provided that the references are clear and specific. However, the final determination of the docketing of a sufficient renewal application is made only after a formal license renewal application has been submitted to the NRC.

For each area of review, the NRC staff should implement the following review procedures.

1.1.3.1 Docketing and Sufficiency of Application

Upon receipt of a tendered application for license renewal, the reviewer should determine whether the applicant has made a reasonable effort to provide the required administrative, technical, and environmental information (Ref. 1). The reviewer should use the review checklist provided in Table 1.1-1 to determine whether the application is reasonably complete and conforms to the requirements outlined in 10 CFR Part 54.

Items I.1 through I.10 in the checklist address administrative information. For the purpose of this review, the reviewer checks the "Yes" column if the required information is included in the application. Item II in the checklist addresses timeliness of the application.

Items II.1 through II.3, III, and IV in the checklist address technical information, the Final Safety Analysis Report (FSAR) supplement, and technical specification changes, respectively. Chapters 2, 3, and 4 of the standard review plan for license renewal (SRP-LR) provide information regarding the technical review. Although the purpose of the docketing and sufficiency review is not to determine the technical adequacy of the application, the reviewer should determine whether the applicant has provided reasonably complete information in the application to address the renewal rule requirements. The reviewer may request assistance from appropriate technical review branches to determine whether the application provides sufficient information to address the items in the checklist so that the staff can begin their technical review. The reviewer should check the "Yes" column for a checklist item if the applicant has provided reasonably complete information in the application to address the checklist item.

Item V of the checklist addresses environmental information. The environmental review staff should review the supplement to the environmental report prepared by the applicant in accordance with the guidelines in NUREG-1555, "Standard Review Plans for Environmental Reviews for Nuclear Power Plants," Supplement 1, "Operating License Renewal" (Ref. 2). The reviewer checks the "Yes" column if the renewal application contains environmental information consistent with the requirements of 10 CFR Part 51.

The application should address each item in the checklist in order to be considered reasonably complete and sufficient. If the reviewer determines that an item in the checklist is not applicable, the reviewer should include a brief statement that the item is not applicable and provide the basis for the statement.

If information in the application for a checklist item is either not provided or not reasonably complete and no justification is provided, the reviewer should check the "No" column for that checklist item. Except for Item VI as discussed in Subsection 1.1.3.2, checking ANY "No," column indicates that the application is not acceptable for docketing as a sufficient renewal application unless the applicant modifies the application to provide the missing or incomplete information.

If the reviewer concludes, and management concurs, that the application is not acceptable for docketing as a sufficient application, the letter (typically preceded by a management call between the staff and the applicant) to the applicant should clearly state that (a) the application is not sufficient and is not acceptable for docketing and (b) the current license will expire at its expiration date. The letter also should include a description of the deficiencies found in the application and offer an opportunity for the applicant to supplement its application to provide the missing or incomplete information. The reviewer should review the supplemented application, if submitted, to determine whether it is acceptable for docketing as a sufficient application.

If the reviewer is able to answer "Yes" to the applicable items in the checklist, the application is acceptable for docketing as a sufficient renewal application. The applicant should be notified by letter that the application is accepted for docketing. Normally, the letter should be issued within 30 days of receipt of a renewal application. A notice of acceptance for docketing of the application and notice of opportunity for a hearing regarding renewal of the license is published in the *Federal Register*.

When the application is acceptable for docketing as a sufficient application, the staff begins its technical review. For license renewal applications, the NRC maintains the docket number of the current operating license for administrative convenience.

1.1.3.2   Timeliness of Application

If a sufficient application is submitted at least 5 years before the expiration of the current operating license, the reviewer checks the "Yes" column for Item VI in the checklist. If the supplemented application, as discussed in Subsection 1.1.3.1, is submitted at least 5 years before the expiration of the current operating license, the reviewer checks the "Yes" column for Item VI in the checklist.

If the reviewer checks the "No" column in Item VI in the checklist, indicating that a sufficient renewal application has not been submitted at least 5 years before the expiration of the current operating license, the letter (typically preceded by a management call between the staff and the applicant) to the applicant should clearly state that (a) the application is not timely, (b) the provisions in 10 CFR 2.109(b) have not been satisfied, and (c) the current license will expire on the expiration date. However, if the application is otherwise determined to be acceptable for docketing, the technical review can begin.

1.1.4   Evaluation Findings

The reviewer determines whether sufficient and adequate information has been provided to satisfy the provisions outlined here. Depending on the results of this review, one of the following conclusions is included in the staff's letter to the applicant:

- On the basis of its review, as discussed above, the staff has determined that the applicant has submitted sufficient information that is acceptable for docketing, in accordance with 10 CFR 54.19, 54.21, 54.22, 54.23, 54.4, and 51.53(c). However, the staff's determination does not preclude the request for additional information as the review proceeds.
- On the basis of its review, as discussed above, the staff has determined that the application is *not acceptable* for docketing as a timely and sufficient renewal application.

1.1.5   Implementation

Except for cases in which the applicant proposes an acceptable alternative method for complying with specified portions of NRC regulations, the methods described herein are used by NRC staff members in their evaluation of conformance with NRC regulations.

1.1.6   References

1. NRC Regulatory Guide 1.188, "Standard Format and Content for Applications to Renew Nuclear Power Plant Operating Licenses," U.S. Nuclear Regulatory Commission, January 2005.

2. NUREG-1555, "Standard Review Plans for Environmental Reviews for Nuclear Power Plants," Supplement 1, "Operating License Renewal," U.S. Nuclear Regulatory Commission, October 1999.

Table 1.1-1 Acceptance Review Checklist for License Renewal Application Acceptability for Docketing

		Yes	No
I.	**General Information**		
1.	Application identifies specific unit(s) applying for license renewal	☐	☐
2.	Filing of renewal application 10 CFR 54.17(a) is in accordance with:		
A.	10 CFR Part 2, Subpart A; 10 CFR 2.101	☐	☐
B.	10 CFR 50.4		
a.	Application is addressed to the Document Control Desk as specified in 10 CFR 50.4(a)	☐	☐
b.	Signed original application and 13 copies are provided to the Document Control Desk. One copy is provided to the appropriate Regional office [10 CFR 50.4(b)(3)]	☐	☐
c.	Form of the application meets the requirements of 10 CFR 50.4(c)	☐	☐
C.	10 CFR 50.30		
a.	Application is filed in accordance with 10 CFR 50.4 [10 CFR 50.30(a)(1)]	☐	☐
b.	Application is submitted under oath or affirmation [10 CFR 50.30(b)]	☐	☐
3.	Applicant is eligible to apply for a license and is not a foreign-owned or foreign-controlled entity [10 CFR 54.17(b)]	☐	☐
4.	Application is not submitted earlier than 20 years before expiration of current license [10 CFR 54.17(c)]	☐	☐
5.	Application states whether it contains applications for other kinds of licenses [10 CFR 54.17(d)]	☐	☐
6.	Information incorporated by reference in the application is contained in other documents previously filed with the Commission, and the references are clear and specific [10 CFR 54.17(e)]	☐	☐
7.	Restricted data or other defense information, if any, is separated from unclassified information in accordance with 10 CFR 50.33(j) [10 CFR 54.17(f)]	☐	☐
8.	If the application contains restricted data, written agreement on the control of accessibility to such information is provided [10 CFR 54.17(g)]	☐	☐
9.	Information specified in 10 CFR 50.33(a) through (e), (h), and (i) is provided or referenced [10 CFR 54.19(a)]:	☐	☐
A.	Name of applicant	☐	☐
B.	Address of applicant	☐	☐
C.	Business description	☐	☐
D.	Citizenship and ownership details	☐	☐
E.	License information	☐	☐

				Yes	No
	F.	Construction or alteration dates		☐	☐
	G.	Regulatory agencies and local publications		☐	☐
10.		Conforming changes, as needed, to the standard indemnity agreement have been submitted (10 CFR 140.92, Appendix B) to account for the proposed change in the expiration date [10 CFR 54.19(b)]		☐	☐

II.   Technical Information

1.   An integrated plant assessment [10 CFR 54.21(a)] is provided, and consists of:

A.   For those SSCs within the scope of license renewal [10 CFR 54.4], identification and listing of those structures and components that are subject to an aging management review (AMR) in accordance with 10 CFR 54.21(a)(1)(i) and (ii)

			Yes	No
	a.	Description of the boundary of the system or structure considered (if applicant initially scoped at the system or structure level). Within this boundary, identification of structures and components subject to an AMR. For commodity groups, description of basis for the grouping	☐	☐
	b.	Lists of structures and components subject to an AMR	☐	☐
B.		Description and justification of methods used to identify structures and components subject to an AMR [10 CFR 54.21(a)(2)]	☐	☐

C.   Demonstration that the effects of aging will be adequately managed for each structure and component identified, so that their intended function(s) will be maintained consistent with the current licensing basis for the period of extended operation [10 CFR 54.21(a)(3)]

			Yes	No
	a.	Description of the intended function(s) of the structures and components	☐	☐
	b.	Identification of applicable aging effects based on materials, environment, operating experience, etc.	☐	☐
	c.	Identification and description of aging management programs	☐	☐
	d.	Demonstration of aging management provided	☐	☐

2.   An evaluation of time-limited aging analyses (TLAAs) is provided, and consists of:

			Yes	No
A.		Listing of plant-specific TLAAs in accordance with the six criteria specified in 10 CFR 54.3 [10 CFR 54.21(c)(1)]	☐	☐
B.		An evaluation of each identified TLAA using one of the three approaches specified in 10 CFR 54.21(c)(1)(i) to (iii)	☐	☐
3.		All plant-specific exemptions granted pursuant to 10 CFR 50.12 and in effect that are based on a TLAA are listed, and evaluations justifying the continuation of these exemptions for the period of extended operation are provided [10 CFR 54.21(c)(2)]	☐	☐

			Yes	No

| | A. | Listing of plant-specific exemptions that are based on TLAAs as defined in 10 CFR 54.3 [10 CFR 54.21(c)(2)] | ☐ | ☐ |
| | B. | An evaluation of each identified exemption justifying the continuation of these exemptions for the period of extended operation [10 CFR 54.21(c)(2)] | ☐ | ☐ |

III. An FSAR supplement [10 CFR 54.21(d)] is provided and contains the following information:

		Yes	No
1.	Summary description of the aging management programs and activities for managing the effects of aging	☐	☐
2.	Summary description of the evaluation of TLAAs	☐	☐

IV. Technical Specification Changes

Any technical specification changes necessary to manage the aging effects during the period of extended operation and their justifications are included in the application [10 CFR 54.22] ☐ ☐

V. Environmental Information

Application includes a supplement to the environmental report that is in accordance with the requirements of Subpart A of 10 CFR Part 51 [10 CFR 54.23] ☐ ☐

VI. Timeliness Provision

The application is sufficient and submitted at least 5 years before expiration of current license [10 CFR 2.109(b)]. If not, application can be accepted for docketing, but the timely renewal provision in 10 CFR 2.109(b) does not apply ☐ ☐

VII. Conclusions Regarding Acceptance of Application for Docketing

The application is reasonably complete and meets the Acceptance Review Checklist criteria I through V and is recommended for docketing ☐ ☐

CHAPTER 2

SCOPING AND SCREENING METHODOLOGY FOR IDENTIFYING STRUCTURES AND COMPONENTS SUBJECT TO AGING MANAGEMENT REVIEW AND IMPLEMENTATION RESULTS

2.1   SCOPING AND SCREENING METHODOLOGY

Review Responsibilities

Primary – Assigned branch

Secondary – None

2.1.1   Areas of Review

This section addresses the scoping and screening methodology for license renewal. As required by 10 CFR 54.21(a)(2), the applicant, in its integrated plant assessment (IPA), is to describe and justify methods used to identify systems, structures, and components (SSCs) subject to an aging management review (AMR). The SSCs subject to AMR are those that perform an intended function, as described on 10 CFR 54.4, and meet two criteria:

1. They perform such functions without moving parts or without a change in configuration or properties, as set forth in 10 CFR 54.21(a)(1)(i) (denoted as "passive" components and structures in this SRP), and

2. They are not subject to replacement based on a qualified life or specified time period, as set forth in 10 CFR 54.21(a)(1)(ii) (denoted as "long-lived" structures and components).

The identification of the SSCs within the scope of license renewal is called "scoping." For those SSCs within the scope of license renewal, the identification of "passive," "long-lived" structures and components that are subject to an AMR is called "screening."

To verify that the applicant has properly implemented its methodology, the staff reviews the implementation results separately, following the guidance in Sections 2.2 through 2.5.

The following areas relating to the applicant's scoping and screening methodology are reviewed.

2.1.1.1   Scoping

The methodology used by the applicant to implement the scoping requirements of 10 CFR 54.4, "Scope," is reviewed.

2.1.1.2   Screening

The methodology used by the applicant to implement the screening requirements of 10 CFR 54.21(a)(1) is reviewed.

2.1.2   Acceptance Criteria

The acceptance criteria for the areas of review are based on the following regulations:

*   10 CFR 54.4(a) as it relates to the identification of plant SSCs within the scope of the rule;
*   10 CFR 54.4(b) as it relates to the identification of the intended functions of plant SSCs determined to be within the scope of the rule; and

- 10 CFR 54.21(a)(1) and (a)(2) as they relate to the methods utilized by the applicant to identify plant structures and components subject to an AMR.

Specific criteria necessary to determine whether the applicant has met the relevant requirements of 10 CFR 54.4(a), 54.4(b), 54.21(a)(1), and 54.21(a)(2) are as follows.

2.1.2.1 Scoping

The scoping methodology used by the applicant should be consistent with the process described in Section 3.0, "Identify the SSCs within the Scope of License Renewal and Their Intended Functions," of NEI 95-10, "Industry Guideline for Implementing the Requirements of 10 CFR Part 54 - The License Renewal Rule" (Ref. 1), or the justification provided by the applicant for any exceptions should provide a reasonable basis for the exception.

2.1.2.2 Screening

The screening methodology used by the applicant should be consistent with the process described in Section 4.1, "Identification of Structures and Components Subject to an Aging Management Review and Intended Functions," of NEI 95-10 (Ref. 1), as referenced by Regulatory Guide 1.188.

2.1.3 Review Procedures

Preparation for the review of the scoping and screening methodology employed by the applicant should include the following:

- Review of the NRC's safety evaluation report (SER) that was issued along with the operating license for the facility. This review is conducted for the purpose of familiarization with the principal design criteria for the facility and its current licensing basis (CLB), as defined in 10 CFR 54.3(a).

- Review of Chapters 1 through 12 of the Updated Final Safety Analysis Report (UFSAR) and the facility's technical specifications for the purposes of familiarization with the facility design and the nomenclature that is applied to SSCs within the facility (including the bases for such nomenclature). During this review, the SSCs should be identified that are relied upon to remain functional during and after design basis events (DBEs), as defined in 10 CFR 50.49(b)(1)(ii), for which the facility was designed, to ensure that the functions described in 10 CFR 54.4(a)(1) are successfully accomplished. This review should also yield information regarding seismic Category I SSCs as defined in Regulatory Guide 1.29, "Seismic Design Classification" (Ref. 2). For a newer plant, this information is typically contained in Section 3.2.1, "Seismic Classification," of the UFSAR consistent with the Standard Review Plan (NUREG-0800) (Ref. 3).

- Review of Chapter 15 (or equivalent) of the UFSAR to identify the anticipated operational occurrences and postulated accidents that are explicitly evaluated in the accident analyses for the facility. During this review, the SSCs that are relied upon to remain functional during and following design basis events (as defined in 10 CFR 50.49(b)(1)) to ensure the functions described in 10 CFR 54.4(a)(1) should be identified.

- The set of design basis events as defined in the rule is not limited to Chapter 15 (or equivalent) of the UFSAR. Examples of design basis events that may not be described in this chapter include external events, such as floods, storms, earthquakes, tornadoes, or hurricanes, and internal events, such as a high-energy line break. Information regarding design basis events as defined in 10 CFR 50.49(b)(1) may be found in any chapter of the facility UFSAR, the Commission's regulations, NRC orders, exemptions, or license conditions within the CLB. These sources should also be reviewed to identify systems, structures, and components that are relied upon to remain functional during and following design basis events (as defined in 10 CFR 50.49(b)(1)) to ensure the functions described in 10 CFR 54.4(a)(1).

- Review of the facility's Probabilistic Risk Analysis (PRA) Summary Report that was prepared by the licensee in response to Generic Letter (GL) 88-20, "Individual Plant Examination for Severe Accident Vulnerabilities - 10 CFR 50.54(f)," dated November 23, 1988 (Ref. 4). This review should yield additional information regarding the impact of the Individual Plant Examination (IPE) on the CLB for the facility. While the LR Rule is "deterministic," the NRC in the statements of consideration (SOC) accompanying the Rule also states that "In license renewal, probabilistic methods may be most useful, on a plant-specific basis, in helping to assess the relative importance of structures and components that are subject to an aging management review by helping to draw attention to specific vulnerabilities (e.g., results of an IPE or IPEEE)" (60 FR 22468). For example, the reviewer should focus on IPE information pertaining to plant changes or modifications that are initiated by the licensee in accordance with the requirements of 10 CFR 50.59 or 10 CFR 50.90.

- Review of the results of the facility's Individual Plant Examination of External Events (IPEEE) study conducted as a follow-up to the IPE performed as a result of GL 88-20 to identify any changes or modifications made to the facility in accordance with the requirements of 10 CFR 50.59 or 10 CFR 50.90.

- Review of the applicant's docketed correspondence related to the following regulations:

    (a) 10 CFR 50.48, "Fire Protection,"

    (b) 10 CFR 50.49, "Environmental Qualification of Electric Equipment Important to Safety for Nuclear Power Plants,"

    (c) 10 CFR 50.61, "Fracture Toughness Requirements for Protection Against Pressurized Thermal Shock Events" [applicable to pressurized water reactor (PWR) plants],

    (d) 10 CFR 50.62, "Requirements for Reduction of Risk from Anticipated Transients without Scram Events for Light-Water-Cooled Nuclear Power Plants," and

    (e) 10 CFR 50.63, "Loss of All Alternating Current Power" (applicable to PWR plants).

Other staff members are reviewing the applicant's scoping and screening results separately following the guidance in Sections 2.2 through 2.5. The reviewer should keep these other staff members informed of findings that may affect their review of the applicant's scoping and screening results. The reviewer should coordinate this sharing of information through the license renewal project manager.

2.1.3.1   Scoping

Once the information delineated above has been gathered, the reviewer reviews the applicant's methodology to determine whether its depth and breadth are sufficiently comprehensive to identify the SSCs within the scope of license renewal, and the structures and components requiring an AMR. Because "[t]he CLB represents the evolving set of requirements and commitments for a specific plant that are modified as necessary over the life of a plant to ensure continuation of an adequate level of safety" (60 FR 22465, May 8, 1995), the regulations, orders, license conditions, exemptions, and technical specifications defining functional requirements for facility SSCs that make up an applicant's CLB should be considered as the initial input into the scoping process. 10 CFR 50.49 defines DBEs as conditions of normal operation, including anticipated operational occurrences, DBAs, external events, and natural phenomena for which the plant must be designed to ensure (1) the integrity of the reactor pressure boundary, (2) the capability to shut down the reactor and maintain it in safe shutdown condition, or (3) the capability to prevent or mitigate the consequences of accidents that could result in potential offsite exposures comparable to those referred to in 10 CFR 50.34(a)(1), 50.67(b)(2), or 100.11, as applicable. Therefore, to determine the safety-related (SR) SSCs that are within the scope of the rule under 10 CFR 54.4 (a)(1), the applicant must identify those SSCs that are relied upon to remain functional during and following these DBEs, consistent with the CLB of the facility. Most licensees have developed lists or databases that identify systems, structures, and components relied on for compliance with other regulations in a manner consistent with the CLB of their facilities. Consistent with the licensing process and regulatory criteria used to develop such lists or databases, licensees should build upon these information sources to satisfy 10 CFR Part 54 requirements.

With respect to technical specifications, the NRC states (60 FR 22467):

> The Commission believes that there is sufficient experience with its policy on technical specifications to apply that policy generically in revising the license renewal rule consistent with the Commission's desire to credit existing regulatory programs. Therefore, the Commission concludes that the technical specification limiting conditions for operation scoping category is unwarranted and has deleted the requirement that identifies systems, structures, and components with operability requirements in technical specifications as being within the scope of the license renewal review.

Therefore, the applicant need not consider its technical specifications and applicable limiting conditions of operation when scoping for license renewal. This is not to say that the events and functions addressed within the applicant's technical specifications can be excluded in determining the SSCs within the scope of license renewal solely on the basis of such an event's inclusion in the technical specifications. Rather, those SSCs governed by an applicant's technical specifications that are relied upon to remain functional during a DBE, as identified within the applicant's UFSAR, applicable NRC regulations, license conditions, NRC orders, and exemptions, need to be included within the scope of license renewal.

For licensee commitments, such as licensee responses to NRC Bulletins, GLs, or enforcement actions, and those documented in staff safety evaluations or licensee event reports, and which make up the remainder of an applicant's CLB, many of the associated SSCs need not be considered under license renewal. Generic communications, safety evaluations, and other similar documents found on the docket are not regulatory requirements, and commitments made by a licensee to address any associated safety concerns are not typically considered to be design requirements. However, any generic communication, safety evaluation, or licensee commitment that specifically identifies or describes a function associated with a system, structure, or component necessary to fulfill the requirement of a particular regulation, order, license condition, and/or exemption may need to be considered when scoping for license renewal. For example, NRC Bulletin 88-11, "Pressurizer Surge Line Thermal Stratification," states:

> The licensing basis according to 10 CFR 50.55a for all PWRs requires that the licensee meet the American Society of Mechanical Engineers Boiler and Pressure Vessel Code Sections III and XI and to reconcile the pipe stresses and fatigue evaluation when any significant differences are observed between measured data and the analytical results for the hypothesized conditions. Staff evaluation indicates that the thermal stratification phenomenon could occur in all PWR surge lines and may invalidate the analyses supporting the integrity of the surge line. The staff's concerns include unexpected bending and thermal striping (rapid oscillation of the thermal boundary interface along the piping inside surface) as they affect the overall integrity of the surge line for its design life (e.g., the increase of fatigue).

Therefore, this bulletin specifically describes conditions that may affect compliance with the requirements associated with 10 CFR 50.55a and functions specifically related to this regulation that must be considered in the scoping process for license renewal.

An applicant may take an approach in scoping and screening that combines similar components from various systems. For example, containment isolation valves from various systems may be identified as a single system for purposes of license renewal.

Staff from branches responsible for systems may be requested to assist in reviewing the plant design basis and intended function(s), as necessary.

The reviewer should verify that the applicant's scoping methods document the actual information sources used (for example, those identified in Table 2.1-1).

Table 2.1-2 contains specific staff guidance on certain subjects of scoping.

2.1.3.1.1   Safety-Related

The applicant's methodology is reviewed to ensure that the SR SSCs are identified to satisfactorily accomplish any of the intended functions identified in 10 CFR 54.4(a)(1). The reviewer must ascertain how, and to what extent, the applicant incorporated the information in the CLB for the facility in its methodology. Specifically, the reviewer should review the application, as well as all other relevant sources of information outlined above, to identify the set of plant-specific conditions of normal operation, DBAs, external events, and natural phenomena for which the plant must be designed to ensure the following functions:

- The integrity of the reactor coolant pressure boundary;
- The capability to shut down the reactor and maintain it in a safe shutdown condition; or
- The capability to prevent or mitigate the consequences of accidents that could result in potential offsite exposure comparable to the guidelines in 10 CFR 50.34(a)(1), 50.67(b)(2), or 100.11, as applicable.

2.1.3.1.2   Nonsafety-Related

The applicant's methodology is reviewed to ensure that nonsafety-related (NSR) SSCs whose failure could prevent satisfactory accomplishment of any of the functions identified in 10 CFR 54.4(a)(1) are identified as being within the scope of license renewal.

The scoping criterion under 10 CFR 54.4(a)(2), in general, is intended to identify those NSR SSCs that support SR functions. More specifically, this scoping criterion requires an applicant to identify all NSR SSCs whose failure could prevent satisfactory accomplishment of any of the functions identified under 10 CFR 54.4(a)(1). Section III.c(iii) of the SOC (60 FR 22467) clarifies the NRC's intent for this requirement in the following statement:

> The inclusion of nonsafety-related systems, structures, and components whose failure could prevent other systems, structures, and components from accomplishing a safety function is intended to provide protection against safety function failure in cases where the safety-related structure or component is not itself impaired by age-related degradation but is vulnerable to failure from the failure of another structure or component that may be so impaired.

In addition, Section III.c(iii) of the SOC provides the following guidance to assist an applicant in determining the extent to which failures must be considered when applying this scoping criterion:

> Consideration of hypothetical failures that could result from system interdependencies that are not part of the current licensing bases and that have not been previously experienced is not required. [...] However, for some license renewal applicants, the Commission cannot exclude the possibility that hypothetical failures that <u>are part of the CLB</u> may require consideration of second-, third-, or fourth-level support systems.

Therefore, to satisfy the scoping criterion under 10 CFR 54.4(a)(2), the applicant must identify those NSR SSCs (including certain second-, third-, or fourth-level support systems) whose failures are considered in the CLB and could prevent the satisfactory accomplishment of an SR function identified under 10 CFR 54.4(a)(1). In order to identify such systems, the applicant should consider those failures identified in (1) the documentation that makes up its CLB, (2) plant-specific operating experience, and (3) industrywide operating experience that is specifically applicable to its facility. The applicant need not consider hypothetical failures that are not part of the CLB, have not been previously experienced, or are not applicable to its facility.

In part, 10 CFR 54.4(a)(2) requires that the applicant consider all NSR SSCs whose failure could prevent satisfactory accomplishment of any of the functions identified in 10 CFR 54.4(a)(1)(i), 10 CFR 54.4(a)(1)(ii), or 10 CFR 54.4(a)(1)(iii) to be within the scope of license renewal. By letters dated December 3, 2001 and March 15, 2002, the NRC issued a staff position to NEI which provided staff guidance for determining what SSCs meet the 10 CFR 54.4(a)(2) criterion. The December 3, 2001 letter, "License Renewal Issue: Scoping of Seismic II/I Piping Systems," provided specific examples of operating experience which identified pipe failure events [summarized in Information Notice (IN) 2001-09, "Main Feedwater System Degradation in Safety-Related ASME Code Class 2 Piping Inside the Containment of a Pressurized Water Reactor"] and the approaches the NRC considers acceptable to determine which piping systems should be included in scope based on the 10 CFR 54.4(a)(2) criterion.

The March 15, 2002 letter, "License Renewal Issue: Guidance on the Identification and Treatment of Structures, Systems, and Components Which Meet 10 CFR 54.4(a)(2)," further described the staff's recommendations for the evaluation of non-piping SSCs to determine which additional NSR SSCs are within scope. The position states that the applicants should not consider hypothetical failures, but rather should base their evaluation on the plant's CLB, engineering judgment and analyses, and relevant operating experience. The paper further describes operating experience as all documented plant-specific and industrywide experience that can be used to determine the plausibility of a failure. Documentation would include NRC generic communications and event reports, plant-specific condition reports, industry reports, such as significant operating experience reports (SOERs), and engineering evaluations.

For example, the safety classification of a pipe at certain locations, such as valves, may change throughout its length in the plant. In these instances, the applicant should identify the SR portion of the pipe as being within the scope of license renewal under 10 CFR 54.4(a)(1). However, the entire pipe run, including associated piping anchors, may have been analyzed as part of the CLB to establish that it could withstand DBE loads. If this is the case, a failure in the pipe run or in the associated piping anchors could render the SR portion of the piping unable to perform its intended function under CLB design conditions. Therefore, the reviewer must verify that the applicant's methodology would include (1) the remaining NSR piping up to its anchors and (2) the associated piping anchors as being within the scope of license renewal under 10 CFR 54.4(a)(2).

In order to comply, in part, with the requirements of 10 CFR 54.4(a)(2), all applicants must include in scope all NSR piping attached directly to SR piping (within scope) up to a defined anchor point consistent with the plant CLB. This anchor point may be served by a true anchor (a device or structure which ensures forces and moments are restrained in three (3) orthogonal directions) or an equivalent anchor, such as a large piece of plant equipment (e.g., a heat exchanger,) determined by an evaluation of the plant-specific piping design (i.e., design documentation, such as piping stress analysis for the facility).

Applicants should be able to define an equivalent anchor consistent with their CLB (e.g., described in the UFSAR or other CLB documentation), which is being credited for the 10 CFR 54.4(a)(2) evaluation, and be able to describe the structures and components that are part of the NSR piping segment boundary up to and including the anchor point or equivalent anchor point within scope of the rule.

There may be isolated cases where an equivalent anchor point for a particular piping segment is not clearly described within the existing CLB information. In those instances the applicant may use a combination of restraints or supports such that the NSR piping and associated structures and components attached to SR piping is included in scope up to a boundary point which encompasses at least two (2) supports in each of three (3) orthogonal directions.

It is important to note that the scoping criterion under 10 CFR 54.4(a)(2) specifically applies to those functions "identified in paragraphs (a)(1)(i), (ii), and (iii)" of 10 CFR 54.4 and does not apply to functions identified in 10 CFR 54.4(a)(3), as discussed below.

2.1.3.1.3 "Regulated Events"

The applicant's methodology is reviewed to ensure that SSCs relied on in safety analyses or plant evaluations to perform functions that demonstrate compliance with the requirements of the fire protection, environmental qualification, pressurized thermal shock (PTS), anticipated transients without scram (ATWS), and station blackout (SBO) regulations are identified. The reviewer should review the applicant's docketed correspondence associated with compliance of the facility with these regulations.

The scoping criteria in 10 CFR 54.4(a)(3) require an applicant to consider *"[a]ll structures, systems, and components relied on in safety analyses or plant evaluations to perform a function that demonstrates compliance with the [specified] Commission regulations. . ."* In addition, Section III.c(iii) (60 FR 22467) of the SOC states that the NRC intended to limit the potential for unnecessary expansion of the review for SSCs that meet the scoping criteria under 10 CFR 54.4(a)(3) and provides additional guidance that qualifies what is meant by *"those SSCs relied on in safety analyses or plant evaluations to perform a function that demonstrates compliance with the Commission regulations"* in the following statement:

> [T]he Commission intends that this [referring to 10 CFR 54.4(a)(3)] scoping category include all SSC whose function is relied upon to demonstrate compliance with these Commission[ ] regulations. An applicant for license renewal should rely on the plant's current licensing bases, actual plant-specific experience, industrywide operating experience, as appropriate, and existing engineering evaluations to determine those SSC that are the initial focus of license renewal review.

Therefore, all SSCs that are relied upon in the plant's CLB (as defined in 10 CFR 54.3), plant-specific experience, industrywide experience (as appropriate), and safety analyses or plant evaluations to perform a function that demonstrates compliance with NRC regulations identified under 10 CFR 54.4(a)(3) are required to be included within the scope of the rule. For example, if an NSR diesel generator is required for safe shutdown under the fire protection plan, the diesel generator and all SSCs specifically relied upon for that generator to comply with NRC regulations shall be included within the scope of license renewal under 10 CFR 54.4(a)(3). Such SSCs may include, but should not be limited to, the cooling water system or systems relied upon for operability, the diesel support pedestal, and any applicable power supply cable specifically relied upon for safe shutdown in the event of a fire.

In addition, the last sentence of the second paragraph in Section III.c(iii) of the SOC provides the following guidance for limiting the application of the scoping criterion under 10 CFR 54.4(a)(3) as it applies to the use of hypothetical failures:

> Consideration of hypothetical failures that could result from system interdependencies, that are not part of the current licensing bases and that have not been previously experienced is not required. (60 FR 22467)

The SOC does not provide any additional guidance relating to the use of hypothetical failures or the need to consider second-, third-, or fourth-level support systems for scoping under 10 CFR 54.4(a)(3). Therefore, in the absence of any guidance, an applicant need not consider hypothetical failures or second-, third-, or fourth-level support systems in determining the SSCs within the scope of the rule under 10 CFR 54.4(a)(3). For example, if an NSR diesel generator is relied upon only to remain functional to demonstrate compliance with the NRC SBO regulation, the applicant need not consider the following SSCs: (1) an alternate/backup cooling water system, (2) non-seismically-qualified building walls, or (3) an overhead segment of non-seismically-qualified piping (in a Seismic II/I configuration). This guidance is not intended to exclude any support system (whether identified by an applicant's CLB, or as indicated from actual plant-specific experience, industrywide experience [as applicable], safety analyses, or plant evaluations) that is specifically relied upon for compliance with the applicable NRC regulation. For example, if analysis of an NSR diesel generator (relied upon to demonstrate compliance with an applicable NRC regulation) specifically relies upon a second cooling system to cool the diesel generator jacket water cooling system for the generator to be operable, then both cooling systems must be included within the scope of the rule under 10 CFR 54.4(a)(3).

The applicant is required to identify the SSCs whose functions are relied upon to demonstrate compliance with the regulations identified in 10 CFR 54.4(a)(3) (that is, whose functions were credited in the analysis or evaluation). Mere mention of an SSC in the analysis or evaluation does not necessarily constitute support of an intended function as required by the regulation.

For environmental qualification, the reviewer verifies that the applicant has indicated that the environmental qualification equipment is the equipment already identified by the licensee under 10 CFR 50.49(b), that is, equipment relied upon in safety analyses or plant evaluations to demonstrate compliance with NRC regulations for environmental qualification (10 CFR 50.49).

The PTS regulation is applicable only to PWRs. If the renewal application is for a PWR and the applicant relies on a Regulatory Guide 1.154 (Ref. 5) analysis to satisfy 10 CFR 50.61, as described in the plant's CLB, the reviewer verifies that the applicant's methodology would include SSCs relied on in that analysis.

For SBO, the reviewer verifies that the applicant's methodology would include those SSCs relied upon during the "coping duration" and "recovery" phase of an SBO event. In addition, because 10 CFR 50.63(c)(1)(ii) and its associated guidance in Regulatory Guide 1.155 include procedures to recover from an SBO that include offsite and onsite power, the offsite power system that is used to connect the plant to the offsite power source should also be included within the scope of the rule. However, the staff's review is based on the plant-specific current licensing basis, regulatory requirements, and offsite power design configurations.

2.1.3.2   Screening

Once the SSCs within the scope of license renewal have been identified, the next step is determining which structures and components are subject to an AMR (i.e., "screening") (Ref. 1). Table 2.1-3 contains specific staff guidance on certain subjects of screening.

2.1.3.2.1   "Passive"

The reviewer reviews the applicant's methodology to ensure that "passive" structures and components are identified as those that perform their intended functions without moving parts or a change in configuration or properties in accordance with 10 CFR 54.21(a)(1)(i). The description of "passive" may also be interpreted to include structures and components that do not display "a change in state." 10 CFR 54.21(a)(1)(i) provides specific examples of structures and components that do or do not meet the criterion. The reviewer verifies that the applicant's screening methodology includes consideration of the intended functions of structures and components consistent with the plant's CLB, as typified in Tables 2.1-4(a) and (b), respectively (Ref. 1).

The license renewal rule focuses on "passive" structures and components because structures and components that have passive functions generally do not have performance and condition characteristics that are as readily observable as those that perform active functions. "Passive" structures and components, for the purpose of the license renewal rule, are those that perform an intended function, as described in 10 CFR 54.4, without moving parts or without a change in configuration or properties (Ref. 2). The description of "passive" may also be interpreted to include structures and components that do not display "a change of state."

Table 2.1-5 provides a list of typical structures and components identifying whether they meet 10 CFR 54.21(a)(1)(i).

10 CFR 54.21(a)(1)(i) explicitly excludes instrumentation, such as pressure transmitters, pressure indicators, and water level indicators, from an AMR. The applicant does not have to identify pressure-retaining boundaries of this instrumentation because 10 CFR 54.21(a)(1)(i) excludes this instrumentation without exception, unlike pumps and valves. Further, instrumentation is sensitive equipment and degradation of its pressure retaining boundary would be readily determinable by surveillance and testing. If an applicant determines that certain structures and components listed in Table 2.1-5 as meeting 10 CFR 54.21(a)(1)(i) do not meet that requirement for its plant, the reviewer reviews the applicant's basis for that determination.

2.1.3.2.2   "Long-Lived"

The applicant's methodology is reviewed to ensure that "long-lived" structures and components are identified as those that are not subject to periodic replacement based on a qualified life or specified time period. Passive structures and components that are not replaced on the basis of a qualified life or specified time period require an AMR.

Replacement programs may be based on vendor recommendations, plant experience, or any means that establishes a specific replacement frequency under a controlled program. Section f(i)(b) of the SOC provides the following guidance for identifying "long-lived" structures and components:

> In sum, a structure or component that is not replaced either (i) on a specified interval based upon the qualified life of the structure or component or (ii) periodically in accordance with a specified time period is deemed by § 54.21(a)(1)(ii) of this rule to be "long-lived," and therefore subject to the § 54.21(a)(3)aging management review [60 FR 22478].

A qualified life does not necessarily have to be based on calendar time. A qualified life based on run time or cycles are examples of qualified life references that are not based on calendar time (Ref. 3).

Structures and components that are replaced on the basis of performance or condition are not generically excluded from an AMR. Rather, performance or condition monitoring may be evaluated later in the IPA as programs to ensure functionality during the period of extended operation. On this topic, Section f(i)(b) of the SOC provides the following guidance:

> It is important to note, however, that the Commission has decided not to generically exclude passive structures and components that are replaced based on performance or condition from an aging management review. Absent the specific nature of the performance or condition replacement criteria and the fact that the Commission has determined that the components with "passive" functions are not as readily monitorable as components with active functions, such generic exclusion is not appropriate. However, the Commission does not intend to preclude a license renewal applicant from providing site-specific

justification in a license renewal application that a replacement program on the basis of performance or condition for a passive structure or component provides reasonable assurance that the intended function of the passive structure or component will be maintained in the period of extended operation. [60 FR 22478]

2.1.4 Evaluation Findings

When the review of the information in the license renewal application is complete, and the reviewer has determined that it is satisfactory and in accordance with the acceptance criteria in Subsection 2.1.2, a statement of the following type should be included in the staff's safety evaluation report:

> On the basis of its review, as discussed above, the staff concludes that there is reasonable assurance that the applicant's methodology for identifying the systems, structures, and components within the scope of license renewal and the structures and components requiring an aging management review is consistent with the requirements of 10 CFR 54.4 and 10 CFR 54.21(a)(1).

2.1.5 Implementation

Except in those cases in which the applicant proposes an acceptable alternative method for complying with specified portions of NRC regulations, the method described herein will be used by the staff in its evaluation of conformance with NRC regulations (Ref. 6-12 as examples).

2.1.6 References

1. NEI 95-10, "Industry Guideline for Implementing the Requirements of 10 CFR Part 54 – The License Renewal Rule," Nuclear Energy Institute, Revision 6.

2. Regulatory Guide 1.29, Rev. 3, "Seismic Design Classification," U.S. Nuclear Regulatory Commission, March 2007.

3. NUREG-0800, "Standard Review Plan for the Review of Safety Analysis Reports for Nuclear Power Plants," U.S. Nuclear Regulatory Commission, March 2007.

4. Generic Letter (GL) 88-20, "Individual Plant Examination for Severe Accident Vulnerabilities-10 CFR 50.54(f)," dated November 23, 1988.

5. Regulatory Guide 1.154, "Format and Content of Plant-Specific Pressurized Thermal Shock Safety Analysis Reports for Pressurized Water Reactors," January 1987.

6. NUREG-1723, "Safety Evaluation Report Related to the License Renewal of Oconee Nuclear Stations, Units 1, 2, and 3," March 2000.

7. Letter to Douglas J. Walters, Nuclear Energy Institute, from Christopher I. Grimes, NRC, dated August 5, 1999.

8. Summary of December 8, 1999, Meeting with the Nuclear Energy Institute (NEI) on License Renewal Issue (LR) 98-12, "Consumables," Project No. 690, January 21, 2000.

9. Letter to William R. McCollum, Jr., Duke Energy Corporation, from Christopher I. Grimes, NRC, dated October 8, 1999.

10. Letter to Alan Nelson, Nuclear Energy Institute, and David Lochbaum, Union of Concerned Scientists, from Christopher I. Grimes, NRC, "License Renewal Issue: Scoping of Seismic II/I Piping Systems," dated December 3, 2001.

11. Letter to Alan Nelson, Nuclear Energy Institute, and David Lochbaum, Union of Concerned Scientists, from Christopher I. Grimes, NRC, "License Renewal Issue: Guidance on the Identification and Treatment of Structures, Systems, and Components Which Meet 10 CFR 54.4(a)(2)," dated March 15, 2002.

12. Letter to Alan Nelson, Nuclear Energy Institute, and David Lochbaum, Union of Concerned Scientists, from Christopher I. Grimes, NRC, "Staff Guidance on Scoping of Equipment Relied on to Meet the Requirements of the Station Blackout (SBO) Rule (10 CFR 50.63) for License Renewal (10 CFR 54.4(a)(3))," dated April 1, 2002.

Table 2.1-1   Sample Listing of Potential Information Sources

Verified databases (databases that are subject to administrative controls to assure and maintain the integrity of the stored data or information)

Master equipment lists (including NSSS vendor listings)

Q-lists

Updated Final Safety Analysis Reports

Piping and instrument diagrams

NRC Orders, Exemptions, or License Conditions for the facility

Design-basis documents

General arrangement or structural outline drawings

Probabilistic risk assessment summary report

Maintenance rule compliance documentation

Design-basis event evaluations (including plant-specific 10 CFR 50.59 evaluation procedures)

Emergency operating procedures

Docketed correspondence

System interaction commitments

Technical specifications

Environmental qualification program documents

Regulatory compliance reports (including Safety Evaluation Reports)

Severe Accident Management Guidelines

Table 2.1-2 Specific Staff Guidance on Scoping

Issue	Guidance
Commodity groups	The applicant may also group like structures and components into commodity groups. Examples of commodity groups are pipe supports and cable trays. The basis for grouping structures and components can be determined by such characteristics as similar function, similar design, similar materials of construction, similar aging management practices, or similar environments. If the applicant uses commodity groups, the reviewer verifies that the applicant has described the basis for the groups.
Complex assemblies	Some structures and components, when combined, are considered a complex assembly (for example, diesel generator starting air skids or heating, ventilating, and air conditioning refrigerant units). For purposes of performing an AMR, it is important to clearly establish the boundaries of review. An applicant should establish the boundaries for such assemblies by identifying each structure and component that make up the complex assembly and determining whether or not each structure and component is subject to an AMR (Ref. 1).
	NEI 95-10, Revision 0, Appendix C, Example 5 (Ref. 10), illustrates how the evaluation boundary for a control room chiller complex assembly might be determined. The control room chillers were purchased as skid-mounted equipment. These chillers are part of the control room chilled water system. There are two (2) control room chillers. Each is a 100% capacity refrigeration unit. The functions of the control room chillers are to provide a reliable source of chilled water at a maximum temperature of 44°F, to provide a pressure boundary for the control room chilled water system, to provide a pressure boundary for the service water system, and to provide a pressure boundary for the refrigerant. All of these functions are considered intended functions. Typically, control room chillers are considered as one functional unit; however, for purposes of evaluating the effects of aging, it is necessary to consider the individual components. Therefore, the boundary of each control room chiller is established as follows:
	1. At the inlet and outlet flanges of the service water system connections on the control room chiller condenser. Connected piping is part of the service water system.
	2. At the inlet and outlet flanges of the control room chilled water system piping connections on the control room chiller evaporator. Connected piping is part of the control room chilled water system.
	3. For electrical power supplies, the boundary is the output terminals on the circuit breakers supplying power to the skid. This includes the cables from the circuit breaker to the skid and applies for 480 VAC and 120 VAC.
	4. The interface for instrument air supplies is at the instrument air tubing connection to the pressure control regulators, temperature controllers and transmitters, and solenoid valves located on the skid. The tubing from the instrument air header to the device on the skid is part of the instrument air system.
	5. The interface with the annunciator system is at the external connection of the contacts of the device on the skid (limit switch, pressure switch, level

Table 2.1-2   Specific Staff Guidance on Scoping

Issue	Guidance
	switch, etc.) that indicates the alarm condition. The cables are part of the annunciator system.
	Based on the boundary established, the following components would be subject to an aging management review: condenser, evaporator, economizer, chiller refrigerant piping, refrigerant expansion orifice, foundations and bolting, electrical cabinets, cables, conduit, trays and supports, valves
Hypothetical failures	For 10 CFR 54.4(a)(2), an applicant should consider those failures identified in (1) the documentation that makes up its CLB, (2) plant-specific operating experience, and (3) industrywide operating experience that is specifically applicable to its facility. The applicant need not consider hypothetical failures that are not part of CLB and that have not been previously experienced.
	For example, an applicant should consider including (1) the portion of a fire protection system identified in the UFSAR that supplies water to the refueling floor that is relied upon in a DBA analysis as an alternate source of cooling water that can be used to mitigate the consequences from the loss of spent fuel pool cooling, (2) a nonsafety-related, non-seismically-qualified building whose intended function as described in the plant's CLB is to protect a tank that is relied upon as an alternate source of cooling water needed to mitigate the consequences of a DBE, and (3) a segment of nonsafety-related piping identified as a Seismic II/I component in the applicant's CLB (Ref. 7).
Cascading	For 10 CFR 54.4(a)(3), an applicant need not consider hypothetical failures or second-, third, or fourth-level support systems. For example, if a nonsafety-related diesel generator is only relied upon to remain functional to demonstrate compliance with the NRC's SBO regulations, an applicant may not need to consider (1) an alternate/backup cooling water system, (2) the diesel generator non-seismically-qualified building walls, or (3) an overhead segment of non-seismically-qualified piping (in a Seismic II/I configuration). An applicant may not exclude any support system (identified by its CLB, actual plant-specific experience, industrywide experience, as applicable, or existing engineering evaluations) that is specifically relied upon for compliance with, or operation within, applicable NRC regulation. For example, if the analysis of a nonsafety-related diesel generator (relied upon to demonstrate compliance with an applicable NRC regulation) specifically relies upon a second cooling system to cool the diesel generator jacket water cooling system for the diesel to be operable, then both cooling systems must be included within the scope of the rule (Ref. 7).

Table 2.1-3 Specific Staff Guidance on Screening

Issue	Guidance
Consumables	Consumables may be divided into the following four categories for the purpose of license renewal: (a) packing, gaskets, component seals, and O-rings; (b) structural sealants; (c) oil, grease, and component filters; and (d) system filters, fire extinguishers, fire hoses, and air packs. The consumables in both categories (a) and (b) are considered as subcomponents and are not explicitly called out in the scoping and screening procedures. Rather, they are implicitly included at the component level (e.g., if a valve is identified as being in scope, a seal in that valve would also be in scope as a subcomponent of that valve). For category (a), the applicant would generally be able to exclude these subcomponents using a clear basis, such as the example of ASME Section III not being relied on for pressure boundary. For category (b), these subcomponents may perform functions without moving parts or a change in configuration, and they are not typically replaced. The applicant's structural AMP should address these items with respect to an AMR program on a plant-specific basis. The consumables in category (c) are usually short-lived and periodically replaced, and can normally be excluded from an AMR on that basis. Likewise, the consumables that fall within category (d) are typically replaced based on performance or condition monitoring that identifies whether these components are at the end of their qualified lives and may be excluded, on a plant-specific basis, from AMR under 10 CFR 54.21(a)(1)(ii). The applicant should identify the standards that are relied on for the replacement as part of the methodology description (for example, NFPA standards for fire protection equipment) (Ref. 8).
Heat exchanger intended functions	Both the pressure boundary and heat transfer functions for heat exchangers should be considered because heat transfer may be a primary safety function of these components. There may be a unique aging effect associated with different materials in the heat exchanger parts that are associated with the heat transfer function and not the pressure boundary function. Normally the programs that effectively manage aging effects of the pressure boundary function can, in conjunction with the procedures for monitoring heat exchanger performance, effectively manage aging effects applicable to the heat transfer function (Ref. 9).
Multiple functions	Structures and components may have multiple functions. The intended functions as delineated in 10 CFR 54.4(b) are to be reviewed for license renewal. For example, a flow orifice that is credited in a plant's accident analysis to limit flow would have two intended functions. One intended function is pressure boundary. The other intended function is to limit flow. The reviewer verifies that the applicant has considered multiple functions in identifying structure- and component-intended functions.

Table 2.1-4(a) Typical "Passive" Structure-Intended Functions

Structures	
Intended Function	**Description**
Direct Flow	Provide spray shield or curbs for directing flow (e.g., safety injection flow to containment sump)
Expansion/Separation	Provide for thermal expansion and/or seismic separation
Fire Barrier	Provide rated fire barrier to confine or retard a fire from spreading to or from adjacent areas of the plant
Flood Barrier	Provide flood protection barrier (internal and external flooding event)
Gaseous Release Path	Provide path for release of filtered and unfiltered gaseous discharge
Heat Sink	Provide heat sink during station blackout or design-basis accidents
HELB Shielding	Provide shielding against high-energy line breaks (HELB)
Missile Barrier	Provide missile barrier (internally or externally generated)
Pipe Whip Restraint	Provide pipe whip restraint
Pressure Relief	Provide over-pressure protection
Shelter, Protection	Provide shelter/protection to safety-related components
Shielding	Provide shielding against radiation
Shutdown Cooling Water	Provide source of cooling water for plant shutdown
Structural Pressure Barrier	Provide pressure boundary or essentially leak-tight barrier to protect public health and safety in the event of any postulated design-basis events.

Table 2.1-4(b) Typical "Passive" Component-Intended Functions

Components	
Intended Function	**Description**
Absorb Neutrons	Absorb neutrons
Electrical Continuity	Provide electrical connections to specified sections of an electrical circuit to deliver voltage, current, or signals
Insulate (electrical)	Insulate and support an electrical conductor
Filter	Provide filtration
Heat Transfer	Provide heat transfer
Leakage Boundary (Spatial)	Nonsafety-related component that maintains mechanical and structural integrity to prevent spatial interactions that could cause failure of safety-related SSCs
Pressure Boundary	Provide pressure-retaining boundary so that sufficient flow at adequate pressure is delivered, or provide fission product barrier for containment pressure boundary, or provide containment isolation for fission product retention
Spray	Convert fluid into spray
Structural Integrity (Attached)	Nonsafety-related component that maintains mechanical and structural integrity to provide structural support to attached safety-related piping and components
Structural Support	Provide structural and/or functional support to safety-related and/or nonsafety-related components
Throttle	Provide flow restriction

Table 2.1-5   Typical Structures, Components, and Commodity Groups, and 10 CFR 54.21(a)(1)(i) Determinations for Integrated Plant Assessment

Item	Category	Structure, Component, or Commodity Grouping	Structure, Component, or Commodity Group Meets 10 CFR 54.21(a)(1)(i) (Yes/No)
1	Structures	Category I Structures	Yes
2	Structures	Primary Containment Structure	Yes
3	Structures	Intake Structures	Yes
4	Structures	Intake Canal	Yes
5	Structures	Other Non-Category I Structures within the Scope of License Renewal	Yes
6	Structures	Equipment Supports and Foundations	Yes
7	Structures	Structural Bellows	Yes
8	Structures	Controlled Leakage Doors	Yes
9	Structures	Penetration Seals	Yes
10	Structures	Compressible Joints and Seals	Yes
11	Structures	Fuel Pool and Sump Liners	Yes
12	Structures	Concrete Curbs	Yes
13	Structures	Offgas Stack and Flue	Yes
14	Structures	Fire Barriers	Yes
15	Structures	Pipe Whip Restraints and Jet Impingement Shields	Yes
16	Structures	Electrical and Instrumentation and Control Penetration Assemblies	Yes
17	Structures	Instrumentation Racks, Frames, Panels, and Enclosures	Yes
18	Structures	Electrical Panels, Racks, Cabinets, and Other Enclosures	Yes
19	Structures	Cable Trays and Supports	Yes
20	Structures	Conduit	Yes
21	Structures	TubeTrack®	Yes
22	Structures	Reactor Vessel Internals	Yes
23	Structures	ASME Class 1 Hangers and Supports	Yes
24	Structures	Non-ASME Class 1 Hangers and Supports	Yes
25	Structures	Snubbers	No

Table 2.1-5 Typical Structures, Components, and Commodity Groups, and 10 CFR 54.21(a)(1)(i) Determinations for Integrated Plant Assessment

Item	Category	Structure, Component, or Commodity Grouping	Structure, Component, or Commodity Group Meets 10 CFR 54.21(a)(1)(i) (Yes/No)
26	Reactor Coolant Pressure Boundary Components (Note: the components of the RCPB are defined by each plant's CLB and site-specific documentation)	ASME Class 1 Piping	Yes
27	Reactor Coolant Pressure Boundary Components	Reactor Vessel	Yes
28	Reactor Coolant Pressure Boundary Components	Reactor Coolant Pumps	Yes (Casing)
29	Reactor Coolant Pressure Boundary Components	Control Rod Drives	No
30	Reactor Coolant Pressure Boundary Components	Control Rod Drive Housing	Yes
31	Reactor Coolant Pressure Boundary Components	Steam Generators	Yes
32	Reactor Coolant Pressure Boundary Components	Pressurizers	Yes
33	Non-Class I Piping Components	Underground Piping	Yes
34	Non-Class I Piping Components	Piping in Low Temperature Demineralized Water Service	Yes
35	Non-Class I Piping Components	Piping in High Temperature Single Phase Service	Yes
36	Non-Class I Piping Components	Piping in Multiple Phase Service	Yes
37	Non-Class I Piping Components	Service Water Piping	Yes

Table 2.1-5   Typical Structures, Components, and Commodity Groups, and 10 CFR 54.21(a)(1)(i) Determinations for Integrated Plant Assessment

Item	Category	Structure, Component, or Commodity Grouping	Structure, Component, or Commodity Group Meets 10 CFR 54.21(a)(1)(i) (Yes/No)
38	Non-Class I Piping Components	Low Temperature Gas Transport Piping	Yes
39	Non-Class I Piping Components	Stainless Steel Tubing	Yes
40	Non-Class I Piping Components	Instrument Tubing	Yes
41	Non-Class I Piping Components	Expansion Joints	Yes
42	Non-Class I Piping Components	Ductwork	Yes
43	Non-Class I Piping Components	Sprinkler Heads	Yes
44	Non-Class I Piping Components	Miscellaneous Appurtenances (Includes fittings, couplings, reducers, elbows, thermowells, flanges, fasteners, welded attachments, etc.)	Yes
45	Pumps	ECCS Pumps	Yes (Casing)
46	Pumps	Service Water and Fire Pumps	Yes (Casing)
47	Pumps	Lube Oil and Closed Cooling Water Pumps	Yes (Casing)
48	Pumps	Condensate Pumps	Yes (Casing)
49	Pumps	Borated Water Pumps	Yes (Casing)
50	Pumps	Emergency Service Water Pumps	Yes (Casing)
51	Pumps	Submersible Pumps	Yes (Casing)
52	Turbines	Turbine Pump Drives (excluding pumps)	Yes (Casing)
53	Turbines	Gas Turbines	Yes (Casing)
54	Turbines	Controls (Actuator and Overspeed Trip)	No
55	Engines	Fire Pump Diesel Engines	No
56	Emergency Diesel Generators	Emergency Diesel Generators	No
57	Heat Exchangers	Condensers	Yes
58	Heat Exchangers	HVAC Coolers (including housings)	Yes

Table 2.1-5 Typical Structures, Components, and Commodity Groups, and 10 CFR 54.21(a)(1)(i) Determinations for Integrated Plant Assessment

Item	Category	Structure, Component, or Commodity Grouping	Structure, Component, or Commodity Group Meets 10 CFR 54.21(a)(1)(i) (Yes/No)
59	Heat Exchangers	Primary Water System Heat Exchangers	Yes
60	Heat Exchangers	Treated Water System Heat Exchangers	Yes
61	Heat Exchangers	Closed Cooling Water System Heat Exchangers	Yes
62	Heat Exchangers	Lubricating Oil System Heat Exchangers	Yes
63	Heat Exchangers	Raw Water System Heat Exchangers	Yes
64	Heat Exchangers	Containment Atmospheric System Heat Exchangers	Yes
65	Miscellaneous Process Components	Gland Seal Blower	No
66	Miscellaneous Process Components	Recombiners	The applicant shall identify the intended function and apply the IPA process to determine if the grouping is active or passive.
67	Miscellaneous Process Components	Flexible Connectors	Yes
68	Miscellaneous Process Components	Strainers	Yes
69	Miscellaneous Process Components	Rupture Disks	Yes
70	Miscellaneous Process Components	Steam Traps	Yes
71	Miscellaneous Process Components	Restricting Orifices	Yes
72	Miscellaneous Process Components	Air Compressor	No
73	Electrical and I&C	Alarm Unit (e.g., fire detection devices)	No
74	Electrical and I&C	Analyzers (e.g., gas analyzers, conductivity analyzers)	No

Table 2.1-5   Typical Structures, Components, and Commodity Groups, and 10 CFR 54.21(a)(1)(i) Determinations for Integrated Plant Assessment

Item	Category	Structure, Component, or Commodity Grouping	Structure, Component, or Commodity Group Meets 10 CFR 54.21(a)(1)(i) (Yes/No)
75	Electrical and I&C	Annunciators (e.g., lights, buzzers, alarms)	No
76	Electrical and I&C	Batteries	No
77	Electrical and I&C	Cables and Connections, Bus, electrical portions of Electrical and I&C Penetration Assemblies, includes fuse holders outside of cabinets of active electrical SCs (e.g., electrical penetration assembly cables and connections, connectors, electrical splices, fuse holders, terminal blocks, power cables, control cables, instrument cables, insulated cables, communication cables, uninsulated ground conductors, transmission conductors, isolated-phase bus, nonsegregated-phase bus, segregated-phase bus, switchyard bus)	Yes
78	Electrical and I&C	Chargers, Converters, Inverters (e.g., converters-voltage/current, converters-voltage/pneumatic, battery chargers/inverters, motor-generator sets)	No
79	Electrical and I&C	Circuit Breakers (e.g., air circuit breakers, molded case circuit breakers, oil-filled circuit breakers)	No
80	Electrical and I&C	Communication Equipment (e.g., telephones, video or audio recording or playback equipment, intercoms, computer terminals, electronic messaging, radios, transmission line traps, and other power-line carrier equipment)	No
81	Electrical and I&C	Electric Heaters	No Yes for a Pressure Boundary if applicable
82	Electrical and I&C	Heat Tracing	No

Table 2.1-5 Typical Structures, Components, and Commodity Groups, and 10 CFR 54.21(a)(1)(i) Determinations for Integrated Plant Assessment

Item	Category	Structure, Component, or Commodity Grouping	Structure, Component, or Commodity Group Meets 10 CFR 54.21(a)(1)(i) (Yes/No)
83	Electrical and I&C	Electrical Controls and Panel Internal Component Assemblies (may include internal devices such as, but not limited to, switches, breakers, indicating lights, etc.) (e.g., main control board, HVAC control board)	No
84	Electrical and I&C	Elements, RTDs, Sensors, Thermocouples, Transducers (e.g., conductivity elements, flow elements, temperature sensors, radiation sensors, watt transducers, thermocouples, RTDs, vibration probes, amp transducers, frequency transducers, power factor transducers, speed transducers, var. transducers, vibration transducers, voltage transducers)	No Yes for a Pressure Boundary if applicable
85	Electrical and I&C	Fuses	No
86	Electrical and I&C	Generators, Motors (e.g., emergency diesel generators, ECCS and emergency service water pump motors, small motors, motor-generator sets, steam turbine generators, combustion turbine generators, fan motors, pump motors, valve motors, air compressor motors)	No
87	Electrical and I&C	High-Voltage Insulators (e.g., porcelain switchyard insulators, transmission line insulators)	Yes
88	Electrical and I&C	Surge Arresters (e.g., switchyard surge arresters, lightning arresters, surge suppressers, surge capacitors, protective capacitors)	No
89	Electrical and I&C	Indicators (e.g., differential pressure indicators, pressure indicators, flow indicators, level indicators, speed indicators, temperature indicators, analog indicators, digital indicators, LED bar graph indicators, LCD indicators)	No

Table 2.1-5   Typical Structures, Components, and Commodity Groups, and 10 CFR 54.21(a)(1)(i) Determinations for Integrated Plant Assessment

Item	Category	Structure, Component, or Commodity Grouping	Structure, Component, or Commodity Group Meets 10 CFR 54.21(a)(1)(i) (Yes/No)
90	Electrical and I&C	Isolators (e.g., transformer isolators, optical isolators, isolation relays, isolating transfer diodes)	No
91	Electrical and I&C	Light Bulbs (e.g., indicating lights, emergency lighting, incandescent light bulbs, fluorescent light bulbs)	No
92	Electrical and I&C	Loop Controllers (e.g., differential pressure indicating controllers, flow indicating controllers, temperature controllers, controllers, speed controllers, programmable logic controller, single loop digital controller, process controllers, manual loader, selector station, hand/auto station, auto/manual station)	No
93	Electrical and I&C	Meters (e.g., ammeters, volt meters, frequency meters, var meters, watt meters, power factor meters, watt-hour meters)	No
94	Electrical and I&C	Power Supplies	No
95	Electrical and I&C	Radiation Monitors (e.g., area radiation monitors, process radiation monitors)	No
96	Electrical and I&C	Recorders (e.g., chart recorders, digital recorders, events recorders)	No
97	Electrical and I&C	Regulators (e.g., voltage regulators)	No
98	Electrical and I&C	Relays (e.g., protective relays, control/logic relays, auxiliary relays)	No
99	Electrical and I&C	Signal Conditioners	No
100	Electrical and I&C	Solenoid Operators	No
101	Electrical and I&C	Solid-State Devices (e.g., transistors, circuit boards, computers)	No

Table 2.1-5   Typical Structures, Components, and Commodity Groups, and 10 CFR 54.21(a)(1)(i) Determinations for Integrated Plant Assessment

Item	Category	Structure, Component, or Commodity Grouping	Structure, Component, or Commodity Group Meets 10 CFR 54.21(a)(1)(i) (Yes/No)
102	Electrical and I&C	Switches (e.g., differential pressure indicating switches, differential pressure switches, pressure indicator switches, pressure switches, flow switches, conductivity switches, level-indicating switches, temperature-indicating switches, temperature switches, moisture switches, position switches, vibration switches, level switches, control switches, automatic transfer switches, manual transfer switches, manual disconnect switches, current switches, limit switches, knife switches)	No
103	Electrical and I&C	Switchgear, Load Centers, Motor Control Centers, Distribution Panel Internal Component Assemblies (may include internal devices such as, but not limited to, switches, breakers, indicating lights, etc.) (e.g., 4.16 kV switchgear, 480V load centers, 480V motor control centers, 250 VDC motor control centers, 6.9 kV switchgear units, 240/125V power distribution panels)	No
104	Electrical and I&C	Transformers (e.g., instrument transformers, load center transformers, small distribution transformers, large power transformers, isolation transformers, coupling capacitor voltage transformers)	No
105	Electrical and I&C	Transmitters (e.g., differential pressure transmitters, pressure transmitters, flow transmitters, level transmitters, radiation transmitters, static pressure transmitters)	No
106	Valves	Hydraulic-Operated Valves	Yes (Bodies)
107	Valves	Explosive Valves	Yes (Bodies)
108	Valves	Manual Valves	Yes (Bodies)

Table 2.1-5 Typical Structures, Components, and Commodity Groups, and 10 CFR 54.21(a)(1)(i) Determinations for Integrated Plant Assessment

Item	Category	Structure, Component, or Commodity Grouping	Structure, Component, or Commodity Group Meets 10 CFR 54.21(a)(1)(i) (Yes/No)
109	Valves	Small Valves	Yes (Bodies)
110	Valves	Motor-Operated Valves	Yes (Bodies)
111	Valves	Air-Operated Valves	Yes (Bodies)
112	Valves	Main Steam Isolation Valves	Yes (Bodies)
113	Valves	Small Relief Valves	Yes (Bodies)
114	Valves	Check Valves	Yes (Bodies)
115	Valves	Safety Relief Valves	Yes (Bodies)
116	Valves	Dampers, louvers, and gravity dampers	Yes (Housings)
117	Tanks	Air Accumulators	Yes
118	Tanks	Discharge Accumulators (Dampers)	Yes
119	Tanks	Boron Acid Storage Tanks	Yes
120	Tanks	Above Ground Oil Tanks	Yes
121	Tanks	Underground Oil Tanks	Yes
122	Tanks	Demineralized Water Tanks	Yes
123	Tanks	Neutron Shield Tank	Yes
124	Fans	Ventilation Fans (includes intake fans, exhaust fans, and purge fans)	Yes (Housings)
125	Fans	Other Fans	Yes (Housings)
126	Miscellaneous	Emergency Lighting	No
127	Miscellaneous	Hose Stations	Yes

2.2   PLANT-LEVEL SCOPING RESULTS

Review Responsibilities

Primary – Assigned branch(es)

Secondary – None

2.2.1   Areas of Review

This section addresses the plant-level scoping results for license renewal. 10 CFR 54.21(a)(1) requires the applicant to identify and list structures and components subject to an aging management review (AMR). These are "passive," "long-lived" structures and components that are within the scope of license renewal. In addition, 10 CFR 54.21(a)(2) requires the applicant to describe and justify the methods used to identify these structures and components. The staff reviews the applicant's methodology separately, following the guidance in Section 2.1.

The applicant should provide a list of all the plant systems and structures, identifying those that are within the scope of license renewal. If the list exists elsewhere, such as in the Updated Final Safety Analysis Report (UFSAR), it is acceptable to merely identify the reference. The license renewal rule does not require the identification of all plant systems and structures within the scope of license renewal. However, providing such a list may make the review more efficient. On the basis of the Design Basis Events (DBEs) considered in the plant's current licensing basis (CLB) and other CLB information relating to nonsafety-related systems and structures and certain regulated events, the applicant would identify those plant-level systems and structures within the scope of license renewal, as defined in 10 CFR 54.4(a). This is "scoping" of the plant-level systems and structures for license renewal. To verify that the applicant has properly implemented its methodology, the staff focuses its review on the implementation results to confirm that there is no omission of plant-level systems and structures within the scope of license renewal.

Examples of plant systems are the reactor coolant, containment spray, standby gas treatment (BWR), emergency core cooling, open and closed cycle cooling water, compressed air, chemical and volume control (PWR), standby liquid control (BWR), main steam, feedwater, condensate, steam generator blowdown (PWR), and auxiliary feedwater systems (PWR).

Examples of plant structures are the primary containment, secondary containment (BWR), control room, auxiliary building, fuel storage building, radwaste building, and ultimate heat sink cooling tower.

Examples of components are the reactor vessel, reactor vessel internals, steam generator (PWR), and light and heavy load-handling cranes. Some applicants may have categorized such components as plant "systems" for their convenience.

After plant-level scoping, the applicant should identify the portions of the system or structure that perform an intended function, as defined in 10 CFR 54.4(b). Then the applicant should identify those structures and components that are "passive" and "long-lived," in accordance with 10 CFR 54.21(a)(1)(i) and (ii). These "passive," "long-lived" structures and components are those that are subject to an AMR. The staff reviews these results separately following the guidance in Sections 2.3 through 2.5.

The applicant has the flexibility to determine the set of systems and structures it considers as within the scope of license renewal, provided that this set includes the systems and structures that the NRC has determined are within the scope of license renewal. Therefore, the reviewer need not review all systems and structures that the applicant has identified to be within the scope of license renewal because the applicant has the option to include more systems and components than those defined to be within the scope of license renewal by 10 CFR 54.4.

The following areas relating to the methodology implementation results for the plant-level systems and structures are reviewed.

2.2.1.1 Systems and Structures within the Scope of License Renewal

The reviewer verifies the applicant's identification of plant-level systems and structures that are within the scope of license renewal.

2.2.2 Acceptance Criteria

The acceptance criteria for the area of review define methods for determining whether the applicant has identified the systems and structures within the scope of license renewal in accordance with NRC regulations in 10 CFR 54.4. For the applicant's implementation of its methodology to be acceptable, the staff should have reasonable assurance that there has been no omission of plant-level systems and structures within the scope of license renewal.

2.2.2.1 Systems and Structures within the Scope of License Renewal

Systems and structures are within the scope of license renewal as delineated in 10 CFR 54.4(a) if they are

- Safety-related systems and structures that are relied upon to remain functional during and following DBEs [as defined in 10 CFR 50.49(b)(1)] to ensure the following functions:
    - The integrity of the reactor coolant pressure boundary,
    - The capability to shut down the reactor and maintain it in a safe shutdown condition, or
    - The capability to prevent or mitigate the consequences of accidents that could result in potential offsite exposure comparable to the guidelines in 10 CFR 50.34(a)(1), 50.67(b)(2), or 100.11, as applicable.
- All nonsafety-related systems and structures whose failure could prevent satisfactory accomplishment of any of the functions identified in 10 CFR 54.4(a)(1) above.
- All systems and structures relied on in safety analyses or plant evaluations to perform a function that demonstrates compliance with NRC regulations for fire protection (10 CFR 50.48), environmental qualification (10 CFR 50.49), PTS (10 CFR 50.61), ATWS (10 CFR 50.62), and SBO (10 CFR 50.63).

2.2.3 Review Procedures

The reviewer verifies the applicant's scoping and screening results. If the reviewer requests additional information from the applicant regarding why a certain system or structure was not identified by the applicant as being within the scope of license renewal for the applicant's plant, the reviewer should provide a focused question, clearly explaining what information is needed, explaining why it is needed, and how it will allow the staff to make its safety finding. In addition, other staff members review the applicant's scoping and screening methodology separately following the guidance in Section 2.1. The reviewer should keep these other staff members informed of findings that may affect their review of the applicant's methodology. The reviewer should coordinate this sharing of information through the license renewal project manager.

For the area of review, the following review procedures are to be followed.

2.2.3.1   Systems and Structures within the Scope of License Renewal

The reviewer determines whether the applicant has properly identified the plant-level systems and structures within the scope of license renewal by reviewing selected systems and structures that the applicant did not identify as being within the scope of license renewal to verify that they do not have any intended functions.

The reviewer should use the plant UFSAR, orders, applicable regulations, exemptions, and license conditions to determine the design basis for the structures, systems, and components (SSCs) (if components are identified as "systems" by the applicant). The design basis determines the intended function(s) of an SSC. Such functions determine whether the SSC is within the scope of license renewal under 10 CFR 54.4.

This section addresses scoping at a system or structure level. Thus, if any portion of a system or structure performs an intended function as defined in 10 CFR 54.4(b), the system or structure is within the scope of license renewal. The review of individual portions of systems and structures that are within the scope of license renewal are addressed separately in Sections 2.3 through 2.5.

The applicant should submit a list of all plant-level systems and structures, identifying those that are within the scope of license renewal (54.4) and subject to aging management review (54.21(a)(1)). The reviewer should sample selected systems and structures that the applicant did not identify as within the scope of license renewal to determine if they perform any intended functions. The following are examples:

- The applicant does not identify the radiation monitoring system as being within the scope of license renewal. The reviewer may review the UFSAR to verify that this particular system does not perform any intended functions at the applicant's plant.

- The applicant does not identify the polar crane as being within the scope of license renewal. The reviewer may review the UFSAR to verify that this particular structure is not "Seismic II over I," denoting a structure that is not seismic Category I interacting with a Seismic Category I structure as described in Position C.2 of Regulatory Guide 1.29, "Seismic Design Classification" (Ref. 1).

- The applicant does not identify the fire protection pump house as within the scope of license renewal. The reviewer may review the plant's commitments to the fire protection regulation (10 CFR 50.48) to verify that this particular structure does not perform any intended functions at the plant.

- The applicant uses the "spaces" approach for scoping electrical equipment and elects to include all electrical equipment onsite to be within the scope of license renewal except for the 525 kV switchyard and the 230 kV transmission lines. The reviewer may review the UFSAR and commitments to the SBO regulation (10 CFR 50.63) to verify that the 525 kV switchyard and the 230 kV transmission lines do not perform any intended functions at the applicant's plant.

Table 2.2-1 contains additional examples based on lessons learned from the review of the initial license renewal applications, including a discussion of the plant-specific determination of whether a system or structure is within the scope of license renewal.

The applicant may choose to group similar components and structures together in commodity groups for separate analyses. If only a portion of a system or structure has an intended function and is addressed separately in a specific commodity group, it is acceptable for an applicant to identify that system or structure as not being within the scope of license renewal. However, for completeness, the applicant should include some reference indicating that the portion of the system or structure with an intended function that is evaluated with the commodity group.

Section 2.1 contains additional guidance on the following:

- Commodity groups
- Complex assemblies
- Hypothetical failure
- Cascading

If the reviewer has reviewed systems and structures in sufficient detail and does not identify any omissions of systems and structures from those within the scope of license renewal, the staff would have reasonable assurance that the applicant has identified the systems and structures within the scope of license renewal.

If the reviewer determines that the applicant has satisfied the criteria described in this review section, the staff would have reasonable assurance that the applicant has identified the systems and structures within the scope of license renewal.

2.2.4 Evaluation Findings

If the reviewer determines that the applicant has provided information sufficient to satisfy the provisions of the SRP-LR, then the staff's evaluation supports conclusions of the following type, to be included in the safety evaluation report:

> On the basis of its review, as discussed above, the staff concludes that there is reasonable assurance that the applicant has appropriately identified the systems and structures within the scope of license renewal in accordance with 10 CFR 54.4.

2.2.5 Implementation

Except in those cases in which the applicant proposes an acceptable alternative method for complying with specific portions of NRC regulations, the method described herein will be used by the staff in its evaluation of conformance with NRC regulations.

2.2.6 References

1. Regulatory Guide 1.29, Rev. 3, "Seismic Design Classifications," U.S. Nuclear Regulatory Commission, March 2007.

Table 2.2-1   Examples of System and Structure Scoping and Basis for Disposition

Example	Disposition
Recirculation cooling water system	One function of the recirculation cooling water system is to remove decay heat from the stored fuel in the spent fuel pool via the spent fuel pool cooling system. However, the spent fuel pool cooling system at the subject facility is not safety-related, and, following a seismic event, the safety-related spent fuel pool structure and spent fuel pool makeup water supplies ensure the adequate removal of decay heat to prevent potential offsite exposures comparable to those described in 10 CFR Part 100. Therefore, the recirculation cooling water system is not within the scope of license renewal based on the spent fuel decay heat removal function.
SBO diesel generator building	The plant's UFSAR indicates that certain structural components of the SBO diesel generator building for the plant are designed to preclude seismic failure and subsequent impact of the structure on the adjacent safety-related emergency diesel generator building. In addition, the UFSAR indicates that certain equipment attached to the roof of the building has been anchored to resist tornado wind loads. Thus, the SBO diesel generator building is within the scope of license renewal.

2.3   SCOPING AND SCREENING RESULTS: MECHANICAL SYSTEMS

Review Responsibilities

Primary – Assigned branch(es)

Secondary – None

2.3.1   Areas of Review

This section addresses the mechanical systems scoping and screening results for license renewal. Typical mechanical systems consist of the following:

- Reactor coolant system (such as reactor vessel and internals, components forming part of coolant pressure boundary, coolant piping system and connected lines, and steam generators).

- Engineered safety features (such as containment spray and isolation systems, standby gas treatment system, emergency core cooling system, and fan cooler system).

- Auxiliary systems (such as new and spent fuel storage, spent fuel cooling and cleanup systems, suppression pool cleanup system, load handling system, open and closed cycle cooling water systems, ultimate heat sink, compressed air system, chemical and volume control system, standby liquid control system, coolant storage/refueling water systems, ventilation systems, diesel generator system, and fire protection system).

- Steam and power conversion system (such as turbines, main and extraction steam, feedwater, condensate, steam generator blowdown, and auxiliary feedwater).

10 CFR 54.21(a)(1) requires an applicant to identify and list structures and components subject to an aging management review (AMR). These are "passive," "long-lived" structures and components that are within the scope of license renewal (WSLR). In addition, 10 CFR 54.21(a)(2) requires an applicant to describe and justify the methods used to identify these structures and components. The staff reviews the applicant's methodology separately following the guidance in Section 2.1. To verify that the applicant has properly implemented its methodology, the staff focuses its review on the implementation results. Such a focus allows the staff to confirm that there is no omission of mechanical system components that are subject to an AMR by the applicant. If the review identifies no omission, the staff has the basis to find that there is reasonable assurance that the applicant has identified the mechanical system components that are subject to an AMR.

An applicant should list all plant-level systems and structures. On the basis of the Design Basis Events (DBEs) considered in the plant's current licensing basis (CLB) and other CLB information relating to nonsafety-related systems and structures and certain regulated events, the applicant should identify those plant-level systems and structures WSLR, as defined in 10 CFR 54.4(a). This is "scoping" of the plant-level systems and structures for license renewal. The staff reviews the applicant's plant-level "scoping" results separately following the guidance in Section 2.2.

For a mechanical system that is within the scope of license renewal, the applicant should identify the portions of the system that perform an intended function, as defined in 10 CFR 54.4(b). The applicant may identify these particular portions of the system in marked-up piping and instrument diagrams (P&IDs) or in other media. This is "scoping" of mechanical components in a system to identify those that are WSLR for a system.

For those identified mechanical components that are WSLR, the applicant must identify those that are "passive" and "long-lived," as required by 10 CFR 54.21(a)(1)(i) and (ii). These "passive," "long-lived" mechanical components are those that are subject to an AMR. This is "screening" of mechanical components in a system to identify those that are "passive" and "long-lived."

The applicant has the flexibility to determine the set of structures and components for which an AMR is performed, provided that this set includes the structures and components for which the NRC has determined that an AMR is required. This is based on the Statements of Consideration (SOC) for the license renewal rule (60 FR 22478). Therefore, the reviewer need not review all components that the applicant has identified as subject to an AMR because the applicant has the option to include more components than those required to be subject to an AMR pursuant to 10 CFR 54.21(a)(1).

2.3.2   Acceptance Criteria

The acceptance criteria for the areas of review define methods for determining whether the applicant has met the requirements of NRC regulations in 10 CFR 54.21(a)(1). For the applicant's implementation of its methodology to be acceptable, the staff should have reasonable assurance that there has been no omission of mechanical system components that are subject to an AMR.

2.3.2.1   Components within the Scope of License Renewal

Mechanical components are WSLR as delineated in 10 CFR 54.4(a) if they are

- Safety-related structures, systems, or components (SSCs) that are relied upon to remain functional during and following DBEs [as defined in 10 CFR 50.49(b)(1)] to ensure the following functions:
    - The integrity of the reactor coolant pressure boundary;
    - The capability to shut down the reactor and maintain it in a safe shutdown condition; or
    - The capability to prevent or mitigate the consequences of accidents that could result in potential offsite exposure comparable to the guidelines in 10 CFR 50.34(a)(1), 10 CFR 50.67(b)(2), or 10 CFR 100.11, as applicable.
- All nonsafety-related SSCs whose failure could prevent satisfactory accomplishment of any of the functions identified in 10 CFR 54.4(a)(1)(i), (ii), or (iii).
- All SSCs relied on in safety analyses or plant evaluations to perform a function that demonstrates compliance with NRC regulations for fire protection (10 CFR 50.48), environmental qualification (10 CFR 50.49), PTS (10 CFR 50.61), ATWS (10 CFR 50.62), and SBO (10 CFR 50.63).

2.3.2.2   Components Subject to an Aging Management Review

Mechanical components are subject to an AMR if they are WSLR and perform an intended function as defined in 10 CFR 54.4(b) without moving parts or a change in configuration or properties ("passive"), and are not subject to replacement based on a qualified life or specified time period ("long-lived") [10 CFR 54.21(a)(1)(i) and (ii)].

2.3.3   Review Procedures

The reviewer verifies the applicant's scoping and screening results. If the reviewer requests additional information from the applicant regarding why a certain component was not identified by the applicant as being WSLR or subject to an AMR for the applicant's plant, the reviewer should provide a focused question that clearly explains what information is needed, why the information is needed, and how the information will allow the staff to make its safety finding. In addition, other staff members review the applicant's scoping and screening methodology separately, following the guidance in Section 2.1. The reviewer should keep these other staff members informed of findings that may affect their review of the applicant's methodology. The reviewer should coordinate this sharing of information through the license renewal project manager.

For each area of review, the following review procedures are to be followed.

2.3.3.1   Components within the Scope of License Renewal

In this step, the staff determines whether the applicant has properly identified the components that are WSLR. The Rule requires applicants to identify components that are WSLR and subject to an AMR. In the past, LRAs have included a table of components that are WSLR; that information need not be submitted with future LRAs. Although a list of WSLR components will be available at plant sites for inspection, the reviewer should determine through sampling of P&IDs, and review of UFSAR and other plant documents, what portion of the components are within scope. The reviewer should check to see if any components exist that the staff believes are within scope but are not identified by the applicant as being subject to an AMR (and request that the applicant provide justification for omitting those components that are "passive" and "long-lived").

The reviewer should use the UFSAR, orders, applicable regulations, exemptions, and license conditions to determine the design basis for the SSCs. The design basis specifies the intended function(s) of the system(s). That intended function is used to determine the components within that system that are relied upon for the system to perform its intended functions.

The reviewer should focus the review on those components that are not identified as being WSLR, especially the license renewal boundary points and major flow paths. The reviewer should verify that the components do not have intended functions. Portions of the system identified as being WSLR by the applicant do not have to be reviewed because the applicant has the option to include more components within the scope than the rule requires.

Further, the reviewer should select system functions described in the UFSAR that are required by 10 CFR 54.4 to verify that components having intended functions were not omitted from the scope of the rule.

For example, if a reviewer verifies that a portion of a system does not perform an intended function, is not identified as being subject to an AMR by the applicant, and is isolated from the portion of the system that is identified as being subject to an AMR by a boundary valve, the reviewer should verify that the boundary valve is subject to an AMR, or that the valve is not necessary for the within-scope portion of the system to perform its intended function. Likewise, the reviewer should identify, to the extend practical, the system functions of the piping runs and components that are identified as not being WSLR to ensure they do not have intended functions that meet the requirements of 10 CFR 54.4.

Section 2.1 contains additional guidance on the following:

- Commodity groups
- Complex assemblies
- Hypothetical failure
- Cascading

If the reviewer has reviewed components in sufficient detail and does not identify any omissions of components WSLR, the reviewer would have reasonable assurance that the applicant has identified the components WSLR for the mechanical systems.

Table 2.3-1 provides examples of mechanical components scoping lessons learned from the review of the initial license renewal applications and the basis for their disposition.

2.3.3.2   Components Subject to an Aging Management Review

In this step, the reviewer determines whether the applicant has properly identified the components subject to an AMR from among those which are WSLR (i.e., those identified in Subsection 2.3.3.1). The reviewer should review selected components that the applicant has identified as WSLR but as not subject to an AMR. The reviewer should verify that the applicant has not omitted, from an AMR, components that perform intended functions without moving parts or without a change in configuration or properties and that are not subject to replacement on the basis of a qualified life or specified time period.

Starting with the boundary verified in Subsection 2.3.3.1, the reviewer should sample components that are WSLR for that system, but were not identified by the applicant as subject to an AMR. Only components that are "passive" and "long-lived" are subject to an AMR. Table 2.1-5 is provided for the reviewer to assist in identifying whether certain components are "passive." The applicant should justify omitting a component from an AMR that is WSLR at their facility and is listed as "passive" on Table 2.1-5. Although Table 2.1-5 is extensive, it may not be all-inclusive. Thus, the reviewer should use other available information sources, such as prior application reviews, to determine whether a component may be subject to an AMR.

For example, an applicant has marked a boundary of a certain system that is WSLR. The marked-up diagram shows that there are pipes, valves, and air compressors within this boundary. The applicant has identified piping and valve bodies as subject to an AMR. Because Table 2.1-5 indicates that air compressors are not subject to an AMR, the reviewer should find the applicant's determination acceptable.

Section 2.1 contains additional guidance on screening the following:

- Consumables
- Heat exchanger-intended functions
- Multiple functions

If the reviewer does not identify any omissions of components from those that are subject to an AMR, the staff would then have reasonable assurance that the applicant has identified the components subject to an AMR for the mechanical systems.

Table 2.3-2 provides examples of mechanical components screening developed from lessons learned during the review of the initial license renewal applications and bases for their disposition.

If the applicant determines that a component is subject to an AMR, the applicant should also identify the component's intended function, as defined in 10 CFR 54.4. Such functions must be maintained by any necessary AMRs. Table 2.3-3 provides examples of mechanical component-intended functions.

2.3.4 Evaluation Findings

If the reviewer determines that the applicant has provided information sufficient to satisfy the provisions of the SRP-LR, then the staff's evaluation supports conclusions of the following type, to be included in the safety evaluation report:

> On the basis of its review, as discussed above, the staff concludes that there is reasonable assurance that the applicant has appropriately identified the mechanical system components within the scope of license renewal, as required by 10 CFR 54.4, and that the applicant has adequately identified the system components subject to an aging management review in accordance with the requirements stated in 10 CFR 54.21(a)(1).

2.3.5 Implementation

Except in those cases in which the applicant proposes an acceptable alternative method for complying with specific portions of NRC regulations, the method described herein will be used by the staff in its evaluation of conformance with NRC regulations.

2.3.6 References

None.

Table 2.3-1 Examples of Mechanical Components Scoping and Basis for Disposition

Example	Disposition
Piping segment that provides structural support	The safety-related/nonsafety-related boundary along a pipe run may occur at a valve location. The nonsafety-related piping segment between this valve and the next seismic anchor provides structural support in a seismic event. This piping segment is WSLR.
Containment heating and ventilation system ductwork downstream of the fusible links providing cooling to the steam generator compartment and reactor vessel annulus	This nonsafety-related ductwork provides cooling to support the applicant's environmental qualification program. However, the failure of the cavity cooling system ductwork will not prevent the satisfactory completion of any critical safety function during and following a design basis event. Thus, this ductwork is not WSLR.
Standpipe installed inside the fuel oil storage tank	The standpipe as described in the applicant's CLB ensures that there is sufficient fuel oil reserve for the emergency diesel generator to operate for the number of days specified in the plant technical specifications following DBEs. Therefore, this standpipe is WSLR.
Insulation on boron injection tank	The temperature is high enough that insulation is not necessary to prevent boron precipitation. The plant technical specifications require periodic verification of the tank temperature. Thus, the insulation is not relied on to ensure the function of the emergency system and is not WSLR.
Pressurizer spray head	The spray head is not credited for the mitigation of any accidents addressed in the UFSAR accident analyses for many plants. The function of the pressurizer spray is to reduce reactor coolant system pressure during normal operating conditions. However, some plants rely on this component for pressure control to achieve cold shutdown during certain fire events. Failure of the spray head should be evaluated in terms of any possible damage to surrounding safety grade components, in addition to the need for spray. Therefore, this component should be evaluated on a plant-specific basis.

Table 2.3-2 Examples of Mechanical Components Screening and Basis for Disposition

Example	Disposition
Diesel engine jacket water heat exchanger and portions of the diesel fuel oil system and starting air system supplied by a vendor on a diesel generator skid	These are "passive," "long-lived" components having intended functions. They are subject to an AMR for license renewal even though the diesel generator is considered "active."
Fuel assemblies	The fuel assemblies are replaced at regular intervals based on the fuel cycle of the plant. They are not subject to an AMR.
Valve internals (such as disk and seat)	10 CFR 54.21(a)(1)(i) excludes valves, other than the valve body, from AMR. The statements of consideration of the license renewal rule provide the basis for excluding structures and components that perform their intended functions with moving parts or with a change in configuration or properties. Although the valve body is subject to an AMR, valve internals are not.

Table 2.3-3   Examples of Mechanical Component-Intended Functions

Component	Intended Function[a]
Piping	Pressure boundary
Valve body	Pressure boundary
Pump casing	Pressure boundary
Orifice	Pressure boundary flow restriction
Heat exchanger	Pressure boundary heat transfer
Reactor vessel internals	Structural support of fuel assemblies, control rods, and incore instrumentation, to maintain core configuration and flow distribution

[a] The component-intended functions are those that support the system-intended functions. For example, a heat exchanger in the spent fuel cooling system has a pressure boundary-intended function, but may not have a heat transfer function. Similarly, not all orifices have flow restriction as an intended function.

2.4 SCOPING AND SCREENING RESULTS: STRUCTURES

Review Responsibilities

Primary – Assigned branch(es)

Secondary – None

2.4.1 Areas of Review

This section addresses the scoping and screening results of structures and structural components for license renewal. Typical structures include the following:

- The primary containment structure;
- Building structures (such as the intake structure, diesel generator building, auxiliary building, and turbine building);
- Component supports (such as cable trays, pipe hangers, elastomer vibration isolators, equipment frames and stanchions, and HVAC ducting supports);
- Nonsafety-related structures whose failure could prevent safety-related structures, systems, and components (SSCs) from performing their intended functions (that is, seismic Category II over I structures).

Typical structural components include the following: liner plates, walls, floors, roofs, foundations, doors, beams, columns, and frames.

10 CFR 54.21(a)(1) requires an applicant to identify and list structures and components subject to an aging management review (AMR). These are "passive," "long-lived" structures and components that are within the scope of license renewal (WSLR). In addition, 10 CFR 54.21(a)(2) requires an applicant to describe and justify the methods used to identify these structures and components. The staff reviews the applicant's methodology separately following the guidance in Section 2.1. To verify that the applicant has properly implemented its methodology, the staff focuses its review on the implementation results. Such a focus allows the staff to confirm that there is no omission of structures that are subject to an AMR by the applicant. If the review identifies no omission, the staff has the basis to find that there is reasonable assurance that the applicant has identified the structural components that are subject to an AMR.

An applicant should list all plant-level systems and structures. On the basis of the Design Basis Events (DBEs) considered in the plant's current licensing basis (CLB) and other CLB information relating to nonsafety-related systems and structures and certain regulated events, the applicant should identify those plant-level systems and structures WSLR, as defined in 10 CFR 54.4(a). This is "scoping" of the plant-level systems and structures for license renewal. The staff reviews the applicant's plant-level "scoping" results separately following the guidance in Section 2.2.

For structures that are WSLR, an applicant must identify the structural components that are "passive" and "long-lived" in accordance with 10 CFR 54.21(a)(1)(i) and (ii). These "passive," "long-lived" structural components are those that are subject to an AMR ("screening"). The applicant's methodology implementation results for identifying structural components subject to an AMR is the area of review.

The applicant has the flexibility to determine the set of structures and components for which an AMR is performed, provided that this set includes the structures and components for which the NRC has determined that an AMR is required. This flexibility is described in the statements of consideration for the License Renewal Rule (60 FR 22478). Therefore, the reviewer should not focus the review on structural components that the applicant has already identified as subject to an AMR because it is an applicant's option to include more structural components than those subject to an AMR, pursuant to 10 CFR 54.21(a)(1). Rather, the reviewer should focus on those structural components that are not included by the applicant as subject to an AMR to ensure that they do not perform an intended function as defined in 10 CFR 54.4(b) or are not "passive" and "long-lived."

2.4.2   Acceptance Criteria

The acceptance criteria for the areas of review define methods for determining whether the applicant has met the requirements of NRC regulations in 10 CFR 54.21(a)(1). For the applicant's implementation of its methodology to be acceptable, the staff should have reasonable assurance that there has been no omission of structural components that are subject to an AMR.

2.4.2.1   Structural Components Subject to an Aging Management Review

Structural components are WSLR as delineated in 10 CFR 54.4(a) if they are

- Safety-related SSCs that are relied upon to remain functional during and following DBEs [as defined in 10 CFR 50.49(b)(1)] to ensure the following functions:
    - The integrity of the reactor coolant pressure boundary;
    - The capability to shut down the reactor and maintain it in a safe shutdown condition; or
    - The capability to prevent or mitigate the consequences of accidents that could result in potential offsite exposure comparable to the guidelines in 10 CFR 50.34(a)(1), 10 CFR 50.67(b)(2), or 10 CFR 100.11, as applicable.
- All nonsafety-related SSCs whose failure could prevent satisfactory accomplishment of any of the functions identified in 10 CFR 54.4(a)(1)(i), (ii), or (iii).
- All SSCs relied on in safety analyses or plant evaluations to perform a function that demonstrates compliance with NRC regulations for fire protection (10 CFR 50.48), environmental qualification (10 CFR 50.49), PTS (10 CFR 50.61), ATWS (10 CFR 50.62), and SBO (10 CFR 50.63).

Structural components are subject to an AMR if they are WSLR and perform an intended function as defined in 10 CFR 54.4(b) without moving parts or a change in configuration or properties ("passive"), and are not subject to replacement based on a qualified life or specified time period ("long-lived") [10 CFR 54.21(a)(1)(i) and (ii)].

2.4.3   Review Procedures

The reviewer verifies the applicant's scoping and screening results. If the reviewer requests additional information from the applicant regarding why a certain structure was not identified by the applicant as being WSLR or subject to an AMR for the applicant's plant, the reviewer should provide a focused question that clearly explains what information is needed, why the information is needed, and how the information will allow the staff to make its safety finding. In addition, other staff members review the applicant's scoping and screening methodology separately following the guidance in Section 2.1. The reviewer should keep these other staff members informed of findings that may affect their review of the applicant's methodology. The reviewer should coordinate this sharing of information through the license renewal project manager.

For each area of review, the following review procedures are to be followed:

2.4.3.1 Structural Components within the Scope of License Renewal

In this step, the staff determines which structures and structural components are WSLR. The Rule requires applicants to identify structures that are subject to an AMR, but not structures that are WSLR. Whereas, in the past, LRAs have included a table of structures that are WSLR, that information need not be submitted with future LRAs. Although that information will be available at plant sites for inspection, the reviewer should determine through sampling of P&IDs and through review of the UFSAR and other plant documents what portion of the components are within scope. The reviewer should check to see if any structures exist that the staff believes are within scope but are not identified by the applicant as being subject to an AMR (and request that the applicant provide justification for omitting those structures that are "passive" and "long-lived").

2.4.3.2 Structural Components Subject to an Aging Management Review

In general, structural components are "passive" and "long-lived." Thus, they are subject to an AMR if they are WSLR. For each of the plant-level structures WSLR, an applicant should identify those structural components that have intended functions. For example, the applicant may identify that its auxiliary building is WSLR. For this auxiliary building, the applicant may identify the structural components of beams, concrete walls, blowout panels, etc., that are subject to an AMR. The applicant should justify omitting a component from an AMR that is WSLR at its facility and is listed as "passive" on Table 2.1-5. Although Table 2.1-5 is extensive, it may not be all-inclusive. Thus, the reviewer should use other available information, such as prior application reviews, to determine whether a component may be subject to an AMR.

As set forth below, the reviewer should focus on individual structures not subject to an AMR, one at a time, to confirm that the structural components that have intended functions have been identified by the applicant. In a few instances, only portions of a particular building are WSLR. For example, a portion of a particular turbine building provides shelter for some safety-related equipment, which is an intended function, and the remainder of this particular building does not have any intended functions. In this case, the reviewer should verify that the applicant has identified the relevant particular portion of the turbine building as being WSLR and subject to an AMR.

The reviewer should use the UFSAR, orders, applicable regulations, exemptions, and license conditions to determine the design basis for the SSCs. The design basis specifies the intended function(s) of the system(s). That intended function is used to determine the components within that system that are relied upon for the system to perform its intended functions.

The reviewer should focus the review on those structural components that have not been identified as being WSLR. For example, for a building WSLR, if an applicant did not identify the building roof as subject to an AMR, the reviewer should verify that the roof has no intended functions, such as a "Seismic II over I" concern in accordance with the plant's CLB. The reviewer need not verify all structural components that have been identified as subject to an

AMR by the applicant because the applicant has the option to include more structural components than the rule requires.

Further, the reviewer should select functions described in the UFSAR to verify that structural components having intended functions were not omitted from the scope of the review. For example, if the UFSAR indicates that a dike within the fire pump house prevents a fuel oil fire from spreading to the electrically driven fire pump, the reviewer should verify that this dike has been identified as being WSLR. Similarly, if a nonsafety-related structure or component is included in the plant's CLB as a part of the safe shutdown path resulting from the resolution of the unresolved safety issue, USI A-46 (Ref. 1), the reviewer should verify that the structure or component has been included WSLR.

The applicant should also identify the intended functions of structural components. Table 2.1-4 provides typical "passive" structural component-intended functions.

The staff has developed additional scoping/screening guidance (Ref. 2). For example, some structural components may be grouped together as a commodity, such as pipe hangers, and some structural components are considered consumable materials, such as sealants. Additional guidance on these and others are contained in Section 2.1 for the following:

- Commodity groups
- Hypothetical failure
- Cascading
- Consumables
- Multiple functions

If the reviewer does not identify any omissions of components from those that are subject to an AMR, the staff would have reasonable assurance that the applicant has identified the components subject to an AMR for the structural systems.

Table 2.4-1 provides examples of structural components scoping/screening lessons learned from the review of initial license renewal applications and the basis for disposition.

If the applicant determines that a structural component may be subject to an AMR, the applicant should also identify the component's intended functions, as defined in 10 CFR 54.4. Such functions must be maintained by any necessary AMPs.

If the reviewer determines that the applicant has satisfied the criteria described in this review section, the staff would have reasonable assurance that the applicant has identified the components that are WSLR and subject to an AMR.

2.4.4   Evaluation Findings

If the reviewer determines that the applicant has provided information sufficient to satisfy the provisions of the SRP-LR, then the staff's evaluation supports conclusions of the following type, to be included in the safety evaluation report:

> On the basis of its review, as discussed above, the staff concludes that there is reasonable assurance that the applicant has appropriately identified the

structural components subject to an aging management review in accordance with the requirements stated in 10 CFR 54.21(a)(1).

2.4.5   Implementation

Except in those cases in which the applicant proposes an acceptable alternative method for complying with specific portions of NRC regulations, the method described herein will be used by the staff in its evaluation of conformance with NRC regulations.

2.4.6   References

1. NUREG-1211, "Regulatory Analysis for Resolution of Unresolved Safety Issue A-46, `Seismic Qualification of Equipment in Operating Plants,'" U.S. Nuclear Regulatory Commission, February 1987.

2. NUREG-0933, "Resolution of Generic Safety Issues," Supplement 32, U.S. Nuclear Regulatory Commission, August 2008.

Table 2.4-1   Examples of Structural Components Scoping/Screening and Basis for Disposition

Example	Disposition
Roof of turbine building	An applicant indicates that degradation or loss of its turbine building roof will not result in the loss of any intended functions. The turbine building contains safety-related SSCs in the basement, which would remain sheltered and protected by several reinforced concrete floors if the turbine building roof were to degrade. Because this roof does not perform an intended function, it is not WSLR.
Post-tensioned containment tendon gallery	The intended function of the post-tensioning system is to impose compressive forces on the concrete containment structure to resist the internal pressure resulting from a DBA with no loss of structural integrity. Although the tendon gallery is not relied on to maintain containment integrity during DBEs, operating experience indicates that water infiltration and high humidity in the tendon gallery can contribute to a significant aging effect on the vertical tendon anchorages that could potentially result in loss of the ability of the post-tensioning system to perform its intended function. However, containment inspections provide reasonable assurance that the tendon anchorages, including those in the gallery, will continue to perform their intended functions. Because the tendon gallery itself does not perform an intended function, it is not WSLR.
Water-stops	Ground water leakage into the auxiliary building could occur as a result of degradation to the water-stops. This leakage may cause flooding of equipment WSLR. (The plant's UFSAR discusses the effects of flooding.) The water-stops perform their functions without moving parts or a change in configuration, and they are not typically replaced. Thus, the water-stops are subject to an AMR. However, they need not be called out explicitly in the scoping/screening results if they are included as parts of structural components that are subject to an AMR.

2.5 SCOPING AND SCREENING RESULTS: ELECTRICAL AND INSTRUMENTATION AND CONTROLS SYSTEMS

Review Responsibilities

Primary – Assigned branch(es)

Secondary – None

2.5.1 Areas of Review

This review plan section addresses the electrical and instrumentation and controls (I&C) scoping and screening results for license renewal. Typical electrical and I&C components that are subject to an aging management review (AMR) for license renewal include electrical cables and connections.

10 CFR 54.21(a)(1) requires an applicant to identify and list structures and components subject to an AMR. These are "passive," "long-lived" structures and components that are within the scope of license renewal (WSLR). In addition, 10 CFR 54.21(a)(2) requires an applicant to describe and justify the methods used to identify these structures and components. The staff reviews the applicant's methodology separately following the guidance in Section 2.1. To verify that the applicant has properly implemented its methodology, the staff focuses its review on the implementation results. Such a focus allows the staff to confirm that there is no omission of electrical and I&C components that are subject to an AMR by the applicant. If the review identifies no omission, the staff has the basis to find that there is reasonable assurance that the applicant has identified the electrical and I&C components that are subject to an AMR.

An applicant should list all plant-level systems and structures. On the basis of the DBEs considered in the plant's CLB and other CLB information relating to nonsafety-related systems and structures and certain regulated events, the applicant would identify those plant-level systems and structures that are WSLR, as defined in 10 CFR 54.4(a). This is "scoping" of the plant-level systems and structures for license renewal. The staff reviews the applicant's plant-level "scoping" results separately following the guidance in Section 2.2.

For an electrical and I&C system that is WSLR, an applicant may not identify the specific electrical and I&C components that are subject to an AMR. For example, an applicant may not "tag" each specific length of cable that is "passive" and "long-lived," and performs an intended function as defined in 10 CFR 54.4(b). Instead, an applicant may use the so-called "plant spaces" approach (Ref. 1), which is explained below. The "plant spaces" approach provides efficiencies in the AMR of electrical equipment located within the same plant space environment.

Under the "plant spaces" approach, an applicant would identify all "passive," "long-lived" electrical equipment within a specified plant space as subject to an AMR, regardless of whether these components perform any intended functions. For example, an applicant could identify all "passive," "long-lived" electrical equipment located within the turbine building ("plant space") as subject to an AMR for license renewal. In the subsequent AMR, the applicant would evaluate the environment of the turbine building to determine the appropriate aging management activities for this equipment. The applicant has options to further refine this encompassing scope on an as-needed basis. For this example, if the applicant identified elevated temperatures in a particular area within the turbine building, the applicant may elect to further refine the scope in this particular area by (1) identifying electrical equipment that is not subject to an AMR and (2) excluding this equipment from the AMR. In this case, the excluded electrical equipment would be reported in the application as not being subject to an AMR.

10 CFR 54.21(a)(1)(i) provides many examples of electrical and I&C components that are not considered to be "passive" and are not subject to an AMR for license renewal. Therefore, the applicant is expected to identify only a few electrical and I&C components, such as electrical penetrations, cables, and connections that are "passive" and subject to an AMR. However, the TLAA evaluation requirements in 10 CFR 54.21(c) apply to environmental qualification of electrical equipment, which is not limited to "passive" components.

An applicant has the flexibility to determine the set of electrical and I&C components for which an AMR is performed, provided that this set includes the electrical and I&C components for which the NRC has determined an AMR is required. This is based on the statements of consideration for the License Renewal Rule (60 FR 22478). Therefore, the reviewer need not review all components that the applicant has identified as subject to an AMR because the applicant has the option to include more components than those required by 10 CFR 54.21(a)(1).

2.5.2   Acceptance Criteria

The acceptance criteria for the areas of review define methods for determining whether the applicant has met the requirements of NRC regulations in 10 CFR 54.21(a)(1). For the applicant's implementation of its methodology to be acceptable, the staff should have reasonable assurance that there has been no omission of electrical and I&C system components that are subject to an AMR.

2.5.2.1   Components within the Scope of License Renewal

Electrical and I&C components are WSLR as delineated in 10 CFR 54.4(a) if they are

- Safety-related SSCs that are relied upon to remain functional during and following DBEs (as defined in 10 CFR 50.49(b)(1)) to ensure the following functions:
    - The integrity of the reactor coolant pressure boundary;
    - The capability to shut down the reactor and maintain it in a safe shutdown condition; or
    - The capability to prevent or mitigate the consequences of accidents that could result in potential offsite exposure comparable to the guidelines in 10 CFR 50.34(a)(1), 10 CFR 50.67(b)(2) or 10 CFR 100.11, as applicable.
- All nonsafety-related SSCs whose failure could prevent satisfactory accomplishment of any of the functions identified in 10 CFR 54.4(a)(1)(i), (ii) or (iii).
- All SSCs relied on in safety analyses or plant evaluations to perform a function that demonstrates compliance with NRC regulations for fire protection (10 CFR 50.48), environmental qualification (10 CFR 50.49), PTS (10 CFR 50.61), ATWS (10 CFR 50.62), and SBO (10 CFR 50.63).

2.5.2.1.1   Components within the Scope of SBO (10 CFR 50.63)

Both the offsite and onsite power systems are relied upon to meet the requirements of the SBO Rule. This includes the following:

- The onsite power system meeting the requirements under 10 CFR 54.4(a)(1) (safety-related systems)

- Equipment that is required to cope with an SBO (e.g., alternate ac power sources) meeting the requirements under 10 CFR 54.4(a)(3)

- The plant system portion of the offsite power system that is used to connect the plant to the offsite power source meeting the requirements under 10 CFR 54.4(a)(3). The electrical distribution equipment out to the first circuit breaker with the offsite distribution system (i.e., equipment in the switchyard). This path typically includes the circuit breakers that connect to the offsite system power transformers (startup transformers), the transformers themselves, the intervening overhead or underground circuits between circuit breaker and transformer and transformer and onsite electrical distribution system, and the associated control circuits and structures. However, the staff's review is based on the plant-specific current licensing basis, regulatory requirements, and offsite power design configurations.

2.5.2.2   Components Subject to an Aging Management Review

Electrical and I&C components are subject to an AMR if they are WSLR and perform an intended function as defined in 10 CFR 54.4(b) without moving parts or without a change in configuration or properties ("passive"), and are not subject to replacement based on a qualified life or specified time period ("long-lived") [10 CFR 54.21(a)(1)(i) and (ii)].

2.5.3   Review Procedures

The reviewer verifies the applicant's scoping and screening results. If the reviewer requests additional information from the applicant regarding why a certain component was not identified by the applicant as being WSLR or subject to an AMR for the applicant's plant, the reviewer should provide a focused question that clearly explains what information is needed, why the information is needed, and how the information will allow the staff to make its safety finding. In addition, other staff members review the applicant's scoping and screening methodology separately following the guidance in Section 2.1. The reviewer should keep these other staff members informed of findings that may affect their review of the applicant's methodology. The reviewer should coordinate this sharing of information through the license renewal project manager.

The reviewer should verify that an applicant has identified in the license renewal application the electrical and I&C components that are subject to an AMR for its plant. The review procedures are presented below and assume that the applicant has performed "scoping" and "screening" of electrical and I&C system components in that sequence. However, the applicant may elect to perform "screening" before "scoping," which is acceptable because, regardless of the sequence, the end result should encompass the electrical and I&C components that are subject to an AMR.

The scope of 10 CFR 50.49 electric equipment to be included within 10 CFR 54.4(a)(3) is that "long-lived" (qualified life of 40 years or greater) equipment already identified by licensees under 10 CFR 50.49(b), which specifies certain electric equipment important to safety. Licensees may rely upon their listing of environmental qualification equipment, as required by 10 CFR 50.49(d), for the purposes of satisfying 10 CFR 54.4(a)(3) with respect to equipment within the scope of 10 CFR 50.49 (60 FR 22466). However, the License Renewal Rule has a requirement (10 CFR 54.21(c)) on the evaluation of TLAAs, including environmental qualification (10 CFR 50.49).

Environmental qualification equipment is not limited to "passive" equipment. The applicant may identify environmental qualification equipment separately for TLAA evaluation and not include such equipment as subject to an AMR under 10 CFR 54.21(a)(1). The environmental qualification equipment identified for TLAA evaluation would include the "passive" environmental qualification equipment subject to an AMR. The TLAA evaluation would ensure that the environmental qualification equipment would be functional for the period of extended operation. The staff reviews the applicant's environmental qualification TLAA evaluation separately following the guidance in Section 4.4.

For each area of review, the following review procedures are to be followed.

2.5.3.1   Components within the Scope of License Renewal

In this step, the staff determines whether the applicant has properly identified the components that are WSLR. The Rule requires that the LRA identify and list components that are WSLR and are subject to an AMR. Whereas, in the past, LRAs have included a table of components that are WSLR, generally that information need not be submitted with future LRAs. Although that information will be available at plant sites for inspection, the reviewer must determine, through sampling of one-line diagrams and through review of UFSAR and other plant documents, what portion of the components are WSLR. The reviewer must check to see if any components exist that the staff believes are within the scope but are not identified by the applicant as being subject to AMR (any request that the applicant provide justification for omitting those components that are "passive" and "long-lived").

The reviewer should use the UFSAR, orders, applicable regulations, exemptions, and license conditions to determine the design basis for the SSCs. The design basis specifies the intended function(s) of the system(s). That intended function is used to determine the components within that system that are required for the system to perform its intended functions.

The applicant may use the "plant spaces" approach in scoping electrical and I&C components for license renewal. In the "plant spaces" approach, an applicant may indicate that all electrical and I&C components located within a particular plant area ("plant space"), such as the containment and auxiliary building, are WSLR. The applicant may also indicate that all electrical and I&C components located within another plant area ("plant space"), such as the warehouse, are not WSLR. Table 2.5-1 contains examples of this "plant spaces" approach and the corresponding review procedures.

The applicant would use the "plant spaces" approach for the subsequent AMR of the electrical and I&C components. The applicant would evaluate the environment of the "plant spaces" to determine the appropriate aging management activities for equipment located there. The applicant has the option to further refine this encompassing scope on an as-needed basis. For example, if the applicant identified elevated temperatures in a particular area within a building ("plant space"), the applicant may elect to identify only those "passive," "long-lived" electrical and I&C components that perform an intended function in this particular area as subject to an AMR. This approach of limiting the "plant spaces" is consistent with the "plant spaces" approach. In this case, the reviewer verifies that the applicant has specifically identified the electrical and I&C components that may be WSLR in these limited "plant spaces." The reviewer should verify that the electrical and I&C components that the applicant has elected to further exclude do not indeed have any intended functions as defined in 10 CFR 54.4(b).

Section 2.1 contains additional guidance on scoping the following:

- Commodity groups
- Complex assemblies
- Scoping events
- Hypothetical failure
- Cascading

If the reviewer does not identify any omissions of components from those that are WSLR, the staff would have reasonable assurance that the applicant has identified the components WSLR for the electrical and I&C systems.

2.5.3.2 Component Subject to an Aging Management Review

In this step, the reviewer determines whether the applicant has properly identified the components subject to an AMR from among those which are WSLR (i.e., those identified in Subsection 2.5.3.1). The reviewer should review selected components that the applicant has identified as being WSLR to verify that the applicant has identified these components as being subject to an AMR if they perform intended functions without moving parts or without a change in configuration or properties and are not subject to replacement on the basis of a qualified life or specified time period. The description of "passive" may also be interpreted to include structures and components that do not display "a change in state."

Only components that are "passive" and "long-lived" are subject to an AMR. Table 2.1-5 lists many typical components and structures, and their associated intended functions, and identifies whether they are "passive." The reviewer should use Table 2.1-5 in identifying whether certain components are "passive." The reviewer should verify that electrical and I&C components identified as "passive" in Table 2.1-5 have been included by the applicant as being subject to an AMR. Although Table 2.1-5 is extensive, it may not be all-inclusive. Thus, the reviewer should use other available information sources, such as prior application reviews, to determine whether a component may be subject to an AMR.

Section 2.1 contains additional guidance on screening the following:

- Consumables
- Multiple intended functions

If the reviewer does not identify any omissions of components from those that are subject to an AMR, the staff would have reasonable assurance that the applicant has identified the components subject to an AMR for the electrical and I&C systems.

2.5.4 Evaluation Findings

If the reviewer determines that the applicant has provided information sufficient to satisfy the provisions of the SRP-LR, then the staff's evaluation supports conclusions of the following type, to be included in the safety evaluation report:

> On the basis of its review, as discussed above, the staff concludes that there is reasonable assurance that the applicant has appropriately identified the electrical and instrumentation and controls system components subject to an aging management review in accordance with the requirements stated in 10 CFR 54.21(a)(1).

2.5.5   Implementation

Except in those cases in which the applicant proposes an acceptable alternative method for complying with specific portions of NRC regulations, the method described herein will be used by the staff in its evaluation of conformance with NRC regulations.

2.5.6   References

1. SAND96-0344, "Aging Management Guideline for Commercial Nuclear Power Plants-Electrical Cable and Terminations," Sandia National Laboratories, September 1996, page 6-11.

Table 2.5-1 Examples of "Plant Spaces" Approach for Electrical and I&C Scoping and Corresponding Review Procedures

Example	Review Procedures
An applicant indicates that all electrical and I&C components on site are WSLR.	This is acceptable, and a staff review is not necessary because all electrical and I&C components are included without exception and would include those required by the rule.
An applicant indicates that all electrical and I&C components located in seven specific buildings (containment, auxiliary building, turbine building, etc.) are WSLR.	The reviewer should review electrical systems and components in areas outside of these seven buildings ("plant spaces"). The reviewer should verify that the applicant has included any direct-buried cables in trenches between these buildings as WSLR if they perform an intended function. The reviewer should also select buildings other than the seven indicated (for example, the radwaste facility) to verify that they do not contain any electrical and I&C components that perform any intended functions.
An applicant indicates that all electrical and I&C components located onsite, except for the 525 kV switchyard, 230 kV transmission lines, radwaste facility, and 44 kV substation, are WSLR.	The reviewer should select the specifically excluded "plant spaces" (that is, the 525 kV switchyard, 230 kV transmission lines, radwaste facility, and 44 kV substation) to verify that they do not contain any electrical and I&C components that perform any intended functions.

CHAPTER 3

AGING MANAGEMENT REVIEW

3.0   INTRODUCTION TO STAFF REVIEW OF AGING MANAGEMENT

The NRC project manager (PM) responsible for the safety review of the license renewal application (LRA) is responsible for assigning to appropriate NRC Office of Nuclear Reactor Regulation (NRR) divisions the review or audit of aging management reviews (AMRs) of systems, structures, and components or aging management programs (AMPs) identified in the applicant's LRA. The PM documents to which organization each AMR or AMP is assigned. The assigned AMRs and AMPs are reviewed per the criteria described in Sections 3.1 through 3.6 of this standard review plan (SRP-LR, NUREG-1800) for review of license renewal applications, as directed by the scope of each of these sections.

Review of the AMPs requires assessment of ten program elements as defined in this SRP-LR. The NRC division assigned the AMP reviews the ten program elements to verify their technical adequacy. For three of the ten program elements (corrective actions, confirmation process, and administrative controls), the NRC division responsible for review of the quality assurance aspects of the application verifies that the applicant has documented a commitment in the FSAR Supplement to expand the scope of its 10 CFR Part 50, Appendix B program to address the associated program elements for each AMP. If the applicant chooses alternate means of addressing these three program elements (i.e., use of a process other than the applicant's 10 CFR Part 50, Appendix B program), the NRC division assigned to review the AMP requests that the Division responsible for quality assurance review the applicant's proposal on a case-specific basis. Table 3.0-1 is a supplement to the FSAR and contains a list of programs that are applicable to each SRP-LR section and sub-section. It also contains the programs that are used to manage the aging effects associated with various systems.

3.0.1   Background on the Types of Reviews

10 CFR 54.21(a)(3) requires that the LRA demonstrate, for systems, structures, and components (SSCs) within the scope of license renewal and subject to an AMR pursuant to 10 CRF 54.21(a)(1), that the effects of aging are adequately managed so that the intended function(s) are maintained consistent with the current licensing basis (CLB) for the period of extended operation. This AMR consists of identifying the material, environment, aging effects, and the AMP(s) credited for managing the aging effects.

Sections 3.1 through 3.6 of this SRP-LR describe how the AMRs and AMPs are reviewed. One method that the applicant may use to conduct its AMRs is to satisfy the NUREG-1801 (GALL Report) recommendations. The applicant may choose to use methodology other than that in the GALL Report to demonstrate compliance with 10 CFR 54.21(a)(3).

The GALL Report is a technical basis document to the SRP-LR, which provides the staff with guidance in reviewing a license renewal application. An applicant may reference the GALL Report in a license renewal application to demonstrate that the programs at the applicant's facility correspond to those reviewed and approved in the GALL Report. The GALL Report (NUREG-1801) should be treated as an approved topical report. However, if an applicant takes credit for a program in the GALL Report, it is incumbent on the applicant to ensure that the conditions and operating experience at the plant is bounded by the conditions and operating experience for which the GALL Report program was evaluated. If these bounding conditions are not met it is incumbent on the applicant to address the additional effects of aging and augment the aging management program(s) in the GALL Report as appropriate.

The staff will verify that the applicant's programs are consistent with those described in the GALL report and/or with plant conditions and operating experience during the performance of an aging management program audit and review. The focus of the balance of the staff review of a license renewal application is on those programs that an applicant has enhanced to be consistent with the GALL Report, those programs for which the applicant has taken an exception to the program described in the GALL Report, and plant-specific programs not described in the GALL Report.

If an applicant takes credit for a program in the GALL Report, it is incumbent on the applicant to ensure that the plant program contains all the elements of the referenced GALL Report program. In addition, the conditions at the plant must be bounded by the conditions for which the GALL Report program was evaluated. The above verifications must be documented onsite in an auditable form. The applicant should include a certification in the license renewal application that the verifications have been completed and are documented onsite in an auditable form.

The GALL Report contains one acceptable way to manage aging effects for license renewal. An applicant may propose alternatives for staff review in its plant-specific license renewal application. Use of the GALL Report is not required, but its use should facilitate both preparation of a license renewal application by an applicant and timely, uniform review by the NRC staff.

In addition, the GALL Report does not address scoping of structures and components for license renewal. Scoping is plant-specific, and the results depend on the plant design and current licensing basis. The inclusion of a certain structure or component in the GALL Report does not mean that this particular structure or component is within the scope of license renewal for all plants. Conversely, the omission of a certain structure or component in the GALL Report does not mean that this particular structure or component is not within the scope of license renewal for any plants.

The GALL Report contains an evaluation of a large number of structures and components that may be in the scope of a typical LRA. The evaluation results documented in the GALL Report indicate that many existing, typical generic aging management programs are adequate to manage aging effects for particular structures or components for license renewal without change. The GALL Report also contains recommendations on specific areas for which generic existing programs should be augmented for license renewal and documents the technical basis for each such determination. In addition, the GALL Report identifies certain SSCs that may or may not be subject to particular aging effects, and for which industry groups are developing generic aging management programs or investigating whether aging management is warranted. To the extent that the ultimate generic resolution of such an issue will need NRC review and approval for plant-specific implementation, as indicated in a plant-specific FSAR supplement, and reflected in the SER associated with a particular LR application, an amendment pursuant to 10 CFR 50.90 will be necessary.

In this SRP-LR, subsection 3.X.2 (where X denotes number 1-6) presents the acceptance criteria describing methods to determine whether the applicant has met the requirements of NRC's regulations in 10 CFR 54.21. Subsection 3.X.3 presents the review procedures to be followed. Some rows (items) in the AMR tables (in Chapters II through VIII of the GALL Report) establish the need to perform "further evaluations." This can be clearly seen in Tables 3.X-1, and the last two columns (for 2010 and 2005, respectively) denoted the Rev2 and Rev1 AMR Line Item references. The acceptance criteria for satisfying these "further evaluations" are found in Subsections 3.X.2.2. The related review procedures are provided in subsections 3.X.3.2.

In Regulatory Guide 1.188, "Standard Format and Content for Applications to Renew Nuclear Power Plant Operating Licenses," the NRC has endorsed an acceptable methodology for applicants to structure license renewal applications. Using the guidance described in the aforementioned Regulatory Guide, the applicant documents in the LRA whether its AMR item is consistent or not consistent with the GALL Report.

A portion of the AMR includes the assessment of the AMPs in the GALL Report. The applicant may choose to use an AMP that is consistent with the GALL Report AMP, or may choose a plant-specific AMP.

If a GALL Report AMP is selected to manage aging, the applicant may take one or more exceptions to specific GALL Report AMP program elements. However, any deviation or exception to the GALL Report AMP should be described and justified. Exceptions are portions of the GALL Report AMP that the applicant does not intend to implement.

In some cases, an applicant may choose an existing plant program that does not currently meet all the program elements defined in the GALL Report AMP. If this is the situation, the applicant makes a commitment to augment the existing program to satisfy the GALL Report AMP elements prior to the period of extended operation. This commitment is an AMP "enhancement."

Enhancements are revisions or additions to existing aging management programs that the applicant commits to implement prior to the period of extended operation. Enhancements include, but are not limited to, those activities needed to ensure consistency with the GALL Report recommendations. Enhancements may expand, but not reduce, the scope of an AMP.

An audit and review is conducted at the applicant's facility to evaluate AMPs that the applicant claims to be consistent with the GALL Report. Reviews are performed to address those AMRs or AMPs related to emergent issues, stated to be not consistent with the GALL Report, or based on an NRC-approved precedent (e.g., AMRs and AMPs addressed in an NRC SER of a previous LRA). As a result of the criteria established in 10 CFR Part 54, the guidance provided in SRP-LR, GALL Report, Regulatory Guide 1.188, and the applicant's exceptions and/or enhancements to a GALL Report AMP, the following types of AMRs and AMPs are audited or reviewed by the NRC staff.

AMRs
- AMR results consistent with the GALL Report
- AMR results for which further evaluation is recommended by the GALL Report
- AMR results not consistent with or not addressed in the GALL Report

AMPs
- Consistent with GALL Report AMPs
- Plant-specific AMPs

FSAR Supplement
- Each LRA AMP will provide an FSAR Supplement which defines changes to the FSAR that will be made as a condition of a renewed license. This FSAR Supplement defines the aging management programs the applicant is crediting to satisfy 10 CFR 54.21(a)(3).
- The FSAR Supplement should also contain a commitment to implement the LRA AMP enhancement prior to the period of extended operation.

3.0.2 Applications with Approved Extended Power Uprates

Extended power uprates (EPU) are licensing actions that some licensees have recently requested the NRC staff to approve. This can affect aging management. In an NRC staff letter to the Advisory Committee on Reactor Safeguards, dated October 26, 2004 (ADAMS Accession ML042790085), the NRC Executive Director for Operation states that "All license renewal applications with an approved EPU will be required to perform an operating experience review and its impact on [aging] management programs for structures, and components before entering the period of extended operation." One way for an applicant with an approved EPU to satisfy this criterion is to document its commitment to performing an operating experience review and its impact on aging management programs for systems, structures, and components (SSCs) before entering the period of extended operation as part of its license renewal application. Such licensee commitments should be documented in the NRC staff's SER, written in support of issuing a renewed license. The staff expects to impose a license condition on any renewed license to ensure that the applicant completes these activities no later than the committed date. EPU impact on SSCs should be part of the license renewal review. If necessary, the PM assigns a responsible group to address EPU.

NOTE: In the Summary of Aging Management review tables, the Component field provides a description of the component, material, and environment combinations for which the GALL Report recommends a specified aging management program to manage the related aging effect/mechanism. To provide SRP-LR tables in a format most useful to the reviewer and industry, the presentation and specific language describing structure/components, materials, aging effects/mechanisms, and supplemental information supporting the aging management programs may vary somewhat between chapters. Tables 3.3-1, 3.4-1, and 3.6-1 exhibit similar language while Tables 3.1-1 and 3.5-1 exhibit language that varies somewhat. Also note that the capitalization and punctuation scheme in these Tables 3.X-1 can seem variable, the SRP-LR component column is an artifact of how material and component information has been consolidated in this report. This is not a technical difference, but one that supports the presentation of the materials in the desired table format. When in doubt, refer back to the specified "Rev2 Item," the AMR Item(s) upon which a given row is based.

Table 3.0-1 FSAR Supplement for Aging Management of Applicable Systems

GALL Chapter	GALL Program	Description of Program	Implementation Schedule*	Applicable GALL Report and SRP-LR Chapter References
XI.E1	Insulation Material for Electrical Cables and Connections Not Subject to 10 CFR 50.49 Environmental Qualification Requirements	The program consists of accessible electrical cables and connections installed in adverse localized environments to be visually inspected at least once every 10 years for cable jacket and connection insulation surface anomalies, such as embrittlement, discoloration, cracking, melting, swelling, or surface contamination, that could indicate incipient conductor insulation aging degradation from temperature, radiation, or moisture.	First inspection for license renewal completed prior to the period of extended operation	GALL VI / SRP 3.6
XI.E2	Insulation Material for Electrical Cables and Connections Not Subject to 10 CFR 50.49 Environmental Qualification Requirements Used in Instrumentation Circuits	The program calls for the review of calibration results or findings of surveillance tests on electrical cables and connections used in circuits with sensitive, high-voltage, low-level current signals, such as radiation monitoring and nuclear instrumentation, to provide an indication of the existence of aging effects based on acceptance criteria related to instrumentation circuit performance. By reviewing the results obtained during normal calibration or surveillance, an applicant may detect severe aging degradation prior to the loss of the cable and connection intended function. The review of calibration results or findings of surveillance tests is performed at least once every 10 years. In cases where cables are not included as part of calibration or surveillance program testing circuit, a proven cable test (such as insulation resistance tests, time domain reflectometry tests, or other testing judged to be effective in determining cable system insulation condition as justified in the application) is performed. The test frequency is based on engineering evaluation and is at least once every 10 years.	First review of calibration results or findings of surveillance test results or cable tests for license renewal completed prior to the period of extended operation	GALL VI / SRP 3.6
XI.E3	Inaccessible Power Cables Not Subject to 10 CFR 50.49 Environmental	The program calls for inaccessible or underground (e.g. in conduit, duct bank, or direct buried) power	First tests or first inspections for license renewal	GALL VI / SRP 3.6

Table 3.0-1   FSAR Supplement for Aging Management of Applicable Systems

GALL Chapter	GALL Program	Description of Program	Implementation Schedule*	Applicable GALL Report and SRP-LR Chapter References
	Qualification Requirements	(greater than or equal to 400 volts) cables exposed to significant moisture, to be tested at least once every 6 years to provide an indication of the condition of the conductor insulation. The specific type of test to be used should be capable of detecting reduced insulation resistance of the cable's insulation system due to wetting or submergence. The applicant can assess the condition of the cable insulation with reasonable confidence using one or more of the following techniques: Dielectric Loss (Dissipation Factor/Power Factor), AC Voltage Withstand, Partial Discharge, Step Voltage, Time Domain Reflectometry, Insulation Resistance and Polarization Index, Line Resonance Analysis, or other testing that is state-of-the-art at the time the tests are performed. One or more tests are used to determine the condition of the cables so they will continue to meet their intended function during the period of extended operation. The inspection frequency for water collection is established and performed based on plant-specific operating experience with cable wetting or submergence in manholes (i.e., the inspection is performed periodically based on water accumulation over time and event driven occurrences such as heavy rain or flooding). The periodic inspection should occur at least annually. The inspection should include direct observation that cables are not wetted or submerged, that cables/splices and cable support structures are intact, and dewatering/drainage systems (i.e., sump pumps) and associated alarms operate properly. In addition, operation of dewatering devices should be inspected and operation verified prior to any known or predicted heavy rain or flooding	completed prior to the period of extended operation	

Table 3.0-1 FSAR Supplement for Aging Management of Applicable Systems

GALL Chapter	GALL Program	Description of Program	Implementation Schedule*	Applicable GALL Report and SRP-LR Chapter References
		events.		
XI.E4	Metal Enclosed Bus	The program calls for the visual inspection of metal enclosed bus (MEB) internal surfaces to detect age-related degradation, including cracks, corrosion, foreign debris, excessive dust buildup, and evidence of moisture intrusion. MEB insulating material is visually inspected for signs of embrittlement, cracking, chipping, melting, swelling, discoloration, or surface contamination, which may indicate overheating or aging degradation. The internal bus insulating supports are visually inspected for structural integrity and signs of cracks. MEB external surfaces are visually inspected for loss of material due to general, pitting, and crevice corrosion. Accessible elastomers (e.g., gaskets, boots, and sealants) are inspected for degradation, including surface cracking, crazing, scuffing, and changes in dimensions (e.g., "ballooning" and "necking"), shrinkage, discoloration, hardening and loss of strength. A sample of accessible bolted connections is inspected for increased resistance of connection by using thermography or by measuring connection resistance using a micro-ohmmeter. These inspections are performed at least once every 10 years.

As an alternative to thermography or measuring connection resistance of accessible bolted connections covered with heat shrink tape, sleeving, insulating boots, etc., the applicant may use visual inspection of insulation material to detect surface anomalies, such as embrittlement, cracking, chipping, melting, discoloration, swelling, or surface contamination. When this alternative visual inspection is used to check bolted connections, the | First inspection for license renewal completed prior to the period of extended operation | GALL VI / SRP 3.6 |

Table 3.0-1 FSAR Supplement for Aging Management of Applicable Systems

GALL Chapter	GALL Program	Description of Program	Implementation Schedule*	Applicable GALL Report and SRP-LR Chapter References
		first inspection is completed prior to the period of extended operation and every 5 years thereafter.		
XI.E5	Fuse Holders	The program consists of fuse holders within the scope of license renewal to be tested at least once every 10 years to provide an indication of the condition of the metallic clamp portion of the fuse holders. Testing may include thermography, contact resistance testing, or other appropriate testing methods.	First tests for license renewal completed prior to the period of extended operation	GALL VI / SRP 3.6
XI.E6	Electrical Cable Connections Not Subject to 10 CFR 50.49 Environmental Qualification Requirements	The program consists of a representative sample of electrical connections within the scope of license renewal, which is tested at least once prior to the period of extended operation to confirm that there are no aging effects requiring management during that period. Testing may include thermography, contact resistance testing, or other appropriate testing methods without removing the connection insulation, such as heat shrink tape, sleeving, insulating boots, etc. The one-time test provides additional confirmation to support industry operating experience that shows that electrical connections have not experienced a high degree of failures and that existing installation and maintenance practices are effective.

As an alternative to thermography or measuring connection resistance of the cable connection sample, for the accessible cable connections that are covered with heat shrink tape, sleeving, insulating boots, etc., the applicant may use visual inspection of insulation materials to detect surface anomalies, such as embrittlement, cracking, chipping, melting, discoloration, swelling, or surface contamination. When this alternative visual inspection is used to check cable connections, the first inspection is | First tests for license renewal completed prior to the period of extended operation | GALL VI / SRP 3.6 |

Table 3.0-1   FSAR Supplement for Aging Management of Applicable Systems

GALL Chapter	GALL Program	Description of Program	Implementation Schedule*	Applicable GALL Report and SRP-LR Chapter References
		completed prior to the period of extended operation and every 5 years thereafter.		
XI.M1	ASME Section XI Inservice Inspection, Subsections IWB, IWC, and IWD	The program consists of periodic volumetric, surface, and/or visual examination of American Society of Mechanical Engineers (ASME) Class 1, 2, and 3 pressure-retaining components, including welds, pump casings, valve bodies, integral attachments, and pressure-retaining bolting for assessment, signs of degradation, and corrective actions. This program is in accordance with the ASME Code Section XI edition and addenda approved in accordance with provisions of 10 CFR 50.55a during the period of extended operation.	Existing program	GALL IV / SRP 3.1 GALL VII / SRP 3.3
XI.M2	Water Chemistry	This program mitigates aging effects of loss of material due to corrosion, cracking due to stress corrosion cracking (SCC) and related mechanisms, and reduction of heat transfer due to fouling in components exposed to a treated water environment. Chemistry programs are used to control water chemistry for impurities (e.g., chloride, fluoride, and sulfate) that accelerate corrosion. This program relies on monitoring and control of water chemistry to keep peak levels of various contaminants below the system-specific limits, based on Electric Power Research Institute (EPRI) guidelines (a) BWRVIP-190 (EPRI 1016579, BWR Water Chemistry Guidelines – 2008 Revision) for BWRs or (b) EPRI 1014986 (PWR Primary Water Chemistry – Revision 6) and EPRI 1016555 (PWR Secondary Water Chemistry – Revision 7) for pressurized water reactors (PWRs).	Existing program	GALL IV / SRP 3.1 GALL V / SRP 3.2 GALL VII / SRP 3.3 GALL VIII / SRP 3.4 GALL III / SRP 3.5

Table 3.0-1   FSAR Supplement for Aging Management of Applicable Systems

GALL Chapter	GALL Program	Description of Program	Implementation Schedule*	Applicable GALL Report and SRP-LR Chapter References
XI.M3	Reactor Head Closure Stud Bolting	This program includes (a) ISI in conformance with the requirements of the ASME Code, Section XI, Subsection IWB, Table IWB-2500-1, and (b) preventive measures to mitigate cracking. The program also relies on recommendations to address reactor head stud bolting degradation as delineated in NUREG-1339 and NRC Regulatory Guide (RG) 1.65.	Existing program	GALL IV / SRP 3.1
XI.M4	BWR Vessel ID Attachment Welds	The program includes (a) inspection and flaw evaluation in conformance with the guidelines of staff-approved BWRVIP-48-A to ensure the long-term integrity and safe operation of boiling water reactor (BWR) vessel internal components.	Existing program	GALL IV / SRP 3.1
XI.M5	BWR Feedwater Nozzle	This program includes (a) enhancing ISI specified in the ASME Code, Section XI, with the recommendation of General Electric (GE) NE-523-A71-0594 to perform periodic ultrasonic testing inspection of critical regions of the BWR feedwater nozzle.	Existing program	GALL IV / SRP 3.1
XI.M6	BWR Control Rod Drive Return Line Nozzle	The program includes mandatory in-service inspection in accordance with ASME Code Section XI, Subsection IWB, Table IWB 2500-1 and augmented ISI examinations in accordance with the applicant's commitments to Generic Letter 80-095 to implement the recommendations in NUREG-0619.	Program should be implemented prior to period of extended operation	GALL IV / SRP 3.1
XI.M7	BWR Stress Corrosion Cracking	The program to manage intergranular stress corrosion cracking (IGSCC) in stainless steel or nickel alloy BWR coolant pressure boundary piping is delineated in NUREG-0313, Rev. 2, and NRC Generic Letter (GL) 88-01 and its Supplement 1. The program includes (a) preventive measures to mitigate IGSCC and (b) inspection and flaw evaluation to monitor IGSCC and its effects.	Existing program	GALL IV / SRP 3.1 GALL V / SRP 3.2 GALL VII / SRP 3.3
XI.M8	BWR Penetrations	The program includes inspection and flaw evaluation	Existing program	

Table 3.0-1 FSAR Supplement for Aging Management of Applicable Systems

GALL Chapter	GALL Program	Description of Program	Implementation Schedule*	Applicable GALL Report and SRP-LR Chapter References
		in conformance with the guidelines of staff-approved boiling water reactor vessel and internals project documents BWRVIP-47-A, BWRVIP-49-A, and BWRVIP-27-A, to ensure the long-term integrity and safe operation of BWR vessel internal components.		GALL IV / SRP 3.1
XI.M9	BWR Vessel Internals	The program includes inspection and flaw evaluation in conformance with the guidelines of applicable and staff-approved BWRVIP documents, and to ensure the long-term integrity and safe operation of BWR vessel internal components.

This program also consists of (1) determination of the susceptibility of cast austenitic stainless steel components, (2) accounting for the synergistic effect of thermal aging and neutron irradiation, and (3) implementing a supplemental examination program, as necessary.

This program also addresses aging degradation of X-750 alloy and precipitation-hardened (PH) martensitic stainless steel (e.g., 15-5 and 17-4 PH steel) materials and martensitic stainless steel (e.g., 403, 410, 431 steel) that are used in BWR vessel internal components. | Existing program | GALL IV / SRP 3.1 |
| XI.M10 | Boric Acid Corrosion | The program consists of (a) visual inspection of external surfaces that are potentially exposed to borated water leakage, (b) timely discovery of leak path and removal of the boric acid residues, (c) assessment of the damage, and (d) follow-up inspection for adequacy. This program is implemented in response to NRC GL 88-05 and recent operating experience. | Existing program | GALL V / SRP 3.2

GALL VI / SRP 3.6

GALL VII / SRP 3.3

GALL VIII / SRP 3.4 |

Table 3.0-1   FSAR Supplement for Aging Management of Applicable Systems

GALL Chapter	GALL Program	Description of Program	Implementation Schedule*	Applicable GALL Report and SRP-LR Chapter References
				GALL III / SRP 3.5
XI.M11B	Cracking of Nickel-Alloy Components and Loss of Material due to Boric Acid-Induced Corrosion in Reactor Coolant Pressure Boundary Components (PWRS only)	This program addresses cracking of nickel-alloy components and loss of material due to boric acid-induced corrosion in susceptible, safety-related components in the vicinity of nickel-alloy reactor coolant pressure boundary components. It provides (a) inspection requirements for the PWR vessel, steam generator, pressurizer components, and piping if they contain the primary water stress corrosion cracking (PWSCC) susceptible materials designated alloys 600/82/182 and (b) inspection requirements for reactor pressure vessel upper heads.	Program should be implemented prior to period of extended operation	GALL IV / SRP 3.1
XI.M12	Thermal Aging Embrittlement of Cast Austenitic Stainless Steel (CASS)	The program consists of the determination of the susceptibility of CASS piping, piping components, and piping elements in PWR emergency core cooling system (ECCS) systems, including interfacing pipe lines to the chemical and volume control system and to the spent fuel pool; and in BWR ECCS systems, including interfacing pipe lines to the suppression chamber and to the drywell and suppression chamber spray system in regard to thermal aging embrittlement based on the casting method, molybdenum content, and ferrite percentage. For potentially susceptible piping, aging management is accomplished either through enhanced volumetric examination or component-specific flaw tolerance evaluation.	Existing program	GALL IV / SRP 3.1 GALL V / SRP 3.2
XI.M16A	PWR Vessel Internals	The program relies on implementation of the EPRI Report No. 1016596 (MRP-227) and EPRI Report No. 1016609 (MRP-228) to manage the aging effects on the reactor vessel internal components. This program is used to manage (a) various forms of	Program should be implemented prior to period of extended operation	GALL IV / SRP 3.1

Table 3.0-1 FSAR Supplement for Aging Management of Applicable Systems

GALL Chapter	GALL Program	Description of Program	Implementation Schedule*	Applicable GALL Report and SRP-LR Chapter References
		cracking, including SCC, PWSCC, irradiation-assisted stress corrosion cracking (IASCC), or cracking due to fatigue/cyclical loading; (b) loss of material induced by wear; (c) loss of fracture toughness due to either thermal aging or neutron irradiation embrittlement; (d) dimensional changes and potential loss of fracture toughness due to void swelling and irradiation growth; and (e) loss of preload due to thermal and irradiation-enhanced stress relaxation or creep.		
XI.M17	Flow-Accelerated Corrosion (FAC)	The program consists of (a) conducting appropriate analysis and baseline inspections, (b) determining the extent of thinning and replacement/repair of components, and (c) performing follow-up inspections to confirm or quantify and take long-term corrective actions. The program relies on implementation of EPRI guidelines of NSAC-202L-R2 or R3.	Existing program	GALL IV / SRP 3.1 GALL V / SRP 3.2 GALL VIII / SRP 3.4
XI.M18	Bolting Integrity	This program focuses on closure bolting for pressure-retaining components and relies on recommendations for a comprehensive bolting integrity program, as delineated in NUREG-1339, and industry recommendations, as delineated in EPRI NP-5769, with the exceptions noted in NUREG-1339 for safety-related bolting. The program also relies on industry recommendations for comprehensive bolting maintenance, as delineated in the EPRI TR-104213. The program generally includes periodic inspection of closure bolting for indications of loss of preload, cracking, and loss of material due to corrosion, rust, etc. The program also includes preventive measures to preclude or minimize loss of preload and cracking.	Existing program	GALL IV / SRP 3.1 GALL V / SRP 3.2 GALL VII / SRP 3.3 GALL VIII / SRP 3.4

Table 3.0-1 FSAR Supplement for Aging Management of Applicable Systems

GALL Chapter	GALL Program	Description of Program	Implementation Schedule*	Applicable GALL Report and SRP-LR Chapter References
		A related aging management program (AMP) XI.M1, "ASME Section XI Inservice Inspection (ISI) Subsections IWB, IWC, and IWD," includes inspections of safety-related and non-safety-related closure bolting and supplements this bolting integrity program. Other related programs, AMPs XI.S1, "ASME Section XI, Subsection IWE"; XI.S3, "ASME Section XI Subsection IWF"; XI.S6, "Structures Monitoring"; XI.S7, "RG 1.127, "Inspection of Water-Control Structures Associated with Nuclear Power Plant"; and XI.M23, "Inspection of Overhead Heavy Load and Light Load (Related to Refueling) Handling Systems," manage the inspection of safety-related and non-safety related structural bolting.		
XI.M19	Steam Generators	This program consists of aging management activities for the steam generator tubes, plugs, sleeves, and secondary side components that are contained within the steam generator in accordance with the plant technical specifications and includes commitments to NEI 97-06.	Existing program	GALL IV / SRP 3.1
XI.M20	Open-Cycle Cooling Water System	This program relies on implementing NRC GL 89-13, which includes (a) surveillance and control of biofouling, (b) tests to verify heat transfer, (c) routine inspection and maintenance program, (d) system walkdown inspection, and (e) review of maintenance, operating, and training practices and procedures. The Open-Cycle Cooling Water System program applies to components constructed of various materials, including steel, stainless steel, aluminum copper alloys, polymeric materials, and concrete.	Existing program	GALL IV / SRP 3.1 GALL V / SRP 3.2 GALL VII / SRP 3.3 GALL VIII / SRP 3.4
XI.M21A	Closed Treated Water Systems	The program includes (a) water treatment, including the use of corrosion inhibitors, to modify the chemical composition of the water such that the function of the	Program should be implemented prior to period of	GALL IV / SRP 3.1

Table 3.0-1   FSAR Supplement for Aging Management of Applicable Systems

GALL Chapter	GALL Program	Description of Program	Implementation Schedule*	Applicable GALL Report and SRP-LR Chapter References
		equipment is maintained and such that the effects of corrosion are minimized; (b) chemical testing of the water to ensure that the water treatment program maintains the water chemistry within acceptable guidelines; and (c) inspections to determine the presence or extent of corrosion and/or cracking.	extended operation	GALL V / SRP 3.2 GALL VII / SRP 3.3 GALL VIII / SRP 3.4
XI.M22	Boraflex Monitoring	The program consists of (a) neutron attenuation testing ("blackness testing") to determine gap formation, (b) sampling for the presence of silica in the spent fuel pool along with boron loss, and (c) monitoring and analysis of criticality to assure that the required 5% sub-criticality margin is maintained. This program is implemented in response to NRC GL 96-04.	Existing program	GALL VII / SRP 3.3
XI.M23	Inspection of Overhead Heavy Load and Light Load Handling Related to Refueling) Handling Systems	The program evaluates the effectiveness of the maintenance monitoring program and the effects of past and future usage on the structural reliability of cranes and hoists. The number and magnitude of lifts made by the hoist or crane are also reviewed. Rails and girders are visually inspected on a routine basis for degradation; functional tests are performed to assure their integrity. These cranes must also comply with the maintenance rule requirements provided in 10 CFR 50.65.	Existing program	GALL VII / SRP 3.3
XI.M24	Compressed Air Monitoring	The program consists of monitoring moisture content and corrosion, and performance of the entire system, including (a) preventive monitoring of water (moisture), and other contaminants to keep within the specified limits and (b) inspection of components for indications of loss of material due to corrosion. This program is in response to NRC GL 88-14 and INPO's Significant Operating Experience Report (SOER) 88-01. It also relies on the ASME OM Guide Part 17 and ISA-S7.0.1-1996 as guidance for testing and monitoring air quality and moisture.	Existing program	GALL VII / SRP 3.3

Table 3.0-1   FSAR Supplement for Aging Management of Applicable Systems

GALL Chapter	GALL Program	Description of Program	Implementation Schedule*	Applicable GALL Report and SRP-LR Chapter References
XI.M25	BWR Reactor Water Cleanup System	This program includes ISI and monitoring and control of reactor coolant water chemistry. Related to the inspection guidelines for RWCU piping welds outboard of the second isolation valve, the program includes measures delineated in NUREG-0313, Revision 2, and NRC GL 88-01 and ISI in conformance with the ASME Section XI.	Existing program	GALL VII, SRP 3.3
XI.M26	Fire Protection	The program includes fire barrier inspections. The fire barrier inspection program requires periodic visual inspection of fire barrier penetration seals, fire barrier walls, ceilings, and floors, and periodic visual inspection and functional tests of fire-rated doors to ensure that their operability is maintained. The program also includes periodic inspection and testing of halon/carbon dioxide fire suppression systems.	Existing program	GALL VII / SRP 3.3
XI.M27	Fire Water System	This program consists of periodic full-flow flush tests, system performance tests to prevent corrosion from biofouling components in the fire protection system, and testing or replacement of sprinklers that have been in place for 50 years. The system is normally maintained at required operating pressure and is monitored such that loss of system pressure is immediately detected and corrective actions initiated. The program relies on the testing of piping and components in the water-based fire protection system in accordance with applicable National Fire Protection Association (NFPA) commitments. In addition, this program can be modified to include (a) portions of the fire protection sprinkler system that are subjected to full flow tests prior to the period of extended operation, and (b) portions of the fire protection system exposed to water are internally visually inspected.	Program should be implemented prior to period of extended operation	GALL VII / SRP 3.3
XI.M29	Aboveground Metallic Tanks	This program includes preventive measures to mitigate corrosion by protecting the external surfaces	Existing program	GALL VII / SRP 3.3 GALL VIII / SRP 3.4

Table 3.0-1 FSAR Supplement for Aging Management of Applicable Systems

GALL Chapter	GALL Program	Description of Program	Implementation Schedule*	Applicable GALL Report and SRP-LR Chapter References
		of steel components, per standard industry practice, with sealant or caulking at the concrete-component interface. Visual inspection during periodic system walkdowns should be sufficient to monitor degradation of the protective paint, coating, calking, or sealant. Program effectiveness is determined by measuring the thickness of the tank bottoms to ensure that significant degradation is not occurring and that the component's intended function is maintained during the period of extended operation.		
XI.M30	Fuel Oil Chemistry	The program relies on a combination of surveillance and maintenance procedures. Monitoring and controlling fuel oil contamination in accordance with the guidelines of American Society for Testing and Materials (ASTM) Standards D1796, D2276, D2709, and D4057 maintains the fuel oil quality. Exposure to fuel oil contaminants, such as water and microbiological organisms, is minimized by periodic cleaning/draining of tanks and by verifying the quality of new oil before its introduction into the storage tanks.	Existing program	GALL VII / SRP 3.3
XI.M31	Reactor Vessel Surveillance	This program, extending the scope of 10 CFR Part 50, Appendix H, "Reactor Vessel Material Surveillance Program Requirements," provides sufficient material data and dosimetry to monitor irradiation embrittlement at the end of the period of extended operation, and to determine the need for operating restrictions on the inlet temperature, neutron spectrum, and neutron flux. If surveillance capsules are not withdrawn during the period of extended operation, operating restrictions are to be established to ensure that the plant is operated under the conditions to which the surveillance capsules were exposed. All capsules in the reactor vessel that are removed and tested must meet the test	The surveillance capsule withdrawal schedule revised before the period of extended operation	GALL IV / SRP 3.1

Table 3.0-1   FSAR Supplement for Aging Management of Applicable Systems

GALL Chapter	GALL Program	Description of Program	Implementation Schedule*	Applicable GALL Report and SRP-LR Chapter References
		procedures and reporting requirements of ASTM E 185-82 to the extent practicable for the configuration of the specimens in the capsule. Any changes to the capsule withdrawal schedule, including spare capsules, must be approved by the NRC prior to implementation. Untested capsules placed in storage must be maintained for future insertion.		
XI.M32	One-Time Inspection	The program consists of a one-time inspection of selected components to verify the system-wide effectiveness of an AMP that is designed to prevent or minimize aging to the extent that it will not cause the loss of intended function during the period of extended operation. This program provides inspections that verify that unacceptable degradation is not occurring. It also may trigger additional actions that ensure the intended functions of affected components are maintained during the period of extended operation. The elements of the program include (a) determination of the sample size of components to be inspected based on an assessment of materials of fabrication, environment, plausible aging effects, and operating experience, (b) identification of the inspection locations in the system or component based on the potential for the aging effect to occur, (c) determination of the examination technique, including acceptance criteria that would be effective in managing the aging effect for which the component is examined, and (d) an evaluation of the need for follow-up examinations to monitor the progression of aging if age-related degradation is found that could jeopardize an intended function	Program should be implemented prior to period of extended operation	GALL IV / SRP 3.1 GALL V / SRP 3.2 GALL VII / SRP 3.3 GALL VIII / SRP 3.4

Table 3.0-1   FSAR Supplement for Aging Management of Applicable Systems

GALL Chapter	GALL Program	Description of Program	Implementation Schedule*	Applicable GALL Report and SRP-LR Chapter References
		before the end of the period of extended operation.		
		This program cannot be used for structures or components with known age-related degradation mechanisms or when the environment in the period of extended operation is not expected to be equivalent to that in the prior 40 years. Periodic inspections should be proposed in these cases.		
XI.M33	Selective Leaching	The program includes a one-time visual inspection coupled with either hardness measurement or other mechanical examination techniques such as destructive testing, scraping, or chipping of selected components that may be susceptible to selective leaching. This is to determine whether loss of materials is occurring and whether the process will affect the ability of the components to perform their intended function for the period of extended operation.	Program should be implemented prior to period of extended operation	GALL IV / SRP 3.1 GALL V / SRP 3.2 GALL VII / SRP 3.3 GALL VIII / SRP 3.4
XI.M35	One-Time Inspection of ASME Code Class 1 Small Bore-Piping	This program augments the existing ASME Code, Section XI requirements and is applicable to small-bore ASME Code Class 1 piping and systems with a nominal pipe size diameter less than 4 inches (NPS<4) and greater than or equal to NPS 1. This program provides a one-time volumetric inspection of a sample of this Class 1 piping. The program includes pipes, fittings, branch connections, and all full and partial penetration (socket) welds. The program includes measures to verify that degradation is not occurring, thereby either confirming that there is no need to manage aging-related degradation or validating the effectiveness of any existing program for the period of extended operation. The one-time inspection program for ASME Code Class 1 small-bore piping includes locations that are susceptible to	Program should be implemented prior to period of extended operation	GALL IV / SRP 3.1

Table 3.0-1   FSAR Supplement for Aging Management of Applicable Systems

GALL Chapter	GALL Program	Description of Program	Implementation Schedule*	Applicable GALL Report and SRP-LR Chapter References
		cracking. This program is applicable to systems that have not experienced cracking of ASME Code Class 1 small-bore piping. This program can also be used for systems that experienced cracking but have implemented design changes to effectively mitigate cracking. (Measure of effectiveness includes (1) the one-time inspection sampling is statistically significant; (2) samples will be selected as described in Element 5; and (3) no repeated failures over an extended period of time.) For systems that have experienced cracking and operating experience indicates design changes have not been implemented to effectively mitigate cracking, periodic inspection is proposed, as managed by a plant-specific AMP. Should evidence of cracking be revealed by a one-time inspection, periodic inspection is also proposed, as managed by a plant-specific AMP.		
XI.M36	External Surfaces Monitoring of Mechanical Components	This program is based on system inspections and walkdowns. This program consists of periodic visual inspections of metallic and polymeric components, such as piping, piping components, ducting, polymeric components, and other components. The program manages aging effects through visual inspection of external surfaces for evidence of loss of material, cracking, and change in material properties. When appropriate for the component and material, manipulation may be used to augment visual inspection to confirm the absence of elastomer hardening and loss of strength.	Existing program	GALL V / SRP 3.2 GALL VII / SRP 3.3 GALL VIII / SRP 3.4
XI.M37	Flux Thimble Tube Inspection	The program inspects for the thinning of flux thimble tube walls, which provides a path for the in-core neutron flux monitoring system detectors and forms part of the reactor coolant system pressure boundary. Flux thimble tubes are subject to loss of	Existing Program	GALL IV / SRP 3.1

Table 3.0-1   FSAR Supplement for Aging Management of Applicable Systems

GALL Chapter	GALL Program	Description of Program	Implementation Schedule*	Applicable GALL Report and SRP-LR Chapter References
		material at certain locations in the reactor vessel where flow-induced fretting causes wear at discontinuities in the path from the reactor vessel instrument nozzle to the fuel assembly instrument guide tube. A nondestructive examination methodology, such as eddy current testing, or other applicant-justified and US NRC-accepted inspection methods are used to monitor flux thimble tube wear. This program implements the recommendations of NRC Bulletin 88-09, "Thimble Tube Thinning in Westinghouse Reactors."		
XI.M38	Inspection of Internal Surfaces in Miscellaneous Piping and Ducting Components	The program consists of inspections of the internal surfaces of metallic piping, piping components, ducting, polymeric components, and other components and any water systems other than open-cycle cooling water, treated water, and fire water that are exposed to environments of air – indoor, uncontrolled; air – outdoor; condensation. These internal inspections are performed during the periodic system and component surveillances or during the performance of maintenance activities when the surfaces are made accessible for visual inspection. The program includes visual inspections to ensure that existing environmental conditions are not causing material degradation that could result in a loss of the component's intended function. For certain materials, such as polymers, physical manipulation or pressurization (e.g., hydrotesting) to detect hardening or loss of strength is used to augment the visual examinations conducted under this program. If visual inspection of internal surfaces is not possible, then the applicant needs to provide a plant-specific program.	Existing program	GALL V / SRP 3.2 GALL VII / SRP 3.3 GALL VIII / SRP 3.4 GALL VI / SRP 3.6
XI.M39	Lubricating Oil Analysis	This program ensures that the oil environment in the mechanical systems is maintained to the required	Existing program	GALL V / SRP 3.2

Table 3.0-1   FSAR Supplement for Aging Management of Applicable Systems

GALL Chapter	GALL Program	Description of Program	Implementation Schedule*	Applicable GALL Report and SRP-LR Chapter References
		quality. The program ensures that oil systems are maintained free of contaminants (primarily water and particulates), thereby preserving an environment that is not conducive to loss of material or reduction of heat transfer. Testing activities include sampling and analysis of lubricating oil for detrimental contaminants. The presence of water or particulates may also indicate in-leakage and corrosion product buildup.		GALL VII / SRP 3.3 GALL VIII / SRP 3.4
XI.M40	Monitoring of Neutron-Absorbing Materials other than Boraflex	This program relies on periodic inspection, testing, monitoring, and analysis of the criticality design to assure that the required 5 percent sub-criticality margin is maintained.	Program should be implemented prior to period of extended operation	GALL VII / SRP 3.3
XI.M41	Buried and Underground Piping and Tanks	This comprehensive program is designed to manage the aging of the external surfaces of buried and underground piping and tanks. It addresses piping and tanks composed of any material, including metallic, polymeric, concrete, and cementitious materials. The program manages aging through preventive, mitigative, and inspection activities. It manages all applicable aging effects, such as loss of material, cracking, and changes in material properties.	Inspections to be completed before the period of extended operation	GALL V / SRP 3.2 GALL VII / SRP 3.3 GALL VIII / SRP 3.4
XI.S1	ASME Section XI, Subsection IWE Inservice Inspection (IWE)	The ASME Section XI, Subsection IWE program consists of periodic visual, surface, and volumetric inspection of pressure-retaining components of steel and concrete containments for signs of degradation, assessment of damage, and corrective actions. The program also includes aging management for the potential loss of material due to corrosion in the inaccessible areas of the BWR Mark I steel containment, and surface examination for the detection of cracking of structural bolting. This	Existing program	GALL II / SRP 3.5

Table 3.0-1   FSAR Supplement for Aging Management of Applicable Systems

GALL Chapter	GALL Program	Description of Program	Implementation Schedule*	Applicable GALL Report and SRP-LR Chapter References
		program is in accordance with ASME Section XI, Subsection IWE, 2004 edition.		
XI.S2	ASME Section XI, Subsection IWL Inservice Inspection (IWL)	The ASME Section XI, Subsection IWL program consists of (a) periodic visual inspection of concrete surfaces for reinforced and prestressed concrete containments, (b) periodic visual inspection and sample tendon testing of unbonded post-tensioning systems for prestressed concrete containments for signs of degradation, assessment of damage, and corrective actions, and testing of the tendon corrosion protection medium and free water. Measured tendon lift-off forces are compared to predicted tendon forces calculated in accordance with RG 1.35.1. This program is in accordance with ASME Section XI, Subsection IWL, 2004 edition.	Existing program	GALL II / SRP 3.5
XI.S3	ASME Section XI, Subsection IWF Inservice inspection (IWF)	This program consists of periodic visual examination of component supports and high-strength structural bolting for signs of degradation, evaluation, and corrective actions. This program is in accordance with ASME Section XI, Subsection IWF, 2004 edition.	Existing program	GALL II / SRP 3.5 GALL III / SRP 3.5
XI.S4	10 CFR Part 50, Appendix J	This program consists of monitoring leakage rates through containment liner/welds, penetrations, fittings, and other access openings to detect degradation of the containment pressure boundary. Corrective actions are taken if leakage rates exceed acceptance criteria. This program is implemented in accordance with 10 CFR Part 50 Appendix J, RG 1.163 and NEI 94-01, Rev. 0.	Existing program	GALL II / SRP 3.5
XI.S5	Masonry Walls	The program consists of inspections, based on IE Bulletin 80-11 and plant-specific monitoring proposed by IN 87-67, for managing loss of material and cracking of masonry walls.	Existing program	GALL III / SRP 3.5
XI.S6	Structures Monitoring	The program consists of periodic inspection and monitoring the condition of structures and structure	Existing program	GALL VII / SRP 3.3

Table 3.0-1   FSAR Supplement for Aging Management of Applicable Systems

GALL Chapter	GALL Program	Description of Program	Implementation Schedule*	Applicable GALL Report and SRP-LR Chapter References
		component supports to ensure that aging degradation leading to loss of intended functions will be detected and that the extent of degradation can be determined. This program is implemented in accordance with NUMARC 93-01, Rev. 2 and RG 1.160, Rev. 2.		GALL II / SRP 3.5 GALL III / SRP 3.5 GALL VI / SRP 3.6
XI.S7	R.G. 1.127, Inspection of Water-Control Structures Associated with Nuclear Power Plants	The program consists of inspection and surveillance programs for dams, slopes, canals, intake structure, and other water-control structures associated with emergency cooling water systems or flood protection based on RG 1.127, Rev. 1. The program also includes structural steel and structural bolting associated with water-control structures, steel or wood piles and sheeting required for the stability of embankments and channel slopes, and miscellaneous steel, such as sluice gates and trash racks.	Existing program	GALL III / SRP 3.5
XI.S8	Protective Coating Monitoring and Maintenance	This program consists of guidance for selection, application, inspection, and maintenance of protective coatings. This program is implemented in accordance with RG 1.54, Rev. 1 or latest revision.	Existing program	GALL III / SRP 3.5
GALL Appendix A	Quality Assurance	The 10 CFR Part 50, Appendix B quality assurance program provides for corrective actions, the confirmation process, and administrative controls for AMPs for license renewal. The scope of this existing program is expanded to include nonsafety-related structures and components that are subject to an AMR for license renewal.	Existing program	GALL VII / SRP 3.3 GALL VIII / SRP 3.4 GALL III / SRP 3.5 GALL VI / SRP 3.6
SRP Appendix A	Plant-Specific AMP	The program should contain information associated with the bases for determining that aging effects will be managed during the period of extended operation.	Program should be implemented prior to period of extended operation	GALL IV / SRP 3.1 GALL V / SRP 3.2 GALL VII / SRP 3.3

Table 3.0-1   FSAR Supplement for Aging Management of Applicable Systems

GALL Chapter	GALL Program	Description of Program	Implementation Schedule*	Applicable GALL Report and SRP-LR Chapter References
				GALL VIII / SRP 3.4
				GALL II-III / SRP 3.5
				GALL VI / SRP 3.6

* An applicant need not incorporate the implementation schedule into its FSAR. However, the reviewer should verify that the applicant has identified and committed in the license renewal application to any future aging management activities to be completed before the period of extended operation. The staff expects to impose a license condition on any renewed license to ensure that the applicant will complete these activities no later than the committed date.

3.1   AGING MANAGEMENT OF REACTOR VESSEL, INTERNALS, AND REACTOR COOLANT SYSTEM

Review Responsibilities

Primary - Branch assigned responsibility by PM as described in SRP-LR Section 3.0 of this SRP-LR.

3.1.1   Areas of Review

This section addresses the aging management review (AMR) and the associated aging management program (AMP) of the reactor vessel, internals, and reactor coolant system. For a recent vintage plant, the information related to the reactor vessel, internals, and reactor coolant system is contained in Chapter 5, "Reactor Coolant System and Connected Systems," of the plant's final safety analysis report (FSAR), consistent with the Standard Review Plan for the Review of Safety Analysis Reports for Nuclear Power Plants (NUREG-0800). For older plants, the location of applicable information is plant-specific because an older plant's FSAR may have predated NUREG-0800.

The reactor vessel, internals, and reactor coolant system includes the reactor vessel and internals. For Boiling Water Reactors (BWRs), this system also includes the reactor coolant recirculation system and portions of other systems connected to the pressure vessel extending to the first isolation valve outside of containment or to the first anchor point. These connected systems include residual heat removal, low-pressure core spray, high-pressure core spray, low-pressure coolant injection, high-pressure coolant injection, reactor core isolation cooling, isolation condenser, reactor coolant cleanup, feedwater, and main steam. For Pressurized Water Reactors (PWRs), the reactor coolant system includes the primary coolant loop, the pressurizer, and the steam generators. For PWRs the reactor coolant system also includes the pressurizer relief tank, which is not an ASME Code Class 1 component. The connected systems for PWRs include the residual heat removal or low pressure injection system, core flood spray or safety injection tank, chemical and volume control system or high-pressure injection system, and sampling system.

The responsible review organization is to review the following license renewal application (LRA) AMR and AMP items assigned to it, per SRP-LR Section 3.0:

AMRs
- AMR results consistent with the GALL Report
- AMR results for which further evaluation is recommended by the GALL Report
- AMR results not consistent with or not addressed in the GALL Report

AMPs
- Consistent with GALL Report AMPs
- Plant-specific AMPs

FSAR Supplement
- The responsible review organization is to review the FSAR Supplement associated with each assigned AMP.

3.1.2 Acceptance Criteria

The acceptance criteria for the areas of review describe methods for determining whether the applicant has met the requirements of the NRC's regulations in 10 CFR 54.21.

3.1.2.1 AMR Results Consistent with the GALL Report

The AMR and the AMPs applicable to the reactor vessel, internals, and reactor coolant system are described and evaluated in Chapter IV of NUREG-1801 (GALL Report).

The applicant's LRA should provide sufficient information so that the reviewer is able to confirm that the specific LRA AMR item and the associated LRA AMP are consistent with the cited GALL Report AMR item. The reviewer should then confirm that the LRA AMR item is consistent with the GALL Report AMR item to which it is compared. When the applicant is crediting a different aging management program than recommended in the GALL Report, the reviewer should confirm that the alternate aging management program is valid to use for aging management and will be capable of managing the effects of aging as adequately as the aging management program recommended by the GALL Report.

3.1.2.2 AMR Results for Which Further Evaluation is Recommended by the GALL Report

The basic acceptance criteria defined in Subsection 3.1.2.1 need to be applied first for all of the AMRs and AMPs reviewed as part of this section. In addition, if the GALL Report AMR item to which the LRA AMR item is compared identifies that further evaluation is recommended, then additional criteria apply as identified by the GALL Report for each of the following aging effect/aging mechanism combinations. Refer to Table 3.1-1, comparing the "Further Evaluation Recommended" and the "Rev2 Item" columns, for the AMR items that reference the following subsections. The 2005 AMR item counterpart is provided in the "Rev1 Item" column.

3.1.2.2.1 Cumulative Fatigue Damage

Fatigue is a time-limited aging analysis (TLAA) as defined in 10 CFR 54.3. TLAAs are required to be evaluated in accordance with 10 CFR 54.21(c)(1). This TLAA is addressed separately in Section 4.3, "Metal Fatigue Analysis," of this SRP-LR.

3.1.2.2.2 Loss of Material due to General, Pitting, and Crevice Corrosion

1. Loss of material due to general, pitting, and crevice corrosion could occur in the steel PWR steam generator upper and lower shell and transition cone exposed to secondary feedwater and steam. The existing program relies on control of water chemistry to mitigate corrosion and Inservice Inspection (ISI) to detect loss of material. The extent and schedule of the existing steam generator inspections are designed to ensure that flaws cannot attain a depth sufficient to threaten the integrity of the welds. However, according to NRC Information Notice (IN) 90-04, the program may not be sufficient to detect pitting and crevice corrosion, if general and pitting corrosion of the shell is known to exist. The GALL Report recommends augmented inspection to manage this aging effect. Furthermore, the GALL Report clarifies that this issue is limited to Westinghouse Model 44 and 51 Steam Generators, where a high-stress region exists at the shell to transition cone weld. Acceptance criteria are described in Branch Technical Position RLSB-1 (Appendix A.1 of this SRP-LR).

2. Loss of material due to general, pitting, and crevice corrosion could occur in the steel PWR steam generator shell assembly exposed to secondary feedwater and steam. The existing program relies on control of secondary water chemistry to mitigate corrosion. However, some applicants have replaced only the bottom part of their recirculating steam generators, generating a cut in the middle of the transition cone, and, consequently, a new transition cone closure weld. The GALL Report recommends volumetric examinations performed in accordance with the requirements of ASME Code Section XI for upper shell-to and lower shell-to transition cones with gross structural discontinuities for managing loss of material due to general, pitting, and crevice corrosion in the welds for Westinghouse Model 44 and 51 Steam Generators, where a high-stress region exists at the shell to transition cone weld.

The new continuous circumferential weld, resulting from cutting the transition cone as discussed above, is a different situation from the SG transition cone welds containing geometric discontinuities. Control of water chemistry does not preclude loss of material due to pitting and crevice corrosion at locations of stagnant flow conditions. The new transition area weld is a field-weld as opposed to having been made in a controlled manufacturing facility, and the surface conditions of the transition weld may result in flow conditions more conducive to initiation of general, pitting, and crevice corrosion than those of the upper and lower transition cone welds. Crediting of the ISI program for the new SG transition cone weld may not be an effective basis for managing loss of material in this weld, as the ISI criteria would only perform a VT-2 visual leakage examination of the weld as part of the system leakage test performed pursuant to ASME Section XI requirements. In addition, ASME Section XI does not require licensees to remove insulation when performing visual examination on non-borated treated water systems. Therefore, the effectiveness of the chemistry control program should be verified to ensure that loss of material due to general, pitting and crevice corrosion is not occurring.

For the new continuous circumferential weld, the GALL Report recommends further evaluation to verify the effectiveness of the chemistry control program. A one-time inspection at susceptible locations is an acceptable method to determine whether an aging effect is not occurring or an aging effect is progressing very slowly, such that the component's intended function will be maintained during the period of extended operation. Furthermore, the GALL Report clarifies that this issue is limited to replacement recirculating steam generators with a new transition cone closure weld.

3.1.2.2.3 Loss of Fracture Toughness due to Neutron Irradiation Embrittlement

1. Neutron irradiation embrittlement is a TLAA to be evaluated for the period of extended operation for all ferritic materials that have a neutron fluence greater than 10^{17} n/cm2 (E >1 MeV) at the end of the license renewal term. Certain aspects of neutron irradiation embrittlement are TLAAs as defined in 10 CFR 54.3. TLAAs are required to be evaluated in accordance with 10 CFR 54.21(c)(1). This TLAA is addressed separately in Section 4.2, "Reactor Vessel Neutron Embrittlement Analysis," of this SRP-LR.

2. Loss of fracture toughness due to neutron irradiation embrittlement could occur in BWR and PWR reactor vessel beltline shell, nozzle, and welds exposed to reactor coolant and neutron flux. A reactor vessel materials surveillance program monitors neutron irradiation embrittlement of the reactor vessel. The reactor vessel surveillance program is plant-specific, depending on matters such as the composition of limiting materials, availability of surveillance capsules, and projected fluence levels. In accordance with

10 CFR Part 50, Appendix H, an applicant is required to submit its proposed withdrawal schedule for approval prior to implementation. Untested capsules placed in storage must be maintained for future insertion. Thus, further staff evaluation is required for license renewal. Specific recommendations for an acceptable AMP are provided in Chapter XI, Section M31 of the GALL Report.

3. Ductility – Reduction in Fracture Toughness is a plant-specific TLAA for Babcock and Wilcox (B&W) reactor internals to be evaluated for the period of extended operation in accordance with the staff's safety evaluation concerning "Demonstration of the Management of Aging Effects for the Reactor Vessel Internals," Babcock and Wilcox Owners Group report number BAW-2248, which is included in BAW-2248A, March 2000. Plant-specific TLAAs are addressed in Section 4.7, "Other Plant-Specific Time-Limited Aging Analyses," of this SRP-LR.

3.1.2.2.4   Cracking due to Stress Corrosion Cracking and Intergranular Stress Corrosion Cracking

1. Cracking due to stress corrosion cracking (SCC) and intergranular stress corrosion cracking (IGSCC) could occur in the stainless steel and nickel alloy BWR top head enclosure vessel flange leak detection lines. The GALL Report recommends that a plant-specific AMP be evaluated because existing programs may not be capable of mitigating or detecting cracking due to SCC and IGSCC. Acceptance criteria are described in Branch Technical Position RLSB-1 (Appendix A.1 of this SRP-LR).

2. Cracking due to SCC and IGSCC could occur in stainless steel BWR isolation condenser components exposed to reactor coolant. The existing program relies on control of reactor water chemistry to mitigate SCC and on ASME Section XI ISI to detect cracking. However, the existing program should be augmented to detect cracking due to SCC and IGSCC. The GALL Report recommends an augmented program to include temperature and radioactivity monitoring of the shell-side water and eddy current testing of tubes to ensure that the component's intended function will be maintained during the period of extended operation. Acceptance criteria are described in Branch Technical Position RLSB-1 (Appendix A.1 of this SRP-LR).

3.1.2.2.5   Crack Growth due to Cyclic Loading

Crack growth due to cyclic loading could occur in reactor vessel shell forgings clad with stainless steel using a high-heat-input welding process. Growth of intergranular separations (underclad cracks) in the heat-affected zone under austenitic stainless steel cladding is a TLAA to be evaluated for the period of extended operation for all the SA-508-Cl-2 forgings where the cladding was deposited with a high heat input welding process. The methodology for evaluating the underclad flaw should be consistent with the flaw evaluation procedure and criterion in the ASME Section XI Code, 2004 edition[1]. See the SRP-LR, Section 4.7, "Other Plant-Specific Time-Limited Aging Analysis," for generic guidance for meeting the requirements of 10 CFR 54.21(c).

---

[1] Refer to the GALL Report, Chapter I, for applicability of other editions of the ASME Code, Section XI.

3.1.2.2.6 Cracking due to Stress Corrosion Cracking

1. Cracking due to SCC could occur in the PWR stainless steel reactor vessel flange leak detection lines and bottom-mounted instrument guide tubes exposed to reactor coolant. The GALL Report recommends further evaluation to ensure that these aging effects are adequately managed. The GALL Report recommends that a plant-specific AMP be evaluated to ensure that this aging effect is adequately managed. Acceptance criteria are described in Branch Technical Position RLSB-1 (Appendix A.1 of this SRP-LR).

2. Cracking due to SCC could occur in Class 1 PWR cast austenitic stainless steel (CASS) reactor coolant system piping, piping components, and piping elements exposed to reactor coolant. The existing program relies on control of water chemistry to mitigate SCC; however, SCC could occur for CASS components that do not meet the NUREG-0313 guidelines with regard to ferrite and carbon content. The GALL Report recommends further evaluation of a plant-specific program for these components to ensure that this aging effect is adequately managed. Acceptance criteria are described in Branch Technical Position RLSB-1 (Appendix A.1 of this SRP-LR).

3.1.2.2.7 Cracking due to Cyclic Loading

Cracking due to cyclic loading could occur in steel and stainless steel BWR isolation condenser components exposed to reactor coolant. The existing program relies on ASME Section XI ISI. However, the existing program should be augmented to detect cracking due to cyclic loading. The GALL Report recommends an augmented program to include temperature and radioactivity monitoring of the shell-side water and eddy current testing of tubes to ensure that the component's intended function will be maintained during the period of extended operation. Acceptance criteria are described in Branch Technical Position RLSB-1 (Appendix A.1 of this SRP-LR).

3.1.2.2.8 Loss of Material due to Erosion

Loss of material due to erosion could occur in steel steam generator feedwater impingement plates and supports exposed to secondary feedwater. The GALL Report recommends further evaluation of a plant-specific AMP to ensure that this aging effect is adequately managed. Acceptance criteria are described in Branch Technical Position RLSB-1 (Appendix A.1 of this SRP-LR).

3.1.2.2.9 Cracking due to Stress Corrosion Cracking and Irradiation-Assisted Stress Corrosion Cracking

Cracking due to SCC and irradiation-assisted stress corrosion cracking (IASCC) could occur in inaccessible locations for stainless steel and nickel-alloy Primary and Expansion PWR reactor vessel internal components. If aging effects are identified in accessible locations, the GALL Report recommends further evaluation of the aging effects in inaccessible locations on a plant-specific basis to ensure that this aging effect is adequately managed. Acceptance criteria are described in Branch Technical Position RLSB-1 (Appendix A.1 of this SRP-LR).

3.1.2.2.10 Loss of Fracture Toughness due to Neutron Irradiation Embrittlement, Change in Dimension due to Void Swelling, Loss of Preload due to Stress Relaxation, or Loss of Material due to Wear

Loss of fracture toughness due to neutron irradiation embrittlement, change in dimension due to void swelling, loss of preload due to stress relaxation, or loss of material due to wear could occur in inaccessible locations for stainless steel and nickel-alloy Primary and Expansion PWR reactor vessel internal components. If aging effects are identified in accessible locations, the GALL Report recommends further evaluation of the aging effects in inaccessible locations on a plant-specific basis to ensure that this aging effect is adequately managed. Acceptance criteria are described in Branch Technical Position RLSB-1 (Appendix A.1 of this SRP-LR).

3.1.2.2.11 Cracking due to Primary Water Stress Corrosion Cracking

1. Foreign operating experience in steam generators with a similar design to that of Westinghouse Model 51 has identified extensive cracking due to primary water stress corrosion cracking (PWSCC) in steam generator (SG) divider plate assemblies fabricated of Alloy 600 and/or the associated Alloy 600 weld materials, even with proper primary water chemistry (EPRI TR-1014982). Cracks have been detected in the stub runner, adjacent to the tubesheet/stub runner weld and with depths of almost a third of the divider plate thickness. Therefore, the water chemistry program may not be effective in managing the aging effect of cracking due to PWSCC in SG divider plate assemblies. This is of particular concern for steam generators where the tube-tubesheet welds are considered structural welds and/or where the divider plate assembly contributes to the mechanical integrity of the tubesheet.

   Although these SG divider plate cracks may not have a significant safety impact in and of themselves, these cracks could impact adjacent items, such as the tubesheet and the channel head, if they propagate to the boundary with these items. For the tubesheet, PWSCC cracks in the divider plate could propagate to the tubesheet cladding with possible consequences to the integrity of the tube/tubesheet welds. For the channel head, the PWSCC cracks in the divider plate could propagate to the SG triple point and potentially affect the pressure boundary of the SG channel head.

   The existing program relies on control of reactor water chemistry to mitigate cracking due to PWSCC. The GALL Report recommends that a plant-specific AMP be evaluated, along with the primary water chemistry program, because the existing primary water chemistry program may not be capable of mitigating cracking due to PWSCC. Acceptance criteria are described in Branch Technical Position RLSB-1 (Appendix A.1 of this SRP-LR).

2. Cracking due to PWSCC could occur in steam generator nickel alloy tube-to-tubesheet welds exposed to reactor coolant. Unless the NRC has approved a redefinition of the pressure boundary in which the tube-to-tubesheet weld is no longer included, the effectiveness of the primary water chemistry program should be verified to ensure cracking is not occurring:

   • For plants with Alloy 600 steam generator tubes that have not been thermally treated and for which an alternate repair criteria such as C*, F* or W* has been permanently approved, the weld is no longer part of the pressure boundary and no plant specific aging management program is required;

- For plants with Alloy 600 steam generator tubes that have not been thermally treated and for which there is no permanently approved alternate repair criteria such as C*, F* or W*, a plant-specific AMP is required;
- For plants with Alloy 600TT steam generator tubes and for which an alternate repair criteria such as H* has been permanently approved, the weld is no longer part of the pressure boundary and no plant specific aging management program is required;
- For plants with Alloy 600TT steam generator tubes and for which there is no alternate repair criteria such as H* permanently approved, a plant-specific AMP is required;
- For plants with Alloy 690TT steam generator tubes with Alloy 690 tubesheet cladding, the water chemistry is sufficient, and no further action or plant-specific aging management program is required;
- For plants with Alloy 690TT steam generator tubes and with Alloy 600 tubesheet cladding, either a plant-specific program or a rationale for why such a program is not needed is required.

The existing program relies on control of reactor water chemistry to mitigate cracking due to PWSCC. The GALL Report recommends that a plant-specific AMP be evaluated, along with the primary water chemistry program, because the existing primary water chemistry program may not be capable of mitigating cracking due to PWSCC. Acceptance criteria are described in Branch Technical Position RLSB-1 (Appendix A.1 of this SRP-LR).

3.1.2.2.12   Cracking due to Fatigue

EPRI 1016596, *Materials Reliability Program: Pressurized Water Reactor Internals Inspection and Evaluation Guidelines* (MRP-227-Rev. 0) identifies cracking due to fatigue as an aging effect that can occur for the lower flange weld in the core support barrel assembly, fuel alignment plate in the upper internals assembly, and core support plate lower support structure in PWR internals designed by Combustion Engineering. The GALL Report recommends that inspection for cracking in this component be performed if acceptable fatigue life cannot be demonstrated by TLAA through the period of extended operation as defined in 10 CFR 54.3.

3.1.2.2.13   Cracking due to Stress Corrosion Cracking and Fatigue

Cracking due to stress corrosion cracking and fatigue could occur in nickel alloy control rod guide tube assemblies, guide tube support pins exposed to reactor coolant, and neutron flux. The GALL Report, AMR Item IV.B2.RP-355, recommends further evaluation of a plant-specific AMP to ensure this aging effect is adequately managed. Acceptance criteria are described in Branch Technical Position RLSB-1 (Appendix A.1 of this SRP-LR).

3.1.2.2.14   Loss of Material due to Wear

Loss of material due to wear could occur in nickel alloy control rod guide tube assemblies, guide tube support pins and in Zircaloy-4 incore instrumentation lower thimble tubes exposed to reactor coolant, and neutron flux. The GALL Report, AMR Items IV.B2.RP-356 and IV.B3.RP-357, recommends further evaluation of a plant-specific AMP to ensure this aging effect is adequately managed. Acceptance criteria are described in Branch Technical Position RLSB-1 (Appendix A.1 of this SRP-LR).

3.1.2.2.15 Quality Assurance for Aging Management of Nonsafety-Related Components

Acceptance criteria are described in Branch Technical Position IQMB-1 (Appendix A.2 of this SRP-LR).

3.1.2.3 AMR Results Not Consistent with or Not Addressed in the GALL Report

Acceptance criteria are described in Branch Technical Position RLSB-1 (Appendix A.1 of this SRP-LR).

3.1.2.4 Aging Management Programs

For those AMPs that will be used for aging management and are based on the program elements of an AMP in the GALL Report, the NRC reviewer performs an audit of aging management programs credited in the LRA to confirm consistency with the GALL AMPs identified in the GALL Report, Chapters X and XI.

If the applicant identifies an exception to any of the program elements of the cited GALL Report AMP, the LRA AMP should include a basis demonstrating how the criteria of 10 CFR 54.21(a)(3) would still be met. The reviewer should then confirm that the LRA AMP with all exceptions would satisfy the criteria of 10 CFR 54.21(a)(3). If, while reviewing the LRA AMP, the reviewer identifies a difference between the LRA AMP and the GALL Report AMP that should have been identified as an exception to the GALL Report AMP, the difference should be reviewed and properly dispositioned. The reviewer should document the disposition of all LRA-defined exceptions and staff-identified differences.

The LRA should identify any enhancements that are needed to permit an existing licensee AMP to be declared consistent with the GALL Report AMP to which the licensee AMP is compared. The reviewer is to confirm both that the enhancement, when implemented, would allow the existing licensee AMP to be consistent with the GALL Report AMP and that the applicant has a commitment in the FSAR Supplement to implement the enhancement prior to the period of extended operation. The reviewer should document the disposition of all enhancements.

If the applicant chooses to use a plant-specific program that is not a GALL AMP, the NRC reviewer should confirm that the plant-specific program satisfies the criteria of Branch Technical Position RLSB-1 (Appendix A.1.2.3 of this SRP-LR).

3.1.2.5 FSAR Supplement

The summary description of the programs and activities for managing the effects of aging for the period of extended operation in the FSAR Supplement should be sufficiently comprehensive, such that later changes can be controlled by 10 CFR 50.59. The description should contain information associated with the bases for determining that aging effects will be managed during the period of extended operation. The description should also contain any future aging management activities, including enhancements and commitments, to be completed before the period of extended operation. Table 3.0-1 of this SRP-LR provides examples of the type of information to be included in the FSAR Supplement. Table 3.1-2 lists the programs that are applicable for this SRP-LR subsection.

3.1.3 Review Procedures

For each area of review, the following review procedures are to be followed.

3.1.3.1 AMR Results Consistent with the GALL Report

The applicant may reference the GALL Report in its LRA, as appropriate, and demonstrate that the AMRs and AMPs at its facility are consistent with those reviewed and approved in the GALL Report. The reviewer should not conduct a re-review of the substance of the matters described in the GALL Report. If the applicant has provided the information necessary to adopt the finding of program acceptability as described and evaluated in the GALL Report, the reviewer should find acceptable the applicant's reference to the GALL Report in its LRA. In making this determination, the reviewer confirms that the applicant has provided a brief description of the system, components, materials, and environment. The reviewer also confirms that the applicant has stated that the applicable aging effects and industry and plant-specific operating experience have been reviewed by the applicant and are evaluated in the GALL Report.

Furthermore, the reviewer should confirm that the applicant has addressed operating experience identified after the issuance of the GALL Report. Performance of this review requires the reviewer to confirm that the applicant has identified those aging effects for the reactor vessel, internals, and reactor coolant system components that are contained in the GALL Report as applicable to its plant.

3.1.3.2 AMR Results for Which Further Evaluation is Recommended by the GALL Report

The basic review procedures defined in Subsection 3.1.3.1 need to be applied first for all of the AMRs and AMPs provided in this section. In addition, if the GALL Report AMR item to which the LRA AMR item is compared identifies that "further evaluation is recommended," then additional criteria apply as identified by the GALL Report for each of the following aging effect/aging mechanism combinations.

3.1.3.2.1 Cumulative Fatigue Damage

Fatigue is a TLAA as defined in 10 CFR 54.3. TLAAs are required to be evaluated in accordance with 10 CFR 54.21(c)(1). The staff reviews the evaluation of this TLAA separately following the guidance in Section 4.3 of this SRP-LR.

3.1.3.2.2 Loss of Material due to General, Pitting, and Crevice Corrosion

1. The GALL Report recommends an augmented program for the management of loss of material due to general, pitting, and crevice corrosion for steel PWR steam generator shell assembly exposed to secondary feedwater and steam. The existing program relies on control of water chemistry to mitigate corrosion and ISI to detect loss of material. The extent and schedule of the existing steam generator inspections are designed to ensure that flaws cannot attain a depth sufficient to threaten the integrity of the welds. However, according to NRC IN 90-04, the program may not be sufficient to detect pitting and crevice corrosion, if general and pitting corrosion of the shell is known to exist. Therefore, the GALL Report recommends augmented inspection to manage this aging effect. Furthermore, the GALL Report clarifies that this issue is limited to Westinghouse Model 44 and 51 Steam Generators where a high-stress region exists at the shell to

transition cone weld. Acceptance criteria are described in Branch Technical Position RLSB-1 (Appendix A.1 of this SRP-LR). Loss of material due to general, pitting, and crevice corrosion could also occur for the steel top head enclosure (without cladding) top head nozzles (vent, top head spray or RCIC, and spare) exposed to reactor coolant. The existing program relies on control of reactor water chemistry to mitigate corrosion. However, control of water chemistry does not preclude loss of material due to pitting and crevice corrosion at locations of stagnant flow conditions. Therefore, the effectiveness of the water chemistry control program should be verified to ensure that corrosion is not occurring. The reviewer verifies on a case-by-case basis that the applicant has proposed a program that will manage loss of material due to general, pitting and crevice corrosion by providing enhanced inspection and supplemental methods to detect loss of material and ensure that the component-intended function will be maintained during the period of extended operation.

2. The GALL Report recommends further evaluation of programs to manage the loss of material due to general, pitting, and crevice corrosion for the new transition cone closure weld generated in the steel PWR replacement recirculating steam generator transition cone shell exposed to secondary feedwater and steam. The existing program relies on control of reactor water chemistry to mitigate corrosion and on ISI to detect loss of material. The reviewer verifies on a case-by-case basis that the applicant has proposed an augmented program that will manage loss of material due to general, pitting, and crevice corrosion and ensure that the component-intended function will be maintained during the period of extended operation.

    The reviewer verifies that the applicant has described the surface condition and the resultant flow near the new transition cone closure weld (e.g., weld crown, ground flush, etc.) and how these parameters could affect the susceptibility of this weld to this aging effect, relative to that of the upper and lower transition welds. Based on this information, the reviewer verifies whether any additional aging management of the new transition weld is necessary. If additional aging management is necessary, the reviewer verifies whether the applicant has described an aging management program of the new transition cone closure weld (including examination frequency and technique) that will be effective in managing an aging effect, such as the loss of material due to general, pitting, and crevice corrosion during the period of extended operation for the new transition cone closure weld.

3.1.3.2.3 Loss of Fracture Toughness due to Neutron Irradiation Embrittlement

1. Neutron irradiation embrittlement is a TLAA as defined in 10 CFR 54.3. TLAAs are required to be evaluated in accordance with 10 CFR 54.21(c)(1). The staff reviews the evaluation of this TLAA following the guidance in Section 4.2 of this SRP-LR.

2. The GALL Report recommends further evaluation of the reactor vessel materials surveillance program for the period of extended operation. Neutron embrittlement of the reactor vessel is monitored by a reactor vessel materials surveillance program. Reactor vessel surveillance program is plant-specific, depending on matters such as the composition of limiting materials, availability of surveillance capsules, and projected fluence levels. In accordance with 10 CFR Part 50, Appendix H, an applicant must submit its proposed withdrawal schedule for approval prior to implementation. Untested capsules placed in storage must be maintained for future insertion. Thus, further staff evaluation is required for license renewal. The reviewer verifies on a case-by-case basis

that the applicant has proposed an adequate reactor vessel materials surveillance program for the period of extended operation. Specific recommendations for an acceptable AMP are provided in Chapter XI, Section M31 of the GALL Report.

3. Ductility – Reduction in Fracture Toughness for Babcock and Wilcox reactor internals is a TLAA as defined in 10 CFR 54.3. TLAAs are required to be evaluated in accordance with 10 CFR 54.21(c)(1). The staff reviews the evaluation of this TLAA following the guidance in Section 4.7 of this SRP-LR consistent with the action item documented in the staff's safety evaluation for MRP-227, Revision 0.

3.1.3.2.4 Cracking due to Stress Corrosion Cracking and Intergranular Stress Corrosion Cracking

1. The GALL Report recommends that a plant-specific AMP be evaluated to manage cracking due to SCC and IGSCC in stainless steel and nickel alloy BWR top head enclosure vessel flange leak detection lines. The reviewer reviews the applicant's proposed program on a case-by-case basis to ensure that an adequate program will be in place for the management of these aging effects.

2. The GALL Report recommends an augmented program to include temperature and radioactivity monitoring of the shell-side water and eddy current testing of tubes for the management of cracking due to SCC and IGSCC of the stainless steel BWR isolation condenser components. The existing program relies on control of reactor water chemistry to mitigate SCC and IGSCC and on ASME Section XI ISI to detect leakage. However, the existing program should be augmented to detect cracking due to SCC and IGSCC. The reviewer reviews the applicant's proposed program on a case-by-case basis to ensure that an adequate program will be in place for the management of these aging effects.

3.1.3.2.5   Crack Growth due to Cyclic Loading

The GALL Report recommends further evaluation of programs to manage crack growth due to cyclic loading in reactor vessel shell forgings clad with stainless steel using a high-heat-input welding process. Growth of intergranular separations (underclad cracks) in the heat affected zone under austenitic stainless steel cladding is a TLAA to be evaluated for the period of extended operation for all the SA-508-Cl-2 forgings where the cladding was deposited with a high heat input welding process. The methodology for evaluating the underclad flaw should be consistent with the current well-established flaw evaluation procedure and criterion in the ASME Section XI Code. The SRP-LR, Section 4.7, "Other Plant-Specific Time-Limited Aging Analysis," provides generic guidance for meeting the requirements of 10 CFR 54.21(c). The staff reviews the evaluation of this TLAA separately following the guidance in Section 4.7 of this SRP-LR.

3.1.3.2.6   Cracking due to Stress Corrosion Cracking

1. The GALL Report recommends that a plant-specific AMP be evaluated to manage cracking due to SCC in stainless steel PWR reactor vessel flange leak detection lines and bottom-mounted instrument guide tubes exposed to reactor coolant. The reviewer reviews the applicant's proposed program on a case-by-case basis to ensure that an adequate program will be in place for the management of these aging effects.

2. The GALL Report recommends that a plant-specific AMP be evaluated to manage cracking due to SCC in CASS PWR Class 1 reactor coolant system piping, piping components, and piping elements exposed to reactor coolant that do not meet the carbon and ferrite content guidelines of NUREG-0313. The reviewer reviews the applicant's proposed program on a case-by-case basis to ensure that an adequate program will be in place for the management of these aging effects.

3.1.3.2.7   Cracking due to Cyclic Loading

The GALL Report recommends an augmented program for the management of cracking due to cyclic loading in steel and stainless steel BWR isolation condenser components. The existing program relies on ASME Section XI ISI for detection. However, the inspection requirements should be augmented to detect cracking due to cyclic loading. An augmented program to include temperature and radioactivity monitoring of the shell-side water and eddy current testing of tubes is recommended to ensure that the component's intended function will be maintained during the period of extended operation. The reviewer verifies on a case-by-case basis that the applicant has proposed an augmented program that will detect cracking and ensure that the component-intended function will be maintained during the period of extended operation.

3.1.3.2.8   Loss of Material due to Erosion

The GALL Report recommends further evaluation of a plant-specific AMP for the management of loss of material due to erosion of steel steam generator feedwater impingement plates and supports exposed to secondary feedwater. The reviewer reviews the applicant's proposed program on a case-by-case basis to ensure that an adequate program will be in place for the management of these aging effects.

3.1.3.2.9   Cracking due to Stress Corrosion Cracking and Irradiation-Assisted Stress Corrosion Cracking

The GALL Report recommends further evaluation of cracking due to SCC and IASCC for inaccessible locations for Primary and Expansion PWR reactor vessel internal components if aging effects are identified for these components in accessible locations. The reviewer reviews the applicant's proposed program on a case-by-case basis to ensure that an adequate program will be in place for the management of these aging effects consistent with the action item documented in the staff's safety evaluation for MRP-227, Revision 0..

3.1.3.2.10   Loss of Fracture Toughness due to Neutron Irradiation Embrittlement; Change in Dimension due to Void Swelling; Loss of Preload due to Stress Relaxation; or Loss of Material due to Wear

The GALL Report recommends further evaluation of loss of fracture toughness due to neutron irradiation embrittlement, change in dimension due to void swelling, loss of preload due to stress relaxation, or loss of material due to wear for inaccessible locations for Primary and Expansion PWR reactor vessel internal components, if aging effects are identified for these components in accessible locations. The reviewer reviews the applicant's proposed program on a case-by-case basis to ensure that an adequate program will be in place for the management of these aging effects consistent with the action item documented in the staff's safety evaluation for MRP-227, Revision 0.

3.1.3.2.11   Cracking due to Primary Water Stress Corrosion Cracking

1. The GALL Report recommends that a plant-specific AMP be evaluated, along with the primary water chemistry program, to manage cracking due to PWSCC in nickel alloy divider plate assemblies made of Alloy 600 and/or the associated Alloy 600 weld materials for steam generators with a similar design to that of Westinghouse Model 51. The effectiveness of the chemistry control program should be verified to ensure that cracking due to PWSCC is not occurring. The reviewer verifies the materials of construction of the applicant's SG divider plate assembly. If these materials are susceptible to cracking, the reviewer verifies that the applicant has evaluated the potential for cracking in the divider plate to propagate into other components (e.g., tubesheet cladding). If propagation into these other components is possible, the reviewer verifies if the applicant has described an inspection program (examination technique and frequency) for ensuring that no cracks are propagating into other items (e.g., tube sheet and channel head) that could challenge the integrity of those items. The reviewer reviews the applicant's proposed program on a case-by-case basis to ensure that an adequate program will be in place for the management of this aging effect.

2. The GALL Report recommends that a plant-specific AMP be evaluated, along with the primary water chemistry program, to manage cracking due to PWSCC in recirculating steam generator nickel alloy tube-to-tubesheet welds exposed to reactor coolant. The effectiveness of the primary water chemistry program should be verified to ensure that cracking due to PWSCC is not occurring. The reviewer verifies the combination of materials of construction of the steam generator tubes and tubesheet cladding and the classification of the tube-to-tubesheet weld. If this combination requires further evaluation, the reviewer reviews the applicant's proposed program on a case-by-case basis to ensure that an adequate program will be in place for the management of this aging effect.

3.1.3.2.12   Cracking due to Fatigue

The GALL Report recommends further evaluation of cracking due to fatigue in the lower flange weld in the core support barrel assembly, fuel alignment plate in the upper internals assembly, and core support plate in the lower support structure in PWR internals designed by Combustion Engineering. The reviewer determines whether a TLAA has been performed for each component, consistent with the action item documented in the staff's safety evaluation for MRP-227, Revision 0. If a TLAA has not been performed, the reviewer determines whether the applicant has performed an evaluation to identify the potential location and extent of fatigue cracking for each component consistent with the action item documented in the staff's safety evaluation for MRP-227, Revision 0.

3.1.3.2.13   Cracking due to Stress Corrosion Cracking and Fatigue

The GALL Report recommends further evaluation of cracking due to stress corrosion cracking and fatigue in the nickel alloy control rod guide tube assemblies, guide tube support pins exposed to reactor coolant, and neutron flux. The reviewer reviews the applicant's proposed program on a case-by-case basis to ensure that an adequate program will be in place for the management of these aging effects consistent with the action item documented in the staff's safety evaluation for MRP-227, Revision 0.

3.1.3.2.14 Loss of Material due to Wear

The GALL Report recommends further evaluation of loss of material due to wear in nickel alloy control rod guide tube assemblies, guide tube support pins and in Zircaloy-4 incore instrumentation lower thimble tubes exposed to reactor coolant, and neutron flux. The reviewer reviews the applicant's proposed program on a case-by-case basis to ensure that an adequate program will be in place for the management of these aging effects consistent with the action item documented in the staff's safety evaluation for MRP-227, Revision 0.

3.1.3.2.15 Quality Assurance for Aging Management of Nonsafety-Related Components

The applicant's AMPs for license renewal should contain the elements of corrective actions, the confirmation process, and administrative controls. Safety-related components are covered by 10 CFR Part 50, Appendix B, which is adequate to address these program elements. However, Appendix B does not apply to nonsafety-related components that are subject to an aging management review for license renewal. Nevertheless, the applicant has the option to expand the scope of its 10 CFR Part 50, Appendix B program to include these components and address the associated program elements. If the applicant chooses this option, the reviewer verifies that the applicant has documented such a commitment in the FSAR Supplement. If the applicant chooses alternative means, the branch responsible for quality assurance should be requested to review the applicant's proposal on a case-by-case basis.

3.1.3.3 AMR Results Not Consistent with or Not Addressed in the GALL Report

The reviewer should confirm that the applicant, in its LRA, has identified applicable aging effects, listed the appropriate combination of materials and environments, and AMPs that will adequately manage the aging effects. The AMP credited by the applicant could be an AMP that is described and evaluated in the GALL Report or a plant-specific program. Review procedures are described in Branch Technical Position RSLB-1 (Appendix A.1 of this SRP-LR).

3.1.3.4 Aging Management Programs

The reviewer confirms that the applicant has identified the appropriate AMPs as described and evaluated in the GALL Report. If the applicant commits to an enhancement to make its LRA AMP consistent with a GALL Report AMP, then the reviewer is to confirm that this enhancement, when implemented, will make the LRA AMP consistent with the GALL Report AMP. If the applicant identifies, in the LRA AMP, an exception to any of the program elements of the GALL Report AMP, the reviewer is to confirm that the LRA AMP with the exception will satisfy the criteria of 10 CFR 54.21(a)(3). If the reviewer identifies a difference, not identified by the LRA, between the LRA AMP and the GALL Report AMP, with which the LRA claims to be consistent, the reviewer should confirm that the LRA AMP with this difference satisfies 10 CFR 54.21(a)(3). The reviewer should document the basis for accepting enhancements, exceptions, or differences. The AMPs evaluated in the GALL Report pertinent to the reactor vessel, internals, and reactor coolant system are summarized in Table 3.1-1 of this SRP-LR. The "Rev 2 Item" (for 2010) and "Rev1 Item" (for 2005 counterpart) columns identify the AMR item numbers in the GALL Report, Chapter IV, presenting detailed information summarized by this row.

3.1.3.5   FSAR Supplement

The reviewer confirms that the applicant has provided in its FSAR supplement information equivalent to that in Table 3.0-1 for aging management of the reactor vessel, internals, and reactor coolant system. Table 3.1-2 lists the AMPs that are applicable for this SRP-LR subsection. The reviewer also confirms that the applicant has provided information for Subsection 3.1.3.3, "AMR Results Not Consistent with or Not Addressed in the GALL Report," equivalent to that in Table 3.0-1.

The staff expects to impose a license condition on any renewed license to require the applicant to update its FSAR to include this FSAR Supplement at the next update required pursuant to 10 CFR 50.71(e)(4). As part of the license conditions until the FSAR update is complete, the applicant may make changes to the programs described in its FSAR Supplement without prior NRC approval, provided that the applicant evaluates each such change and finds it acceptable pursuant to the criteria set forth in 10 CFR 50.59. If the applicant updates the FSAR to include the final FSAR supplement before the license is renewed, no condition will be necessary.

As noted in Table 3.0-1, an applicant need not incorporate the implementation schedule into its FSAR. However, the reviewer should confirm that the applicant has identified and committed in the license renewal application to any future aging management activities, including enhancements and commitments to be completed before entering the period of extended operation. The staff expects to impose a license condition on any renewed license to ensure that the applicant will complete these activities no later than the committed date.

3.1.4   Evaluation Findings

If the reviewer determines that the applicant has provided information sufficient to satisfy the provisions of this section, then an evaluation finding similar to the following text should be included in the staff's safety evaluation report:

> On the basis of its review, as discussed above, the staff concludes that the applicant has demonstrated that the aging effects associated with the reactor vessel, internals, and reactor coolant system components will be adequately managed so that the intended functions will be maintained consistent with the CLB for the period of extended operation, as required by 10 CFR 54.21(a)(3).
>
> The staff also reviewed the applicable FSAR Supplement program summaries and concludes that they adequately describe the AMPs credited for managing aging of the reactor vessel, internals and reactor coolant system, as required by 10 CFR 54.21(d).

3.1.5   Implementation

Except in those cases in which the applicant proposes an acceptable alternative method for complying with specified portions of the NRC's regulations, the method described herein will be used by the staff in its evaluation of conformance with NRC regulations.

3.1.6   References

1. NUREG-0800, "Standard Review Plan for the Review of Safety Analysis Reports for Nuclear Power Plants," U.S. Nuclear Regulatory Commission, March 2007.

2. NUREG-1801, "Generic Aging Lessons Learned (GALL) Report," U.S. Nuclear Regulatory Commission, Revision 2, 2010.

3. NEI 97-06, "Steam Generator Program Guidelines," Revision 2, Nuclear Energy Institute, September 2005.

4. NEI 95-10, "Industry Guideline for Implementing the Requirements of 10 CFR Part 54 – The License Renewal Rule," Nuclear Energy Institute, Revision 6.

5. NRC Information Notice 90-04, "Cracking of the Upper Shell-to-Transition Cone Girth Welds in Steam Generators," U.S. Nuclear Regulatory Commission, January 26, 1990.

6. NUREG-0313, Rev. 2, "Technical Report on Material Selection and Processing Guidelines for BWR Coolant Pressure Boundary Piping," U.S. Nuclear Regulatory Commission, January 1988.

7. EPRI 1013706, "PWR Steam Generator Examination Guidelines, Rev. 7," Electric Power Research Institute, October 2007.

8. NRC Regulatory Guide 1.121, "Bases for Plugging Degraded PWR Steam Generator Tubes (for Comment)," U.S. Nuclear Regulatory Commission, May 1976.

9. NRC Generic Letter 95-05, "Voltage-Based Repair Criteria for Westinghouse Steam Generator Tubes Affected by Outside Diameter Stress Corrosion Cracking," U.S. Nuclear Regulatory Commission, August 3, 1995.

10. NRC Information Notice 90-10, "Primary Water Stress Corrosion Cracking (PWSCC) of Inconel 600," U.S. Nuclear Regulatory Commission, February 23, 1990.

11. NRC Information Notice 90-30, "Ultrasonic Inspection Techniques for Dissimilar Metal Welds," U.S. Nuclear Regulatory Commission, May 1, 1990.

12. NRC Generic Letter 89-08, "Erosion/Corrosion-Induced Pipe Wall Thinning," May 2, 1989.

13. NSAC-202L-R3, "Recommendations for an Effective Flow-accelerated Corrosion Program," Electric Power Research Institute, April 1999.

14. NRC Information Notice 96-11, "Ingress of Demineralizer Resins Increase Potential for Stress Corrosion Cracking of Control Rod Drive Mechanism Penetrations," February 14, 1996.

15. BWRVIP-190 (EPRI 1016579), *BWR Vessel and Internals Project: BWR Water Chemistry Guidelines-2008 Revision*, Electric Power Research Institute, Palo Alto, CA, October 2008

16. EPRI NP-5769, "Degradation and Failure of Bolting in Nuclear Power Plants," Volumes 1 and 2, Electric Power Research Institute, Palo Alto, CA, April 1988.

17. EPRI 1014986, "PWR Primary Water Chemistry Guidelines," Revision 6, Volumes 1 and 2, Electric Power Research Institute, Palo Alto, CA, December 2007.

18. NRC Generic Letter 88-01, "NRC Position on IGSCC in BWR Austenitic Stainless Steel Piping," January 25, 1988.

19. NRC Generic Letter 97-01, "Degradation of Control Rod Drive Mechanism Nozzle and Other Vessel Closure Head Penetrations," April 1, 1997.

20. NRC Information Notice 97-46, "Unisolable Crack in High-Pressure Injection Piping," July 9, 1997.

21. NRC Regulatory Guide 1.99, Rev. 2, "Radiation Embrittlement of Reactor Vessel Materials," May 1988.

22. NUREG-0619, "BWR Feedwater Nozzle and Control Rod Drive Return Line Nozzle Cracking," U.S. Nuclear Regulatory Commission, November 1980.

23. NUREG-1339, "Resolution of Generic Safety Issue 29: Bolting Degradation or Failure in Nuclear Power Plants," Richard E. Johnson, U.S. Nuclear Regulatory Commission, June 1990.

24. EPRI TR-104213, "Bolted Joint Maintenance & Application Guide, Electric Power Research Institute," Palo Alto, CA, December 1995.

25. NEI letter dated Dec. 11, 1998, Dave Modeen to Gus Lainas, "Responses to NRC Requests for Additional Information (RAIs) on GL 97-01."

26. EPRI 1016555, "PWR Secondary Water Chemistry Guidelines–Revision 7," Electric Power Research Institute, Palo Alto, CA, February 2009.

27. NRC Information Notice 91-19, "Steam Generator Feedwater Distribution Piping Damage," March 12, 1991.

28. EPRI 1016596, "Materials Reliability Program: Pressurized Water Reactor Internals Inspection and Evaluation Guidelines," MRP-227, Revision 0, Electric Power Research Institute, Palo Alto, CA, December 2008.

29. Topical Report, "Demonstration of the Management of Aging Effects for the Reactor Vessel Internals," BAW-2248A, March 2000.

30. EPRI TR-1014982, "Divider Plate Cracking in Steam Generators - Results of Phase 1: Analysis of Primary Water Stress Corrosion Cracking and Mechanical Fatigue in the Alloy 600 Stub Runner to Divider Plate Weld Material," Electric Power Research Institute, Palo Alto, CA, June 2007.

ID	Type	Component	Aging Effect/Mechanism	Aging Management Programs	Further Evaluation Recommended	Rev2 Item	Rev1 Item
1	BWR/ PWR	High strength, low-alloy steel top head closure stud assembly exposed to air with potential for reactor coolant leakage	Cumulative fatigue damage due to fatigue	Fatigue is a TLAA evaluated for the period of extended operation (See SRP, Sec 4.3 "Metal Fatigue," for acceptable methods to comply with 10 CFR 54.21(c)(1))	Yes, TLAA (See subsection 3.1.2.2.1)	IV.A1.RP-201 IV.A2.RP-54	N/A IV.A2-4(R-73)
2	PWR	Nickel alloy tubes and sleeves exposed to reactor coolant and secondary feedwater/steam	Cumulative fatigue damage due to fatigue	Fatigue is a TLAA evaluated for the period of extended operation (See SRP, Sec 4.3 "Metal Fatigue," for acceptable methods to comply with 10 CFR 54.21(c)(1))	Yes, TLAA (See subsection 3.1.2.2.1)	IV.D1.R-46 IV.D2.R-46	IV.D1-21(R-46) IV.D2-15(R-46)
3	BWR/ PWR	Stainless steel or nickel alloy reactor vessel internal components exposed to reactor coolant and neutron flux	Cumulative fatigue damage due to fatigue	Fatigue is a TLAA evaluated for the period of extended operation (See SRP, Sec 4.3 "Metal Fatigue," for acceptable methods to comply with 10 CFR 54.21(c)(1))	Yes, TLAA (See subsection 3.1.2.2.1)	IV.B1.R-53 IV.B2.RP-303 IV.B3.RP-339 IV.B4.R-53 IV.B3.RP-389 IV.B3.RP-390 IV.B3.RP-391	IV.B1-14(R-53) IV.B2-31(R-53) IV.B3-24(R-53) IV.B4-37(R-53) N/A N/A N/A
4	BWR/ PWR	Steel pressure vessel support skirt and attachment welds	Cumulative fatigue damage due to fatigue	Fatigue is a TLAA evaluated for the period of extended operation (See SRP, Sec 4.3 "Metal Fatigue," for acceptable methods to comply with 10 CFR 54.21(c)(1))	Yes, TLAA (See subsection 3.1.2.2.1)	IV.A1.R-70 IV.A2.R-70	IV.A1-6(R-70) IV.A2-20(R-70)

Table 3.1-1 Summary of Aging Management Programs for Reactor Vessel, Internals, and Reactor Coolant System Evaluated in Chapter IV of the GALL Report

ID	Type	Component	Aging Effect/Mechanism	Aging Management Programs	Further Evaluation Recommended	Rev2 Item	Rev1 Item
5	PWR	Steel, stainless steel, or steel (with stainless steel or nickel alloy cladding) steam generator components, pressurizer relief tank components or piping components or bolting	Cumulative fatigue damage due to fatigue	Fatigue is a TLAA evaluated for the period of extended operation (See SRP, Sec 4.3 "Metal Fatigue," for acceptable methods to comply with 10 CFR 54.21(c)(1))	Yes, TLAA (See subsection 3.1.2.1)	IV.C2.R-13 IV.C2.R-18 IV.D1.R-33 IV.D2.R-33	IV.C2-23(R-13) IV.C2-10(R-18) IV.D1-11(R-33) IV.D2-10(R-33)
6	BWR	Steel (with or without nickel-alloy or stainless steel cladding), or stainless steel; or nickel alloy reactor coolant pressure boundary components: piping, piping components, and piping elements exposed to reactor coolant	Cumulative fatigue damage due to fatigue	Fatigue is a TLAA evaluated for the period of extended operation, and for Class 1 components environmental effects on fatigue are to be addressed. (See SRP, Sec 4.3 "Metal Fatigue," for acceptable methods to comply with 10 CFR 54.21(c)(1))	Yes, TLAA (See subsection 3.1.2.1)	IV.C1.R-220	IV.C1-15(R-220)
7	BWR	Steel (with or without nickel-alloy or stainless steel cladding), or stainless steel; or nickel alloy reactor vessel components: flanges; nozzles; penetrations; safe ends; thermal sleeves;vessel shells, heads and welds exposed to reactor coolant	Cumulative fatigue damage due to fatigue	Fatigue is a TLAA evaluated for the period of extended operation, and for Class 1 components environmental effects on fatigue are to be addressed. (See SRP, Sec 4.3 "Metal Fatigue," for acceptable methods to comply with 10 CFR 54.21(c)(1))	Yes, TLAA (See subsection 3.1.2.1)	IV.A1.R-04	IV.A1-7(R-04)

Table 3.1-1 Summary of Aging Management Programs for Reactor Vessel, Internals, and Reactor Coolant System Evaluated in Chapter IV of the GALL Report

ID	Type	Component	Aging Effect/Mechanism	Aging Management Programs	Further Evaluation Recommended	Rev2 Item	Rev1 Item
8	PWR	Steel (with or without nickel-alloy or stainless steel cladding), or nickel alloy steam generator components exposed to reactor coolant	Cumulative fatigue damage due to fatigue	Fatigue is a TLAA evaluated for the period of extended operation, and for Class 1 components environmental effects on fatigue are to be addressed. (See SRP, Sec 4.3 "Metal Fatigue," for acceptable methods to comply with 10 CFR 54.21(c)(1)	Yes, TLAA (See subsection 3.1.2.2.1)	IV.D1.R-221 IV.D2.R-222	IV.D1-8(R-221) IV.D2-3(R-222)
9	PWR	Steel (with or without nickel-alloy or stainless steel cladding), stainless steel; nickel alloy RCPB piping; flanges; nozzles & safe ends; pressurizer shell heads & welds; heater sheaths & sleeves; penetrations; thermal sleeves exposed to reactor coolant	Cumulative fatigue damage due to fatigue	Fatigue is a TLAA evaluated for the period of extended operation, and for Class 1 components environmental effects on fatigue are to be addressed. (See SRP, Sec 4.3 "Metal Fatigue," for acceptable methods to comply with 10 CFR 54.21(c)(1)	Yes, TLAA (See subsection 3.1.2.2.1)	IV.C2.R-223	IV.C2-25(R-223)
10	PWR	Steel (with or without nickel-alloy or stainless steel cladding), stainless steel; nickel alloy reactor vessel flanges; nozzles; penetrations; pressure housings; safe ends; thermal sleeves; vessel shells, heads and welds exposed to reactor coolant	Cumulative fatigue damage due to fatigue	Fatigue is a TLAA evaluated for the period of extended operation, and for Class 1 components environmental effects on fatigue are to be addressed. (See SRP, Sec 4.3 "Metal Fatigue," for acceptable methods to comply with 10 CFR 54.21(c)(1)	Yes, TLAA (See subsection 3.1.2.2.1)	IV.A2.R-219	IV.A2-21(R-219)

Table 3.1-1 Summary of Aging Management Programs for Reactor Vessel, Internals, and Reactor Coolant System Evaluated in Chapter IV of the GALL Report

ID	Type	Component	Aging Effect/Mechanism	Aging Management Programs	Further Evaluation Recommended	Rev2 Item	Rev1 Item
11	BWR	Steel or stainless steel pump and valve closure bolting exposed to high temperatures and thermal cycles	Cumulative fatigue damage due to fatigue	Fatigue is a TLAA evaluated for the period of extended operation; check ASME Code limits for allowable cycles (less than 7000 cycles) of thermal stress range. (SRP Sec 4.3 "Metal Fatigue," for acceptable methods to comply with 10 CFR 54.21(c)(1))	Yes, TLAA (See subsection 3.1.2.2.1)	IV.C1.RP-44	IV.C1-11(R-28)
12	PWR	Steel steam generator components: upper and lower shells, transition cone; new transition cone closure weld exposed to secondary feedwater or steam	Loss of material due to general, pitting, and crevice corrosion	Chapter XI.M1, "ASME Section XI Inservice Inspection, Subsections IWB, IWC, and IWD," and Chapter XI.M2, "Water Chemistry," and, for Westinghouse Model 44 and 51 S/G, if corrosion of the shell is found, additional inspection procedures are developed	Yes, detection of aging effects is to be evaluated (See subsection 3.1.2.2.2.1 and 3.1.2.2.2.2)	IV.D1.RP-368	IV.D1-12(R-34)
13	BWR/ PWR	Steel (with or without stainless steel cladding) reactor vessel beltline shell, nozzles, and welds exposed to reactor coolant and neutron flux	Loss of fracture toughness due to neutron irradiation embrittlement	TLAA is to be evaluated in accordance with Appendix G of 10 CFR Part 50 and RG 1.99. The applicant may choose to demonstrate that the materials of the nozzles are not controlling for the TLAA evaluations	Yes, TLAA (See subsection 3.1.2.3.1)	IV.A1.R-62 IV.A1.R-67 IV.A2.R-81 IV.A2.R-84	IV.A1-13(R-62) IV.A1-4(R-67) IV.A2-16(R-81) IV.A2-23(R-84)

Table 3.1-1 Summary of Aging Management Programs for Reactor Vessel, Internals, and Reactor Coolant System Evaluated in Chapter IV of the GALL Report

ID	Type	Component	Aging Effect/Mechanism	Aging Management Programs	Further Evaluation Recommended	Rev2 Item	Rev1 Item
14	BWR/ PWR	Steel (with or without cladding) reactor vessel beltline shell, nozzles, and welds; safety injection nozzles	Loss of fracture toughness due to neutron irradiation embrittlement	Chapter XI.M31, "Reactor Vessel Surveillance"	Yes, plant specific or integrated surveillance program (See subsection 3.1.2.2.3.2)	IV.A1.RP-227 IV.A2.RP-228 IV.A2.RP-229	IV.A1-14(R-63) IV.A2-17(R-82) IV.A2-24(R-86)
15	PWR	Stainless steel and nickel alloy reactor vessel internal components exposed to reactor coolant and neutron flux	Reduction in ductility and fracture toughness due to neutron irradiation	Ductility - Reduction in Fracture Toughness is a TLAA to be evaluated for the period of extended operation. See the SRP, Section 4.7, "Other Plant-Specific TLAAs," for acceptable methods for meeting the requirements of 10 CFR 54.21(c)(1).	Yes, TLAA (See subsection 3.1.2.2.3.3)	IV.B4.RP-376	N/A
16	BWR	Stainless steel and nickel alloy top head enclosure vessel flange leak detection line	Cracking due to stress corrosion cracking, intergranular stress corrosion cracking	A plant-specific aging management program is to be evaluated because existing programs may not be capable of mitigating or detecting crack initiation and growth due to SCC in the vessel flange leak detection line	Yes, plant-specific (See subsection 3.1.2.2.4.1)	IV.A1.R-61	IV.A1-10(R-61)
17	BWR	Stainless steel isolation condenser components exposed to reactor coolant	Cracking due to stress corrosion cracking, intergranular stress corrosion cracking	Chapter XI.M1, "ASME Section XI Inservice Inspection, Subsections IWB, IWC, and IWD" for Class 1 components, and Chapter XI.M2, "Water Chemistry" for BWR water, and a plant-specific verification program	Yes, detection of aging effects is to be evaluated (See subsection 3.1.2.2.4.2)	IV.C1.R-15	IV.C1-4(R-15)

Table 3.1-1 Summary of Aging Management Programs for Reactor Vessel, Internals, and Reactor Coolant System Evaluated in Chapter IV of the GALL Report

ID	Type	Component	Aging Effect/Mechanism	Aging Management Programs	Further Evaluation Recommended	Rev2 Item	Rev1 Item
18	PWR	Reactor vessel shell fabricated of SA508-Cl 2 forgings clad with stainless steel using a high-heat-input welding process exposed to reactor coolant	Crack growth due to cyclic loading	Growth of intergranular separations is a TLAA evaluated for the period of extended operation. The Standard Review Plan, Section 4.7, "Other Plant-Specific Time-Limited Aging Analysis," provides guidance for meeting the requirements of 10 CFR 54.21(c).	Yes, TLAA (See subsection 3.1.2.5)	IV.A2.R-85	IV.A2-22(R-85)
19	PWR	Stainless steel reactor vessel closure head flange leak detection line and bottom-mounted instrument guide tubes (external to reactor vessel)	Cracking due to stress corrosion cracking	A plant-specific aging management program is to be evaluated	Yes, plant-specific (See subsection 3.1.2.6.1)	IV.A2.R-74 IV.A2.RP-154	IV.A2-5(R-74) IV.A2-1(RP-13)
20	PWR	Cast austenitic stainless steel Class 1 piping, piping components, and piping elements exposed to reactor coolant	Cracking due to stress corrosion cracking	Chapter XI.M2, "Water Chemistry" and, for CASS components that do not meet the NUREG-0313 guidelines, a plant specific aging management program	Yes, plant-specific (See subsection 3.1.2.6.2)	IV.C2.R-05	IV.C2-3(R-05)
21	BWR	Steel and stainless steel isolation condenser components exposed to reactor coolant	Cracking due to cyclic loading	Chapter XI.M1, "ASME Section XI Inservice Inspection, Subsections IWB, IWC, and IWD" for Class 1 components The ISI program is to be augmented by a plant-specific verification program	Yes, detection of aging effects is to be evaluated (See subsection 3.1.2.7)	IV.C1.R-225	IV.C1-5(R-225)

Table 3.1-1 Summary of Aging Management Programs for Reactor Vessel, Internals, and Reactor Coolant System Evaluated in Chapter IV of the GALL Report

ID	Type	Component	Aging Effect/Mechanism	Aging Management Programs	Further Evaluation Recommended	Rev2 Item	Rev1 Item
22	PWR	Steel steam generator feedwater impingement plate and support exposed to secondary feedwater	Loss of material due to erosion	A plant-specific aging management program is to be evaluated	Yes, plant-specific (See subsection 3.1.2.2.8)	IV.D1.R-39	IV.D1-13(R-39)
23	PWR	Stainless steel or nickel alloy PWR reactor vessel internal components (inaccessible locations) exposed to reactor coolant and neutron flux	Cracking due to stress corrosion cracking, and irradiation-assisted stress corrosion cracking	Chapter XI.M16A, "PWR Vessel Internals," and Chapter XI.M2, "Water Chemistry"	Yes, if accessible Primary, Expansion or Existing program components indicate aging effects that need management (See subsection 3.1.2.2.9)	IV.B2.RP-268 IV.B3.RP-309 IV.B4.RP-238	N/A N/A N/A
24	PWR	Stainless steel or nickel alloy PWR reactor vessel internal components (inaccessible locations) exposed to reactor coolant and neutron flux	Loss of fracture toughness due to neutron irradiation embrittlement; or changes in dimension due to void swelling; or loss of preload due to thermal and irradiation enhanced stress relaxation; or loss of material due to wear	Chapter XI.M16A, "PWR Vessel Internals"	Yes, if accessible Primary, Expansion or Existing program components indicate aging effects that need management (See subsection 3.1.2.2.10)	IV.B2.RP-269 IV.B3.RP-311 IV.B4.RP-239	N/A N/A N/A
25	PWR	Steel (with nickel-alloy cladding) or nickel alloy steam generator primary side components: divider plate and tube-to-tube sheet welds exposed to reactor coolant	Cracking due to primary water stress corrosion cracking	Chapter XI.M2, "Water Chemistry"	Yes, plant-specific (See subsections 3.1.2.2.11.1 and 3.1.2.2.11.2)	IV.D1.RP-367 IV.D1.RP-385 IV.D2.RP-185	IV.D1-6(RP-21) N/A IV.D2-4(R-35)

Table 3.1-1 Summary of Aging Management Programs for Reactor Vessel, Internals, and Reactor Coolant System Evaluated in Chapter IV of the GALL Report

ID	Type	Component	Aging Effect/Mechanism	Aging Management Programs	Further Evaluation Recommended	Rev2 Item	Rev1 Item
26	PWR	Stainless steel Combustion Engineering core support barrel assembly: lower flange weld exposed to reactor coolant and neutron flux; Upper internals assembly: fuel alignment plate (applicable to plants with core shrouds assembled with full height shroud plates) exposed to reactor coolant and neutron flux; Lower support structure: core support plate (applicable to plants with a core support plate) exposed to reactor coolant and neutron flux	Cracking due to fatigue	Chapter XI.M16A, "PWR Vessel Internals," and Chapter XI.M2, "Water Chemistry," if fatigue life cannot be confirmed by TLAA	Yes, evaluate to determine the potential locations and extent of fatigue cracking (See subsection 3.1.2.2.12)	IV.B3.RP-333 IV.B3.RP-338 IV.B3.RP-343	N/A
27	PWR	Nickel alloy Westinghouse control rod guide tube assemblies, guide tube support pins exposed to reactor coolant and neutron flux	Cracking due to stress corrosion cracking and fatigue	A plant-specific aging management program is to be evaluated	Yes, plant-specific (See subsection 3.1.2.2.13)	IV.B2.RP-355	N/A

Table 3.1-1 Summary of Aging Management Programs for Reactor Vessel, Internals, and Reactor Coolant System Evaluated in Chapter IV of the GALL Report

ID	Type	Component	Aging Effect/Mechanism	Aging Management Programs	Further Evaluation Recommended	Rev2 Item	Rev1 Item
28	PWR	Nickel alloy Westinghouse control rod guide tube assemblies, guide tube support pins, and Zircaloy-4 Combustion Engineering incore instrumentation thimble tubes exposed to reactor coolant and neutron flux	Loss of material due to wear	A plant-specific aging management program is to be evaluated	Yes, plant-specific (See subsection 3.1.2.2.14)	IV.B2.RP-356 IV.B3.RP-357	N/A N/A
29	BWR	Nickel alloy core shroud and core plate access hole cover (welded covers) exposed to reactor coolant	Cracking due to stress corrosion cracking, intergranular stress corrosion cracking, irradiation-assisted stress corrosion cracking	Chapter XI.M1, "ASME Section XI Inservice Inspection, Subsections IWB, IWC, and IWD," and Chapter XI.M2, "Water Chemistry," and for BWRs with a crevice in the access hole covers, augmented inspection using UT or other acceptable techniques	No	IV.B1.R-94	IV.B1-5(R-94)
30	BWR	Stainless steel or nickel alloy penetration: drain line exposed to reactor coolant	Cracking due to stress corrosion cracking, intergranular stress corrosion cracking, cyclic loading	Chapter XI.M1, "ASME Section XI Inservice Inspection, Subsections IWB, IWC, and IWD," and Chapter XI.M2, "Water Chemistry"	No	IV.A1.RP-371	IV.A1-5(R-69)
31	BWR	Steel and stainless steel isolation condenser components exposed to reactor coolant	Loss of material due to general (steel only), pitting, and crevice corrosion	Chapter XI.M1, "ASME Section XI Inservice Inspection, Subsections IWB, IWC, and IWD," and Chapter XI.M2, "Water Chemistry"	No	IV.C1.RP-39	IV.C1-6(R-16)

ID	Type	Component	Aging Effect/Mechanism	Aging Management Programs	Further Evaluation Recommended	Rev2 Item	Rev1 Item
32	PWR	Stainless steel, nickel alloy, or CASS reactor vessel internals, core support structure, exposed to reactor coolant and neutron flux	Cracking, or loss of material due to wear	Chapter XI.M1, "ASME Section XI Inservice Inspection, Subsections IWB, IWC, and IWD"	No	IV.B2.RP-382 IV.B3.RP-382 IV.B4.RP-382	IV.B2-26(R-142) IV.B3-22(R-170) IV.B4-42(R-179)
33	PWR	Stainless steel, steel with stainless steel cladding Class 1 reactor coolant pressure boundary components exposed to reactor coolant	Cracking due to stress corrosion cracking	Chapter XI.M1, "ASME Section XI Inservice Inspection, Subsections IWB, IWC, and IWD" for ASME components, and Chapter XI.M2, "Water Chemistry"	No	IV.C2.R-09 IV.C2.R-217 IV.C2.R-30 IV.C2.RP-344 IV.D1.RP-232	IV.C2-5(R-09) IV.C2-20(R-217) IV.C2-27(R-30) IV.C2-2(R-07) IV.D1-1(R-07)
34	PWR	Stainless steel, steel with stainless steel cladding pressurizer relief tank (tank shell and heads, flanges, nozzles) exposed to treated borated water >60°C (>140°F)	Cracking due to stress corrosion cracking	Chapter XI.M1, "ASME Section XI Inservice Inspection, Subsections IWB, IWC, and IWD" for ASME components, and Chapter XI.M2, "Water Chemistry"	No	IV.C2.RP-231	IV.C2-22(R-14)
35	PWR	Stainless steel, steel with stainless steel cladding reactor coolant system cold leg, hot leg, surge line, and spray line piping and fittings exposed to reactor coolant	Cracking due to cyclic loading	Chapter XI.M1, "ASME Section XI Inservice Inspection, Subsections IWB, IWC, and IWD" for Class 1 components	No	IV.C2.R-56	IV.C2-26(R-56)

Table 3.1-1 Summary of Aging Management Programs for Reactor Vessel, Internals, and Reactor Coolant System Evaluated in Chapter IV of the GALL Report

ID	Type	Component	Aging Effect/Mechanism	Aging Management Programs	Further Evaluation Recommended	Rev2 Item	Rev1 Item
36	PWR	Steel, stainless steel pressurizer integral support exposed to air with metal temperature up to 288°C (550°F)	Cracking due to cyclic loading	Chapter XI.M1, "ASME Section XI Inservice Inspection, Subsections IWB, IWC, and IWD" for Class 1 components	No	IV.C2.R-19	IV.C2-16(R-19)
37	PWR	Steel reactor vessel flange	Loss of material due to wear	Chapter XI.M1, "ASME Section XI Inservice Inspection, Subsections IWB, IWC, and IWD" for Class 1 components	No	IV.A2.R-87	IV.A2-25(R-87)
38	BWR/ PWR	Cast austenitic stainless steel Class 1 pump casings, and valve bodies and bonnets exposed to reactor coolant >250 deg-C (>482 deg-F)	Loss of fracture toughness due to thermal aging embrittlement	Chapter XI.M1, "ASME Section XI Inservice Inspection, Subsections IWB, IWC, and IWD" for Class 1 components. For pump casings and valve bodies, screening for susceptibility to thermal aging is not necessary.	No	IV.C1.R-08 IV.C2.R-08	IV.C1-3(R-08) IV.C2-6(R-08)
39	BWR/ PWR	Steel, stainless steel, or steel with stainless steel cladding Class 1 piping, fittings and branch connections < NPS 4 exposed to reactor coolant	Cracking due to stress corrosion cracking, intergranular stress corrosion cracking (for stainless steel only), and thermal, mechanical, and vibratory loading	Chapter XI.M1, "ASME Section XI Inservice Inspection, Subsections IWB, IWC, and IWD" for Class 1 components, Chapter XI.M2, "Water Chemistry," and XI.M35, "One-Time Inspection of ASME Code Class 1 Small-bore Piping"	No	IV.C1.RP-230 IV.C2.RP-235	IV.C1-1(R-03) IV.C2-1(R-02)

Table 3.1-1 Summary of Aging Management Programs for Reactor Vessel, Internals, and Reactor Coolant System Evaluated in Chapter IV of the GALL Report

ID	Type	Component	Aging Effect/Mechanism	Aging Management Programs	Further Evaluation Recommended	Rev2 Item	Rev1 Item
40	PWR	Steel with stainless steel or nickel alloy cladding; or stainless steel pressurizer components exposed to reactor coolant	Cracking due to cyclic loading	Chapter XI.M1, "ASME Section XI Inservice Inspection, Subsections IWB, IWC, and IWD" for Class 1 components, and Chapter XI.M2, "Water Chemistry"	No	IV.C2.R-58	IV.C2-18(R-58)
	PWR	Nickel alloy core support pads; core guide lugs exposed to reactor coolant	Cracking due to primary water stress corrosion cracking	Chapter XI.M1, "ASME Section XI Inservice Inspection, Subsections IWB, IWC, and IWD" for Class 1 components, and Chapter XI.M2, "Water Chemistry"	No	IV.A2.RP-57	IV.A2-12(R-88)
41	BWR	Nickel alloy core shroud and core plate access hole cover (mechanical covers) exposed to reactor coolant	Cracking due to stress corrosion cracking, intergranular stress corrosion cracking, irradiation-assisted stress corrosion cracking	Chapter XI.M1, "ASME Section XI Inservice Inspection, Subsections IWB, IWC, and IWD" for Class 1 components, and Chapter XI.M2, "Water Chemistry"	No	IV.B1.R-95	IV.B1-4(R-95)
42	PWR	Steel with stainless steel or nickel alloy cladding or stainless steel primary side components; steam generator upper and lower heads, and tube sheet weld; or pressurizer components exposed to reactor coolant	Cracking due to stress corrosion cracking, primary water stress corrosion cracking	Chapter XI.M1, "ASME Section XI Inservice Inspection, Subsections IWB, IWC, and IWD" for Class 1 components, and Chapter XI.M2, "Water Chemistry"	No	IV.C2.R-25 IV.D2.RP-47	IV.C2-19(R-25) IV.D2-4(R-35)

Table 3.1-1 Summary of Aging Management Programs for Reactor Vessel, Internals, and Reactor Coolant System Evaluated in Chapter IV of the GALL Report

ID	Type	Component	Aging Effect/Mechanism	Aging Management Programs	Further Evaluation Recommended	Rev2 Item	Rev1 Item
43	BWR	Stainless steel and nickel-alloy reactor vessel internals exposed to reactor coolant	Loss of material due to pitting and crevice corrosion	Chapter XI.M1, "ASME Section XI Inservice Inspection, Subsections IWB, IWC, and IWD" for Class 1 components, and Chapter XI.M2, "Water Chemistry"	No	IV.B1.RP-26	IV.B1-15(RP-26)
44	PWR	Steel steam generator secondary manways and handholds (cover only) exposed to air with leaking secondary-side water and/or steam	Loss of material due to erosion	Chapter XI.M1, "ASME Section XI Inservice Inspection, Subsections IWB, IWC, and IWD" for Class 2 components	No	IV.D2.R-31	IV.D2-5(R-31)
45	PWR	Nickel alloy and steel with nickel-alloy cladding reactor coolant pressure boundary components exposed to reactor coolant	Cracking due to primary water stress corrosion cracking	Chapter XI.M1, "ASME Section XI ISI, IWB, IWC, IWD," and Chapter XI.M2, "Water Chemistry," and, for nickel-alloy, Chapter XI.M11B, "Cracking of Nickel-Alloy Components and Loss of Material Due to Boric Acid-induced Corrosion in RCPB Components (PWRs Only)"	No	IV.A2.R-90 IV.A2.RP-186 IV.A2.RP-59 IV.C2.RP-156 IV.C2.RP-159 IV.C2.RP-37 IV.D1.RP-36 IV.D2.RP-36	IV.A2-18(R-90) IV.A2-9(R-75) IV.A2-19(R-89) IV.C2-24(RP-22) IV.C2-13(RP-31) IV.C2-21(R-06) IV.D1-4(R-01) IV.D2-2(R-01)

Table 3.1-1 Summary of Aging Management Programs for Reactor Vessel, Internals, and Reactor Coolant System Evaluated in Chapter IV of the GALL Report

ID	Type	Component	Aging Effect/Mechanism	Aging Management Programs	Further Evaluation Recommended	Rev2 Item	Rev1 Item
46	PWR	Stainless steel, nickel-alloy, nickel-alloy welds and/or buttering control rod drive head penetration pressure housing or nozzles safe ends and welds (inlet, outlet, safety injection) exposed to reactor coolant	Cracking due to stress corrosion cracking, primary water stress corrosion cracking	Chapter XI.M1, "ASME Section XI ISI, IWB, IWC & IWD," and Chapter XI.M2, "Water Chemistry," and, for nickel-alloy, Chapter XI.M11B, "Cracking of Nickel-Alloy Components and Loss of Material Due to Boric Acid-induced corrosion in RCPB Components (PWRs Only)"	No	IV.A2.RP-234	IV.A2-15(R-83)
47	PWR	Stainless steel, nickel-alloy control rod drive head penetration pressure housing exposed to reactor coolant	Cracking due to stress corrosion cracking, primary water stress corrosion cracking	Chapter XI.M1, "ASME Section XI ISI, IWB, IWC & IWD," and Chapter XI.M2, "Water Chemistry"	No	IV.A2.RP-55	IV.A2-11(R-76)
48	PWR	Steel external surfaces: reactor vessel top head, reactor vessel bottom head, reactor coolant pressure boundary piping or components adjacent to dissimilar metal (Alloy 82/182) welds exposed to air with borated water leakage	Loss of material due to boric acid corrosion	Chapter XI.M10, "Boric Acid Corrosion," and Chapter XI.M11B, "Cracking of Nickel-Alloy Components and Loss of Material Due to Boric Acid-Induced Corrosion in RCPB Components (PWRs Only)"	No	IV.A2.RP-379 IV.C2.RP-380	IV.A2-13(R-17) IV.C2-9(R-17)

Table 3.1-1 Summary of Aging Management Programs for Reactor Vessel, Internals, and Reactor Coolant System Evaluated in Chapter IV of the GALL Report

ID	Type	Component	Aging Effect/Mechanism	Aging Management Programs	Further Evaluation Recommended	Rev2 Item	Rev1 Item
49	PWR	Steel reactor coolant pressure boundary external surfaces or closure bolting exposed to air with borated water leakage	Loss of material due to boric acid corrosion	Chapter XI.M10, "Boric Acid Corrosion"	No	IV.A2.R-17 IV.C2.R-17 IV.C2.RP-167 IV.D1.R-17 IV.D2.R-17	IV.A2-13(R-17) IV.C2-9(R-17) N/A IV.D1-3(R-17) IV.D2-1(R-17)
50	BWR/PWR	Cast austenitic stainless steel Class 1 piping, piping component, and piping elements and control rod drive pressure housings exposed to reactor coolant >250 deg-C (>482 deg-F)	Loss of fracture toughness due to thermal aging embrittlement	Chapter XI.M12, "Thermal Aging Embrittlement of Cast Austenitic Stainless Steel (CASS)"	No	IV.A2.R-77 IV.C1.R-52 IV.C2.R-52	IV.A2-10(R-77) IV.C1-2(R-52) IV.C2-4(R-52)
51	PWR	Stainless steel or nickel-alloy Babcock & Wilcox reactor internal components exposed to reactor coolant and neutron flux	Cracking due to stress corrosion cracking, irradiation-assisted stress corrosion cracking, or fatigue	Chapter XI.M16A, "PWR Vessel Internals," and Chapter XI.M2, "Water Chemistry"	No	IV.B4.RP-236 IV.B4.RP-241 IV.B4.RP-244 IV.B4.RP-245 IV.B4.RP-246 IV.B4.RP-247 IV.B4.RP-248 IV.B4.RP-254 IV.B4.RP-256 IV.B4.RP-261 IV.B4.RP-262 IV.B4.RP-352 IV.B4.RP-375	N/A IV.B4-7(R-125) IV.B4-7(R-125) IV.B4-13(R-194) IV.B4-12(R-196) IV.B4-13(R-194) IV.B4-12(R-196) IV.B4-25(R-210) IV.B4-25(R-210) IV.B4-32(R-203) IV.B4-32(R-203) N/A N/A

Table 3.1-1 Summary of Aging Management Programs for Reactor Vessel, Internals, and Reactor Coolant System Evaluated in Chapter IV of the GALL Report

ID	Type	Component	Aging Effect/Mechanism	Aging Management Programs	Further Evaluation Recommended	Rev2 Item	Rev1 Item
52	PWR	Stainless steel or nickel-alloy Combustion Engineering reactor internal components exposed to reactor coolant and neutron flux	Cracking due to stress corrosion cracking, irradiation-assisted stress corrosion cracking, or fatigue	Chapter XI.M16A, "PWR Vessel Internals," and Chapter XI.M2, "Water Chemistry"	No	IV.B3.RP-306 IV.B3.RP-312 IV.B3.RP-313 IV.B3.RP-314 IV.B3.RP-316 IV.B3.RP-320 IV.B3.RP-322 IV.B3.RP-323 IV.B3.RP-324 IV.B3.RP-325 IV.B3.RP-327 IV.B3.RP-328 IV.B3.RP-329 IV.B3.RP-330 IV.B3.RP-334 IV.B3.RP-335 IV.B3.RP-342 IV.B3.RP-358	N/A IV.B3-2(R-149) N/A IV.B3-9(R-162) IV.B3-9(R-162) IV.B3-9(R-162) N/A N/A N/A N/A IV.B3-15(R-155) IV.B3-15(R-155) IV.B3-15(R-155) IV.B3-23(R-167) IV.B3-23(R-167) IV.B3-23(R-167) N/A N/A
53	PWR	Stainless steel or nickel-alloy Westinghouse reactor internal components exposed to reactor coolant and neutron flux	Cracking due to stress corrosion cracking, irradiation-assisted stress corrosion cracking, or fatigue	Chapter XI.M16A, "PWR Vessel Internals," and Chapter XI.M2, "Water Chemistry"	No	IV.B2.RP-265 IV.B2.RP-271 IV.B2.RP-273 IV.B2.RP-275 IV.B2.RP-276 IV.B2.RP-278 IV.B2.RP-280 IV.B2.RP-282 IV.B2.RP-286 IV.B2.RP-289 IV.B2.RP-291 IV.B2.RP-293 IV.B2.RP-294 IV.B2.RP-298 IV.B2.RP-301 IV.B2.RP-346 IV.B2.RP-387	N/A IV.B2-10(R-125) IV.B2-10(R-125) IV.B2-6(R-128) IV.B2-8(R-120) IV.B2-8(R-120) IV.B2-8(R-120) IV.B2-8(R-120) IV.B2-16(R-133) IV.B2-20(R-130) IV.B2-24(R-138) IV.B2-24(R-138) IV.B2-24(R-138) IV.B2-28(R-118) IV.B2-40(R-112) N/A N/A

Table 3.1-1 Summary of Aging Management Programs for Reactor Vessel, Internals, and Reactor Coolant System Evaluated in Chapter IV of the GALL Report

ID	Type	Component	Aging Effect/Mechanism	Aging Management Programs	Further Evaluation Recommended	Rev2 Item	Rev1 Item
54	PWR	Stainless steel bottom mounted instrument system flux thimble tubes (with or without chrome plating) exposed to reactor coolant and neutron flux	Loss of material due to wear	Chapter XI.M16A, "PWR Vessel Internals," and Chapter XI.M37, "Flux Thimble Tube Inspection"	No	IV.B2.RP-284	IV.B2-12(R-143) IV.B2-13(R-145)
55	PWR	Stainless steel thermal shield assembly, thermal shield flexures exposed to reactor coolant and neutron flux	Cracking due to fatigue; Loss of material due to wear	Chapter XI.M16A, "PWR Vessel Internals"	No	IV.B2.RP-302	N/A
56	PWR	Stainless steel or nickel-alloy Combustion Engineering reactor internal components exposed to reactor coolant and neutron flux	Loss of fracture toughness due to neutron irradiation embrittlement; or changes in dimension due to void swelling; or loss of preload due to thermal and irradiation enhanced stress relaxation; or loss of material due to wear	Chapter XI.M16A, "PWR Vessel Internals"	No	IV.B3.RP-307 IV.B3.RP-315 IV.B3.RP-317 IV.B3.RP-318 IV.B3.RP-319 IV.B3.RP-326 IV.B3.RP-331 IV.B3.RP-332 IV.B3.RP-336 IV.B3.RP-359 IV.B3.RP-360 IV.B3.RP-361 IV.B3.RP-362 IV.B3.RP-363 IV.B3.RP-364 IV.B3.RP-365 IV.B3.RP-366	N/A IV.B3-7(R-165) IV.B3-7(R-165) IV.B4-8(R-163) IV.B3-9(R-162) N/A N/A IV.B3-17(R-156) IV.B3-22(R-170) N/A N/A N/A N/A N/A N/A N/A N/A

Table 3.1-1 Summary of Aging Management Programs for Reactor Vessel, Internals, and Reactor Coolant System Evaluated in Chapter IV of the GALL Report

ID	Type	Component	Aging Effect/Mechanism	Aging Management Programs	Further Evaluation Recommended	Rev2 Item	Rev1 Item
58	PWR	Stainless steel or nickel-alloy Babcock & Wilcox reactor internal components exposed to reactor coolant and neutron flux	Loss of fracture toughness due to neutron irradiation embrittlement; or changes in dimension due to void swelling; or loss of preload due to thermal and irradiation enhanced stress relaxation; or loss of material due to wear	Chapter XI.M16A, "PWR Vessel Internals"	No	IV.B4.RP-237 IV.B4.RP-240 IV.B4.RP-242 IV.B4.RP-243 IV.B4.RP-249 IV.B4.RP-250 IV.B4.RP-251 IV.B4.RP-252 IV.B4.RP-253 IV.B4.RP-258 IV.B4.RP-259 IV.B4.RP-260	N/A IV.B4-1(R-128) IV.B4-4(R-183) IV.B4-1(R-128) IV.B4-12(R-196) IV.B4-12(R-196) IV.B4-15(R-190) IV.B4-16(R-188) IV.B4-21(R-191) IV.B4-4(R-183) IV.B4-31(R-205) IV.B4-31(R-205)
59	PWR	Stainless steel or nickel-alloy Westinghouse reactor internal components exposed to reactor coolant and neutron flux	Loss of fracture toughness due to neutron irradiation embrittlement; or changes in dimension due to void swelling; or loss of preload due to thermal and irradiation enhanced stress relaxation; or loss of material due to wear	Chapter XI.M16A, "PWR Vessel Internals"	No	IV.B2.RP-267 IV.B2.RP-270 IV.B2.RP-272 IV.B2.RP-274 IV.B2.RP-281 IV.B2.RP-285 IV.B2.RP-287 IV.B2.RP-288 IV.B2.RP-290 IV.B2.RP-292 IV.B2.RP-295 IV.B2.RP-296 IV.B2.RP-297 IV.B2.RP-299 IV.B2.RP-300 IV.B2.RP-345 IV.B2.RP-354 IV.B2.RP-386 IV.B2.RP-388	N/A IV.B2-1(R-124) IV.B2-6(R-128) IV.B2-6(R-128) IV.B2-9(R-122) IV.B2-14(R-137) IV.B2-17(R-135) IV.B2-18(R-132) IV.B2-21(R-140) IV.B2-21(R-140) IV.B2-22(R-141) N/A N/A IV.B2-34(R-115) IV.B2-33(R-108) N/A N/A N/A N/A
60	BWR	Steel piping, piping components, and piping elements exposed to reactor coolant	Wall thinning due to flow-accelerated corrosion	Chapter XI.M17, "Flow-Accelerated Corrosion"	No	IV.C1.R-23	IV.C1-7(R-23)

Table 3.1-1 Summary of Aging Management Programs for Reactor Vessel, Internals, and Reactor Coolant System Evaluated in Chapter IV of the GALL Report

ID	Type	Component	Aging Effect/Mechanism	Aging Management Programs	Further Evaluation Recommended	Rev2 Item	Rev1 Item
61	PWR	Steel steam generator steam nozzle and safe end, feedwater nozzle and safe end, AFW nozzles and safe ends exposed to secondary feedwater/steam	Wall thinning due to flow-accelerated corrosion	Chapter XI.M17, "Flow-Accelerated Corrosion"	No	IV.D1.R-37 IV.D2.R-38	IV.D1-5(R-37) IV.D2-7(R-38)
62	PWR	High-strength, low alloy steel, or stainless steel closure bolting; stainless steel control rod drive head penetration flange bolting exposed to air with reactor coolant leakage	Cracking due to stress corrosion cracking	Chapter XI.M18, "Bolting Integrity"	No	IV.A2.R-78 IV.C2.R-11 IV.D1.R-10	IV.A2-6(R-78) IV.C2-7(R-11) IV.D1-2(R-10)
63	BWR	Steel or stainless steel closure bolting exposed to air with reactor coolant leakage	Loss of material due to general (steel only), pitting, and crevice corrosion or wear	Chapter XI.M18, "Bolting Integrity"	No	IV.C1.RP-42	IV.C1-12(R-26)
64	PWR	Steel closure bolting exposed to air – indoor uncontrolled	Loss of material due to general, pitting, and crevice corrosion	Chapter XI.M18, "Bolting Integrity"	No	IV.C2.RP-166	N/A
65	PWR	Stainless steel control rod drive head penetration flange bolting exposed to air with reactor coolant leakage	Loss of material due to wear	Chapter XI.M18, "Bolting Integrity"	No	IV.A2.R-79	IV.A2-7(R-79)

Table 3.1-1 Summary of Aging Management Programs for Reactor Vessel, Internals, and Reactor Coolant System Evaluated in Chapter IV of the GALL Report

ID	Type	Component	Aging Effect/Mechanism	Aging Management Programs	Further Evaluation Recommended	Rev2 Item	Rev1 Item
66	PWR	High-strength, low alloy steel, or stainless steel closure bolting; stainless steel control rod drive head penetration flange bolting exposed to air with reactor coolant leakage	Loss of preload due to thermal effects, gasket creep, and self-loosening	Chapter XI.M18, "Bolting Integrity"	No	IV.A2.R-80 IV.C2.R-12	IV.A2-8(R-80) IV.C2-8(R-12)
67	BWR/PWR	Steel or stainless steel closure bolting exposed to air – indoor with potential for reactor coolant leakage	Loss of preload due to thermal effects, gasket creep, and self-loosening	Chapter XI.M18, "Bolting Integrity"	No	IV.C1.RP-43 IV.D1.RP-46 IV.D2.RP-46	IV.C1-10(R-27) IV.D1-10(R-32) IV.D2-6(R-32)
68	PWR	Nickel alloy steam generator tubes exposed to secondary feedwater or steam	Changes in dimension ("denting") due to corrosion of carbon steel tube support plate	Chapter XI.M19, "Steam Generators," and Chapter XI.M2, "Water Chemistry"	No	IV.D1.R-43 IV.D2.R-226	IV.D1-19(R-43) IV.D2-13(R-226)
69	PWR	Nickel alloy steam generator tubes and sleeves exposed to secondary feedwater or steam	Cracking due to outer diameter stress corrosion cracking and intergranular attack	Chapter XI.M19, "Steam Generators," and Chapter XI.M2, "Water Chemistry"	No	IV.D1.R-47 IV.D1.R-48 IV.D2.R-47 IV.D2.R-48	IV.D1-23(R-47) IV.D1-22(R-48) IV.D2-17(R-47) IV.D2-16(R-48)
70	PWR	Nickel alloy steam generator tubes, repair sleeves, and tube plugs exposed to reactor coolant	Cracking due to primary water stress corrosion cracking	Chapter XI.M19, "Steam Generators," and Chapter XI.M2, "Water Chemistry"	No	IV.D1.R-40 IV.D1.R-44 IV.D2.R-40 IV.D2.R-44	IV.D1-18(R-40) IV.D1-20(R-44) IV.D2-12(R-40) IV.D2-14(R-44)

ID	Type	Component	Aging Effect/Mechanism	Aging Management Programs	Further Evaluation Recommended	Rev2 Item	Rev1 Item
71	PWR	Steel, chrome plated steel, stainless steel, nickel alloy steam generator U-bend supports including anti-vibration bars exposed to secondary feedwater or steam	Cracking due to stress corrosion cracking or other mechanism(s); loss of material due general (steel only), pitting, and crevice corrosion	Chapter XI.M19, "Steam Generators," and Chapter XI.M2, "Water Chemistry"	No	IV.D1.RP-226 IV.D1.RP-384	IV.D1-15(RP-15) IV.D1-14(RP-14)
72	PWR	Steel steam generator tube support plate, tube bundle wrapper, supports and mounting hardware exposed to secondary feedwater or steam	Loss of material due to erosion, general, pitting, and crevice corrosion, ligament cracking due to corrosion	Chapter XI.M19, "Steam Generators," and Chapter XI.M2, "Water Chemistry"	No	IV.D1.R-42 IV.D1.RP-161 IV.D2.R-42 IV.D2.RP-162	IV.D1-17(R-42) IV.D1-9(RP-16) IV.D2-11(R-42) N/A
73	PWR	Nickel alloy steam generator tubes and sleeves exposed to phosphate chemistry in secondary feedwater or steam	Loss of material due to wastage and pitting corrosion	Chapter XI.M19, "Steam Generators," and Chapter XI.M2, "Water Chemistry"	No	IV.D1.R-50	IV.D1-25(R-50)
74	PWR	Steel steam generator upper assembly and separators including feedwater inlet ring and support exposed to secondary feedwater or steam	Wall thinning due to flow-accelerated corrosion	Chapter XI.M19, "Steam Generators," and Chapter XI.M2, "Water Chemistry"	No	IV.D1.RP-49	IV.D1-26(R-51)
75	PWR	Steel steam generator tube support lattice bars exposed to secondary feedwater or steam	Wall thinning due to flow-accelerated corrosion and general corrosion	Chapter XI.M19, "Steam Generators," and Chapter XI.M2, "Water Chemistry"	No	IV.D1.RP-48	IV.D1-16(R-41)

ID	Type	Component	Aging Effect/Mechanism	Aging Management Programs	Further Evaluation Recommended	Rev2 Item	Rev1 Item
76	PWR	Steel, chrome plated steel, stainless steel, nickel alloy steam generator U-bend supports including anti-vibration bars exposed to secondary feedwater or steam	Loss of material due to fretting	Chapter XI.M19, "Steam Generators"	No	IV.D1.RP-225	IV.D1-15(RP-15)
77	PWR	Nickel alloy steam generator tubes and sleeves exposed to secondary feedwater or steam	Loss of material due to wear and fretting	Chapter XI.M19, "Steam Generators"	No	IV.D1.RP-233 IV.D2.RP-233	IV.D1-24(R-49) IV.D2-18(R-49)
78	PWR	Nickel alloy steam generator components such as, secondary side nozzles (vent, drain, and instrumentation) exposed to secondary feedwater or steam	Cracking due to stress corrosion cracking	Chapter XI.M2, "Water Chemistry," and Chapter XI.M32, "One-Time Inspection," or Chapter XI.M1, "ASME Section XI Inservice Inspection, Subsections IWB, IWC, and IWD."	No	IV.D2.R-36	IV.D2-9(R-36)
79	BWR	Stainless steel; steel with nickel-alloy or stainless steel cladding; and nickel-alloy reactor coolant pressure boundary components exposed to reactor coolant	Loss of material due to pitting and crevice corrosion	Chapter XI.M2, "Water Chemistry," and Chapter XI.M32, "One-Time Inspection"	No	IV.C1.RP-158	IV.C1-14(RP-27)

Table 3.1-1 Summary of Aging Management Programs for Reactor Vessel, Internals, and Reactor Coolant System Evaluated in Chapter IV of the GALL Report

ID	Type	Component	Aging Effect/Mechanism	Aging Management Programs	Further Evaluation Recommended	Rev2 Item	Rev1 Item
80	PWR	Stainless steel or steel with stainless steel cladding pressurizer relief tank: tank shell and heads, flanges, nozzles (none-ASME Section XI components) exposed to treated borated water >60°C (>140°F)	Cracking due to stress corrosion cracking	Chapter XI.M2, "Water Chemistry," and Chapter XI.M32, "One-Time Inspection"	No	IV.C2.RP-383	N/A
81	PWR	Stainless steel pressurizer spray head exposed to reactor coolant	Cracking due to stress corrosion cracking	Chapter XI.M2, "Water Chemistry," and Chapter XI.M32, "One-Time Inspection"	No	IV.C2.RP-41	IV.C2-17(R-24)
82	PWR	Nickel alloy pressurizer spray head exposed to reactor coolant	Cracking due to stress corrosion cracking, primary water stress corrosion cracking	Chapter XI.M2, "Water Chemistry," and Chapter XI.M32, "One-Time Inspection"	No	IV.C2.RP-40	IV.C2-17(R-24)
83	PWR	Steel steam generator shell assembly exposed to secondary feedwater or steam	Loss of material due to general, pitting, and crevice corrosion	Chapter XI.M2, "Water Chemistry," and Chapter XI.M32, "One-Time Inspection"	No	IV.D1.RP-372 IV.D2.RP-153	N/A IV.D2-8(R-224)
84	BWR	Steel top head enclosure (without cladding) top head nozzles (vent, top head spray or RCIC, and spare) exposed to reactor coolant	Loss of material due to general, pitting, and crevice corrosion	Chapter XI.M2, "Water Chemistry," and Chapter XI.M32, "One-Time Inspection"	No	IV.A1.RP-50	IV.A1-11(R-59)

Table 3.1-1 Summary of Aging Management Programs for Reactor Vessel, Internals, and Reactor Coolant System Evaluated in Chapter IV of the GALL Report

ID	Type	Component	Aging Effect/Mechanism	Aging Management Programs	Further Evaluation Recommended	Rev2 Item	Rev1 Item
85	BWR	Stainless steel, nickel-alloy, and steel with nickel-alloy or stainless steel cladding reactor vessel flanges, nozzles, penetrations, safe ends, vessel shells, heads and welds exposed to reactor coolant	Loss of material due to pitting and crevice corrosion	Chapter XI.M2, "Water Chemistry," and Chapter XI.M32, "One-Time Inspection"	No	IV.A1.RP-157	IV.A1-8(RP-25)
86	PWR	Stainless steel steam generator primary side divider plate exposed to reactor coolant	Cracking due to stress corrosion cracking	Chapter XI.M2, "Water Chemistry"	No	IV.D1.RP-17	IV.D1-7(RP-17)
87	PWR	Stainless steel or nickel-alloy PWR reactor internal components exposed to reactor coolant and neutron flux	Loss of material due to pitting and crevice corrosion	Chapter XI.M2, "Water Chemistry"	No	IV.B2.RP-24 IV.B3.RP-24 IV.B4.RP-24	IV.B2-32(RP-24) IV.B3-25(RP-24) IV.B4-38(RP-24)
88	PWR	Stainless steel; steel with nickel-alloy or stainless steel cladding; and nickel-alloy reactor coolant pressure boundary components exposed to reactor coolant	Loss of material due to pitting and crevice corrosion	Chapter XI.M2, "Water Chemistry"	No	IV.A2.RP-28 IV.C2.RP-23	IV.A2-14(RP-28) IV.C2-15(RP-23)
89	PWR	Steel piping, piping components, and piping elements exposed to closed cycle cooling water	Loss of material due to general, pitting, and crevice corrosion	Chapter XI.M21A, "Closed Treated Water Systems"	No	IV.C2.RP-221	IV.C2-14(RP-10)

Table 3.1-1 Summary of Aging Management Programs for Reactor Vessel, Internals, and Reactor Coolant System Evaluated in Chapter IV of the GALL Report

ID	Type	Component	Aging Effect/Mechanism	Aging Management Programs	Further Evaluation Recommended	Rev2 Item	Rev1 Item
90	PWR	Copper alloy piping, piping components, and piping elements exposed to closed cycle cooling water	Loss of material due to pitting, crevice, and galvanic corrosion	Chapter XI.M21A, "Closed Treated Water Systems"	No	IV.C2.RP-222	IV.C2-11(RP-11)
91	BWR	High-strength low alloy steel closure head stud assembly exposed to air with potential for reactor coolant leakage	Cracking due to stress corrosion cracking; loss of material due to general, pitting, and crevice corrosion, or wear (BWR)	Chapter XI.M3, "Reactor Head Closure Stud Bolting"	No	IV.A1.RP-165 IV.A1.RP-51	N/A IV.A1-9(R-60)
92	PWR	High-strength low alloy steel closure head stud assembly exposed to air with potential for reactor coolant leakage	Cracking due to stress corrosion cracking; loss of material due to general, pitting, and crevice corrosion, or wear (PWR)	Chapter XI.M3, "Reactor Head Closure Stud Bolting"	No	IV.A2.RP-52 IV.A2.RP-53	IV.A2-2(R-71) IV.A2-3(R-72)
93	PWR	Copper alloy >15% Zn or > 8% Al piping, piping components, and piping elements exposed to closed cycle cooling water	Loss of material due to selective leaching	Chapter XI.M33, "Selective Leaching "	No	IV.C2.RP-12	IV.C2-12(RP-12)
94	BWR	Stainless steel and nickel alloy vessel shell attachment welds exposed to reactor coolant	Cracking due to stress corrosion cracking, intergranular stress corrosion cracking	Chapter XI.M4, "BWR Vessel ID Attachment Welds," and Chapter XI.M2, "Water Chemistry"	No	IV.A1.R-64	IV.A1-12(R-64)
95	BWR	Steel (with or without stainless steel cladding) feedwater nozzles exposed to reactor coolant	Cracking due to cyclic loading	Chapter XI.M5, "BWR Feedwater Nozzle"	No	IV.A1.R-65	IV.A1-3(R-65)

Table 3.1-1 Summary of Aging Management Programs for Reactor Vessel, Internals, and Reactor Coolant System Evaluated in Chapter IV of the GALL Report

ID	Type	Component	Aging Effect/Mechanism	Aging Management Programs	Further Evaluation Recommended	Rev2 Item	Rev1 Item
96	BWR	Steel (with or without stainless steel cladding) control rod drive return line nozzles exposed to reactor coolant	Cracking due to cyclic loading	Chapter XI.M6, "BWR Control Rod Drive Return Line Nozzle"	No	IV.A1.R-66	IV.A1-2(R-66)
97	BWR	Stainless steel and nickel alloy piping, piping components, and piping elements greater than or equal to 4 NPS; nozzle safe ends and associated welds	Cracking due to stress corrosion cracking, intergranular stress corrosion cracking	Chapter XI.M7, "BWR Stress Corrosion Cracking," and Chapter XI.M2, "Water Chemistry"	No	IV.A1.R-68 IV.C1.R-20 IV.C1.R-21	IV.A1-1(R-68) IV.C1-9(R-20) IV.C1-8(R-21)
98	BWR	Stainless steel or nickel alloy penetrations: instrumentation and standby liquid control exposed to reactor coolant	Cracking due to stress corrosion cracking, intergranular stress corrosion cracking, cyclic loading	Chapter XI.M8, "BWR Penetrations," and Chapter XI.M2, "Water Chemistry"	No	IV.A1.RP-369	IV.A1-5(R-69)
99	BWR	Cast austenitic stainless steel; PH martensitic stainless steel; martensitic stainless steel; X-750 alloy reactor internal components exposed to reactor coolant and neutron flux	Loss of fracture toughness due to thermal aging and neutron irradiation embrittlement	Chapter XI.M9, "BWR Vessel Internals"	No	IV.B1.RP-182 IV.B1.RP-200 IV.B1.RP-219 IV.B1.RP-220	N/A N/A IV.B1-11(R-101) IV.B1-9(R-103)
100	BWR	Stainless steel reactor vessel internals components (jet pump wedge surface) exposed to reactor coolant	Loss of material due to wear	Chapter XI.M9, "BWR Vessel Internals"	No	IV.B1.RP-377	N/A

Table 3.1-1 Summary of Aging Management Programs for Reactor Vessel, Internals, and Reactor Coolant System Evaluated in Chapter IV of the GALL Report

ID	Type	Component	Aging Effect/Mechanism	Aging Management Programs	Further Evaluation Recommended	Rev2 Item	Rev1 Item
101	BWR	Stainless steel steam dryers exposed to reactor coolant	Cracking due to flow-induced vibration	Chapter XI.M9, "BWR Vessel Internals" for steam dryer	No	IV.B1.RP-155	IV.B1-16(RP-18)
102	BWR	Stainless steel fuel supports and control rod drive assemblies control rod drive housing exposed to reactor coolant	Cracking due to stress corrosion cracking, intergranular stress corrosion cracking	Chapter XI.M9, "BWR Vessel Internals," and Chapter XI.M2, "Water Chemistry"	No	IV.B1.R-104	IV.B1-8(R-104)
103	BWR	Stainless steel and nickel alloy reactor internal components exposed to reactor coolant and neutron flux	Cracking due to stress corrosion cracking, intergranular stress corrosion cracking, irradiation-assisted stress corrosion cracking	Chapter XI.M9, "BWR Vessel Internals," and Chapter XI.M2, "Water Chemistry"	No	IV.B1.R-100 IV.B1.R-105 IV.B1.R-92 IV.B1.R-93 IV.B1.R-96 IV.B1.R-97 IV.B1.R-98 IV.B1.R-99	IV.B1-13(R-100) IV.B1-10(R-105) IV.B1-1(R-92) IV.B1-6(R-93) IV.B1-2(R-96) IV.B1-3(R-97) IV.B1-17(R-98) IV.B1-7(R-99)
104	BWR	X-750 alloy reactor vessel internal components exposed to reactor coolant and neutron flux	Cracking due to intergranular stress corrosion cracking	Chapter XI.M9, "BWR Vessel Internals" for core plate, and Chapter XI.M2, "Water Chemistry"	No	IV.B1.RP-381	N/A
105	BWR/ PWR	Steel piping, piping components and piping element exposed to concrete	None	None, provided 1) attributes of the concrete are consistent with ACI 318 or ACI 349 (low water-to-cement ratio, low permeability, and adequate air entrainment) as cited in NUREG-1557, and 2) plant OE indicates no degradation of the concrete	No, if conditions are met.	IV.E.RP-353	IV.E-6(RP-01)

Table 3.1-1 Summary of Aging Management Programs for Reactor Vessel, Internals, and Reactor Coolant System Evaluated in Chapter IV of the GALL Report

ID	Type	Component	Aging Effect/Mechanism	Aging Management Programs	Further Evaluation Recommended	Rev2 Item	Rev1 Item
106	BWR/ PWR	Nickel alloy piping, piping components and piping element exposed to air – indoor, uncontrolled, or air with borated water leakage	None	None	NA - No AEM or AMP	IV.E.RP-03 IV.E.RP-378	IV.E-1(RP-03) N/A
107	BWR/ PWR	Stainless steel piping, piping components and piping element exposed to gas, concrete, air with borated water leakage, air – indoors, uncontrolled	None	None	NA - No AEM or AMP	IV.E.RP-04 IV.E.RP-05 IV.E.RP-06 IV.E.RP-07	IV.E-2(RP-04) IV.E-3(RP-05) IV.E-4(RP-06) IV.E-5(RP-07)

Table 3.1-2 Aging Management Programs Recommended for Reactor Vessel, Internals, and Reactor Coolant System

GALL Report Chapter/AMP	Program Name
Chapter X.M1	Metal Fatigue of Reactor Coolant Pressure Boundary
Chapter XI.M1	ASME Section XI Inservice Inspection, Subsections IWB, IWC, and IWD
Chapter XI.M2	Water Chemistry
Chapter XI.M3	Reactor Head Closure Stud Bolting
Chapter XI.M4	BWR Vessel ID Attachment Welds
Chapter XI.M5	BWR Feedwater Nozzle
Chapter XI.M6	BWR Control Rod Drive Return Line Nozzle
Chapter XI.M7	BWR Stress Corrosion Cracking
Chapter XI.M8	BWR Penetrations
Chapter XI.M9	BWR Vessel Internals
Chapter XI.M10	Boric Acid Corrosion
Chapter XI.M11B	Cracking of Nickel-Alloy Components and Loss of Material Due to Boric Acid-induced Corrosion in Reactor Coolant Pressure Boundary Components (PWRs only)
Chapter XI.M12	Thermal Aging of Cast Austenitic Stainless Steel (CASS)
Chapter XI.M16A	PWR Vessel Internals
Chapter XI.M17	Flow-Accelerated Corrosion
Chapter XI.M18	Bolting Integrity
Chapter XI.M19	Steam Generators
Chapter XI.M21A	Closed Treated Water Systems
Chapter XI.M31	Reactor Vessel Surveillance
Chapter XI.M32	One-Time Inspection
Chapter XI.M33	Selective Leaching
Chapter XI.M35	One-Time Inspection of ASME Code Class 1 Small Bore-Piping
Chapter XI.M37	Flux Thimble Tube Inspection
Appendix for GALL	Quality Assurance for Aging Management Programs
SRP-LR Appendix A	Plant-specific AMP

3.2 AGING MANAGEMENT OF ENGINEERED SAFETY FEATURES

Review Responsibilities

Primary - Branch assigned responsibility by PM as described in SRP-LR Section 3.0 of this SRP-LR.

3.2.1 Areas of Review

This section addresses the aging management review (AMR) and the associated aging management program (AMP) of the engineered safety features. For a recent vintage plant, the information related to the engineered safety features is contained in Chapter 6, "Engineered Safety Features," of the plant's FSAR, consistent with the "Standard Review Plan for the Review of Safety Analysis Reports for Nuclear Power Plants" (NUREG-0800). The engineered safety features contained in this review plan section are generally consistent with those contained in NUREG-0800 except for the refueling water, control room habitability, and residual heat removal systems. For older plants, the location of applicable information is plant-specific because an older plant's Final Safety Analysis Report (FSAR) may have predated NUREG-0800.

The engineered safety features consist of containment spray, standby gas treatment [Boiling Water Reactors (BWRs)], containment isolation components, and emergency core cooling systems.

The responsible review organization is to review the following license renewal application (LRA) AMR and AMP items assigned to it, per SRP-LR Section 3.0:

AMRs

- AMR results consistent with the GALL Report
- AMR results for which further evaluation is recommended by the GALL Report
- AMR results not consistent with or not addressed in the GALL Report

AMPs

- Consistent with GALL Report AMPs
- Plant-specific AMPs

FSAR Supplement

- The responsible review organization is to review the FSAR Supplement associated with each assigned AMP.

3.2.2 Acceptance Criteria

The acceptance criteria for the areas of review describe methods for determining whether the applicant has met the requirements of the NRC's regulations in 10 CFR 54.21.

3.2.2.1 AMR Results Consistent with the GALL Report

The AMR and the AMPs applicable to the engineered safety features are described and evaluated in Chapter V of NUREG-1801 (GALL Report).

The applicant's LRA should provide sufficient information so that the NRC reviewer is able to confirm that the specific LRA AMR item and the associated LRA AMP are consistent with the cited GALL Report AMR item. The reviewer should then confirm that the LRA AMR item is consistent with the GALL Report AMR item to which it is compared.

When the applicant is crediting a different aging management program than recommended in the GALL Report, the reviewer should confirm that the alternate aging management program is valid to use for aging management and will be capable of managing the effects of aging as adequately as the aging management program recommended by the GALL Report.

3.2.2.2   AMR Results for Which Further Evaluation is Recommended by the GALL Report

The basic acceptance criteria defined in Subsection 3.2.2.1 need to be applied first for all of the AMRs and AMPs reviewed as part of this section. In addition, if the GALL Report AMR item to which the LRA AMR item is compared identifies that "further evaluation is recommended," then additional criteria apply as identified by the GALL Report for each of the following aging effect/aging mechanism combinations. Refer to Table 3.2-1, comparing the "Further Evaluation Recommended" and the "Rev2 Item" columns, for the AMR items that reference the following subsections. The 2005 AMR item counterpart is provided in the "Rev1 Item" column.

3.2.2.2.1   Cumulative Fatigue Damage

Fatigue is a time-limited aging analysis (TLAA) as defined in 10 CFR 54.3. TLAAs are required to be evaluated in accordance with 10 CFR 54.21(c). This TLAA is addressed separately in Section 4.3, "Metal Fatigue Analysis," of this SRP-LR.

3.2.2.2.2   Loss of Material due to Cladding Breach

Loss of material due to cladding breach could occur for PWR steel pump casings with stainless steel cladding exposed to treated borated water. The GALL Report references NRC Information Notice 94-63, Boric Acid Corrosion of Charging Pump Casings Caused by Cladding Cracks, and recommends further evaluation of a plant-specific AMP to ensure that the aging effect is adequately managed. Acceptance criteria are described in Branch Technical Position RLSB-1 (Appendix A.1 of this SRP-LR).

3.2.2.2.3   Loss of Material due to Pitting and Crevice Corrosion

1. Loss of material due to pitting and crevice corrosion could occur in partially encased stainless steel tanks exposed to raw water due to cracking of the perimeter seal from weathering. The GALL Report recommends further evaluation to ensure that the aging effect is adequately managed. The GALL Report recommends that a plant-specific AMP be evaluated because moisture and water can egress under the tank if the perimeter seal is degraded. Acceptance criteria are described in Branch Technical Position RSLB-1 (Appendix A.1 of this SRP-LR).

2. Loss of material due to pitting and crevice corrosion could occur for stainless steel piping, piping components, piping elements, and tanks exposed to outdoor air. The possibility of pitting and crevice corrosion also extends to components exposed to air which has recently been introduced into buildings, i.e., components near intake vents. Pitting and crevice corrosion is only known to occur in environments containing sufficient

halides (primarily chlorides) and in which condensation or deliquescence is possible. Condensation or deliquescence should generally be assumed to be possible. Applicable outdoor air environments (and associated indoor air environments) include, but are not limited to, those within approximately 5 miles of a saltwater coastline, those within 1/2 mile of a highway which is treated with salt in the wintertime, those areas in which the soil contains more than trace chlorides, those plants having cooling towers where the water is treated with chlorine or chlorine compounds, and those areas subject to chloride contamination from other agricultural or industrial sources. This item is applicable for the environments described above.

GALL AMP XI.M36, "External Surfaces Monitoring," is an acceptable method to manage the aging effect. The applicant may demonstrate that this item is not applicable by describing the outdoor air environment present at the plant and demonstrating that external pitting or crevice corrosion is not expected. The GALL Report recommends further evaluation to determine whether an aging management program is needed to manage this aging effect based on the environmental conditions applicable to the plant and requirements applicable to the components.

3.2.2.2.4   Loss of Material due to Erosion

Loss of material due to erosion could occur in the stainless steel high-pressure safety injection (HPSI) pump miniflow recirculation orifice exposed to treated borated water. The GALL Report recommends a plant-specific AMP be evaluated for erosion of the orifice due to extended use of the centrifugal HPSI pump for normal charging. The GALL Report references Licensee Event Report (LER) 50-275/94-023 for evidence of erosion. Further evaluation is recommended to ensure that the aging effect is adequately managed. Acceptance criteria are described in Branch Technical Position RSLB-1 (Appendix A.1 of this SRP-LR).

3.2.2.2.5   Loss of Material due to General Corrosion and Fouling that Leads to Corrosion

Loss of material due to general corrosion and fouling that leads to corrosion can occur for steel drywell and suppression chamber spray system nozzle and flow orifice internal surfaces exposed to air - indoor uncontrolled. This could result in plugging of the spray nozzles and flow orifices. This aging mechanism and effect will apply since the spray nozzles and flow orifices are occasionally wetted, even though the majority of the time this system is on standby. The wetting and drying of these components can accelerate corrosion and fouling. The GALL Report recommends further evaluation of a plant-specific AMP to ensure that the aging effect is adequately managed. Acceptance criteria are described in Branch Technical Position RSLB-1 (Appendix A.1 of this SRP-LR).

3.2.2.2.6   Cracking due to Stress Corrosion Cracking

Cracking due to stress corrosion cracking could occur for stainless steel piping, piping components, piping elements and tanks exposed to outdoor air. The possibility of cracking also extends to components exposed to air which has recently been introduced into buildings, i.e., components near intake vents. Cracking is only known to occur in environments containing sufficient halides (primarily chlorides) and in which condensation or deliquescence is possible. Condensation or deliquescence should generally be assumed to be possible. Applicable outdoor air environments (and associated indoor air environments) include, but are not limited to, those within approximately 5 miles of a saltwater coastline, those within 1/2 mile of a highway which is treated with salt in the wintertime, those areas in which the soil contains more than trace chlorides, those plants having cooling towers where the water is treated with chlorine or chlorine compounds, and those areas subject to chloride contamination from other agricultural or industrial sources. This item is applicable for the environments described above.

GALL AMP XI.M36, "External Surfaces Monitoring," is an acceptable method to manage the aging effect. The applicant may demonstrate that this item is not applicable by describing the outdoor air environment present at the plant and demonstrating that external chloride stress corrosion cracking is not expected. The GALL Report recommends further evaluation to determine whether an aging management program is needed to manage this aging effect based on the environmental conditions applicable to the plant and requirements applicable to the components.

3.2.2.2.7   Quality Assurance for Aging Management of Nonsafety-Related Components

Acceptance criteria are described in Branch Technical Position IQMB-1 (Appendix A.2 of this SRP-LR.)

3.2.2.3   AMR Results Not Consistent with or Not Addressed in the GALL Report

Acceptance criteria are described in Branch Technical Position RSLB-1 (Appendix A.1 of this SRP-LR).

3.2.2.4   Aging Management Programs

For those AMPs that will be used for aging management and are based on the program elements of an AMP in the GALL Report, the NRC reviewer performs an audit of aging management programs credited in the LRA to confirm consistency with the GALL AMPs identified in the GALL Report, Chapters X and XI.

If the applicant identifies an exception to any of the program elements of the cited GALL Report AMP, the LRA AMP should include a basis demonstrating how the criteria of 10 CFR 54.21(a)(3) would still be met. The NRC reviewer should then confirm that the LRA AMP with all exceptions would satisfy the criteria of 10 CFR 54.21(a)(3). If, while reviewing the LRA AMP, the reviewer identifies a difference between the LRA AMP and the GALL Report AMP that should have been identified as an exception to the GALL Report AMP, the difference should be reviewed and properly dispositioned. The reviewer should document the disposition of all LRA-defined exceptions and staff-identified differences.

The LRA should identify any enhancements that are needed to permit an existing AMP to be declared consistent with the GALL Report AMP to which the LRA AMP is compared. The reviewer is to confirm both that the enhancement, when implemented, would allow the existing plant AMP to be consistent with the GALL Report AMP and also that the applicant has a commitment in the FSAR Supplement to implement the enhancement prior to the period of extended operation. The reviewer should review and document the disposition of all enhancements.

If the applicant chooses to use a plant-specific program that is not a GALL AMP, the NRC reviewer should confirm that the plant-specific program satisfies the criteria of Branch Technical Position RLSB-1 (Appendix A.1.2.3 of this SRP-LR).

3.2.2.5   FSAR Supplement

The summary description of the programs and activities for managing the effects of aging for the period of extended operation in the FSAR Supplement should be sufficiently comprehensive, such that later changes can be controlled by 10 CFR 50.59. The description should contain information associated with the bases for determining that aging effects will be managed during the period of extended operation. The description should also contain any future aging management activities, including enhancements and commitments, to be completed before the period of extended operation. Table 3.0-1 of this SRP-LR provides examples of the type of information to be included in the FSAR Supplement. Table 3.2-2 lists the programs that are applicable for this SRP-LR subsection.

3.2.3   Review Procedures

For each area of review, the following review procedures are to be followed:

3.2.3.1   AMR Results Consistent with the GALL Report

The applicant may reference the GALL Report in its LRA, as appropriate, and demonstrate that the AMRs and AMPs at its facility are consistent with those reviewed and approved in the GALL Report. The reviewer should not conduct a re-review of the substance of the matters described in the GALL Report. If the applicant has provided the information necessary to adopt the finding of program acceptability as described and evaluated in the GALL Report, the reviewer should find acceptable the applicant's reference to the GALL Report in its LRA. In making this determination, the reviewer confirms that the applicant has provided a brief description of the system, components, materials, and environment. The reviewer also confirms that the applicant has stated that the applicable aging effects and industry and plant-specific operating experience have been reviewed by the applicant and are evaluated in the GALL Report.

Furthermore, the reviewer should confirm that the applicant has addressed operating experience identified after the issuance of the GALL Report. Performance of this review requires the reviewer to confirm that the applicant has identified those aging effects for the engineered safety features system components that are contained in the GALL Report as applicable to its plant.

3.2.3.2   AMR Results for Which Further Evaluation is Recommended by the GALL Report

The basic review procedures defined in Subsection 3.2.3.1 need to be applied first to all of the AMRs and AMPs provided in this section. In addition, if the GALL Report AMR item to which the LRA AMR item is compared identifies that "further evaluation is recommended," then additional criteria apply as identified by the GALL Report for each of the following aging effect/aging mechanism combinations.

3.2.3.2.1   Cumulative Fatigue Damage

Fatigue is a TLAA as defined in 10 CFR 54.3. TLAAs are required to be evaluated in accordance with 10 CFR 54.21(c). The staff reviews the evaluation of this TLAA separately, following the guidance in Section 4.3 of this SRP-LR.

3.2.3.2.2   Loss of Material due to Cladding Breach

The GALL Report recommends further evaluation of programs to manage loss of material due to cladding breach for PWR steel charging pump casings with stainless steel cladding. The GALL Report references NRC Information Notice 94-63, Boric Acid Corrosion of Charging Pump Casings Caused by Cladding Cracks, and recommends further evaluation of a plant-specific program to ensure that the aging effect is adequately managed. The reviewer reviews the applicant's proposed program on a case-by-case basis to ensure that an adequate program will be in place for the management of general corrosion of these components.

3.2.3.2.3   Loss of Material due to Pitting and Crevice Corrosion

1. The GALL Report recommends further evaluation of programs to manage the loss of material due to pitting and crevice corrosion for partially encased stainless steel tanks exposed to raw water. The GALL Report specifically recommends that the program address the bottom of partially encased stainless steel tanks because moisture and water can egress under the tank due to cracking of the perimeter seal from weathering. The reviewer reviews the applicant's proposed program on a case-by-case basis to ensure that an adequate program will be in place for the management of loss of material due to pitting and crevice corrosion of these components

2. The GALL Report recommends further evaluation to manage loss of material due to pitting and crevice corrosion of stainless steel piping, piping components, piping elements, and tanks exposed to outdoor air environments containing sufficient halides (primarily chlorides) and in which condensation or deliquescence is possible. The possibility of pitting and crevice corrosion also extends to components exposed to air which has recently been introduced into buildings, i.e., components near intake vents.

    The reviewer should determine whether an adequate program is credited to manage the aging effect based on the applicable environmental conditions. Pitting and crevice corrosion is only known to occur in environments containing sufficient halides (primarily chlorides) and in which condensation or deliquescence is possible. Condensation or deliquescence should generally be assumed to be possible. Applicable outdoor air environments (and associated indoor air environments) include, but are not limited to, those within approximately 5 miles of a saltwater coastline, those within 1/2 mile of a highway which is treated with salt in the wintertime, those areas in which the soil contains more than trace chlorides, those plants having cooling towers where the water is treated with chlorine or chlorine compounds, and those areas subject to chloride contamination from other agricultural or industrial sources. This item is applicable for the environments described above. The use of GALL AMP XI.M36, "External Surfaces Monitoring," is an acceptable method to manage the aging effect.

3.2.3.2.4   Loss of Material due to Erosion

The GALL Report recommends further evaluation of programs to manage loss of material due to erosion of the stainless steel high pressure safety injection pump miniflow orifice. The reviewer reviews the applicant's proposed program on a case-by-case basis to ensure that an adequate program will be in place to manage this aging effect.

3.2.3.2.5 Loss of Material due to General Corrosion and Fouling that Leads to Corrosion

The GALL Report recommends further evaluation of programs to manage loss of material due to general corrosion and fouling that leads to corrosion for steel drywell and suppression chamber spray system spray nozzles and orifices exposed to air - indoor uncontrolled. This is necessary to prevent the plugging of spray nozzles and spargers of the BWR drywell and suppression chamber spray system. The reviewer reviews the applicant's proposed program on a case-by-case basis to ensure that an adequate program will be in place for the management of loss of material due to general corrosion and fouling of these components.

3.2.3.2.6 Cracking due to Stress Corrosion Cracking

The GALL Report recommends further evaluation to manage cracking due to stress corrosion cracking of stainless steel piping, piping components, piping elements, and tanks exposed to outdoor air environments containing sufficient halides (primarily chlorides) and in which condensation or deliquescence is possible. The possibility of cracking also extends to components exposed to air which has recently been introduced into buildings, i.e., components near intake vents.

The reviewer should determine whether an adequate program is credited to manage the aging effect based on the applicable environmental conditions. Cracking is only known to occur in environments containing sufficient halides (primarily chlorides) and in which condensation or deliquescence is possible. Condensation or deliquescence should generally be assumed to be possible. Applicable outdoor air environments (and associated indoor air environments) include, but are not limited to, those within approximately 5 miles of a saltwater coastline, those within 1/2 mile of a highway which is treated with salt in the wintertime, those areas in which the soil contains more than trace chlorides, those plants having cooling towers where the water is treated with chlorine or chlorine compounds, and those areas subject to chloride contamination from other agricultural or industrial sources. This item is applicable for the environments described above. The use of GALL AMP XI.M36, "External Surfaces Monitoring," is an acceptable method to manage the aging effect.

3.2.3.2.7 Quality Assurance for Aging Management of Nonsafety-Related Components

The applicant's AMPs for license renewal should contain the elements of corrective actions, the confirmation process, and administrative controls. Safety-related components are covered by 10 CFR Part 50, Appendix B, which is adequate to address these program elements. However, Appendix B does not apply to nonsafety-related components that are subject to an aging management review for license renewal. Nevertheless, the applicant has the option to expand the scope of its 10 CFR Part 50, Appendix B program to include these components and address the associated program elements. If the applicant chooses this option, the reviewer verifies that the applicant has documented such a commitment in the FSAR Supplement. If the applicant chooses alternative means, the branch responsible for quality assurance should be requested to review the applicant's proposal on a case-by-case basis.

3.2.3.3 AMR Results Not Consistent with or Not Addressed in the GALL Report

The reviewer should confirm that the applicant, in its LRA, has identified applicable aging effects, listed the appropriate combination of materials and environments, and AMPs that will adequately manage the aging effects. The AMP credited by the applicant could be an AMP that is described and evaluated in the GALL Report or a plant-specific program. Review procedures are described in Branch Technical Position RSLB-1 (Appendix A.1 of this SRP-LR).

3.2.3.4   Aging Management Programs

The reviewer confirms that the applicant has identified the appropriate AMPs as described and evaluated in the GALL Report. If the applicant commits to an enhancement to make its LRA AMP consistent with a GALL Report AMP, then the reviewer is to confirm that this enhancement, when implemented, will make the LRA AMP consistent with the GALL Report AMP. If the applicant identifies, in the LRA AMP, an exception to any of the program elements of the GALL Report AMP, the reviewer is to confirm that the LRA AMP with the exception will satisfy the criteria of 10 CFR 54.21(a)(3). If the reviewer identifies a difference, not identified by the LRA, between the LRA AMP and the GALL Report AMP, with which the LRA claims to be consistent, the reviewer should confirm that the LRA AMP with this difference satisfies 10 CFR 54.21(a)(3). The reviewer should document the basis for accepting enhancements, exceptions or differences. The AMPs evaluated in the GALL Report pertinent to the engineered safety features components are summarized in Table 3.2-1 of this SRP-LR. The "Rev2 Item" (for 2010) and "Rev1 Item" (for 2005 counterpart) columns identify the AMR item numbers in the GALL Report, Chapter V, presenting detailed information summarized by this row.

Table 3.2-1 of this SRP-LR may identify a plant-specific aging management program. If the applicant chooses to use a plant-specific program that is not a GALL AMP, the NRC reviewer should confirm that the plant-specific program satisfies the criteria of Branch Technical Position RLSB-1 (Appendix A.1.2.3 of this SRP-LR).

3.2.3.5   FSAR Supplement

The reviewer confirms that the applicant has provided in its FSAR supplement information equivalent to that in Table 3.0-1 for aging management of the engineered safety features. Table 3.2-2 lists the AMPs that are applicable for this SRP-LR subsection. The reviewer also confirms that the applicant has provided information for Subsection 3.2.3.3, "AMR Results Not Consistent With or Not Addressed in the GALL Report," equivalent to that in Table 3.0-1.

The staff expects to impose a license condition on any renewed license to require the applicant to update its FSAR to include this FSAR Supplement at the next update required pursuant to 10 CFR 50.71(e)(4). As part of the license condition until the FSAR update is complete, the applicant may make changes to the programs described in its FSAR Supplement without prior NRC approval, provided that the applicant evaluates each such change and finds it acceptable pursuant to the criteria set forth in 10 CFR 50.59. If the applicant updates the FSAR to include the final FSAR supplement before the license is renewed, no condition will be necessary.

As noted in Table 3.0-1, an applicant need not incorporate the implementation schedule into its FSAR. However, the reviewer should confirm that the applicant has identified and committed in the LRA to any future aging management activities, including enhancements and commitments, to be completed before entering the period of extended operation. The staff expects to impose a license condition on any renewed license to ensure that the applicant will complete these activities no later than the committed date.

3.2.4 Evaluation Findings

If the reviewer determines that the applicant has provided information sufficient to satisfy the provisions of this section, then an evaluation finding similar to the following text should be included in the staff's safety evaluation report:

> On the basis of its review, as discussed above, the staff concludes that the applicant has demonstrated that the aging effects associated with the engineered safety features systems components will be adequately managed so that the intended functions will be maintained consistent with the CLB for the period of extended operation, as required by 10 CFR 54.21(a)(3).

> The staff also reviewed the applicable FSAR Supplement program summaries and concludes that they adequately describe the AMPs credited for managing aging of the engineered safety features systems, as required by 10 CFR 54.21(d).

3.2.5 Implementation

Except in those cases in which the applicant proposes an acceptable alternative method for complying with specified portions of the NRC's regulations, the method described herein will be used by the staff in its evaluation of conformance with NRC regulations.

3.2.6 References

1. NUREG-0800, "Standard Review Plan for the Review of Safety Analysis Reports for Nuclear Power Plants, LWR Edition" U.S. Nuclear Regulatory Commission, March 2007.

2. NUREG-1801, "Generic Aging Lessons Learned (GALL) Report," U.S. Nuclear Regulatory Commission, Revision 2, 2010.

3. NEI 95-10, "Industry Guideline for Implementing the Requirements of 10 CFR Part 54 – The License Renewal Rule," Nuclear Energy Institute, Revision 6.

Table 3.2-1   Summary of Aging Management Programs for Engineered Safety Features Evaluated in Chapter V of the GALL Report

ID	Type	Component	Aging Effect/Mechanism	Aging Management Programs	Further Evaluation Recommended	Rev2 Item	Rev1 Item
1	BWR/PWR	Stainless steel, Steel Piping, piping components, and piping elements exposed to Treated water (borated)	Cumulative fatigue damage due to fatigue	Fatigue is a time-limited aging analysis (TLAA) to be evaluated for the period of extended operation. See the SRP, Section 4.3 "Metal Fatigue," for acceptable methods for meeting the requirements of 10 CFR 54.21(c)(1).	Yes, TLAA (See subsection 3.2.2.2.1)	V.D1.E-13 V.D2.E-10	V.D1-27(E-13) V.D2-32(E-10)
2	PWR	Steel (with stainless steel cladding) Pump casings exposed to Treated water (borated)	Loss of material due to cladding breach	A plant-specific aging management program is to be evaluated Reference NRC Information Notice 94-63, "Boric Acid Corrosion of Charging Pump Casings Caused by Cladding Cracks."	Yes, verify that plant-specific program addresses clad breach (See subsection 3.2.2.2.2)	V.D1.EP-49	V.D1-32(EP-49)
3	PWR	Stainless steel Partially-encased tanks with breached moisture barrier exposed to Raw water	Loss of material due to pitting and crevice corrosion	A plant-specific aging management program is to be evaluated for pitting and crevice corrosion of tank bottom because moisture and water can egress under the tank due to cracking of the perimeter seal from weathering.	Yes, plant-specific (See subsection 3.2.2.3.1)	V.D1.E-01	V.D1-15(E-01)

Table 3.2-1 Summary of Aging Management Programs for Engineered Safety Features Evaluated in Chapter V of the GALL Report

ID	Type	Component	Aging Effect/Mechanism	Aging Management Programs	Further Evaluation Recommended	Rev2 Item	Rev1 Item
4	BWR/PWR	Stainless steel Piping, piping components, and piping elements; tanks exposed to Air – outdoor	Loss of material due to pitting and crevice corrosion	Chapter XI.M36, "External Surfaces Monitoring of Mechanical Components"	Yes, environmental conditions need to be evaluated (See subsection 3.2.2.3.2)	V.B.EP-107 V.C.EP-107 V.D1.EP-107 V.D2.EP-107	N/A N/A N/A N/A
5	PWR	Stainless steel Orifice (miniflow recirculation) exposed to Treated water (borated)	Loss of material due to erosion	A plant-specific aging management program is to be evaluated for erosion of the orifice due to extended use of the centrifugal HPSI pump for normal charging. See LER 50-275/94-023 for evidence of erosion.	Yes, plant-specific (See subsection 3.2.2.2.4)	V.D1.E-24	V.D1-14(E-24)
6	BWR	Steel Drywell and suppression chamber spray system (internal surfaces): flow orifice; spray nozzles exposed to Air – indoor, uncontrolled (Internal)	Loss of material due to general corrosion; fouling that leads to corrosion	A plant-specific aging management program is to be evaluated	Yes, plant-specific (See subsection 3.2.2.2.5)	V.D2.EP-113	V.D2-1(E-04)
7	BWR/PWR	Stainless steel Piping, piping components, and piping elements; tanks exposed to Air – outdoor	Cracking due to stress corrosion cracking	Chapter XI.M36, "External Surfaces Monitoring of Mechanical Components"	Yes, environmental conditions need to be evaluated (See subsection 3.2.2.6)	V.B.EP-103 V.C.EP-103 V.D1.EP-103 V.D2.EP-103	N/A N/A N/A N/A
8	PWR	Aluminum, Copper alloy (>15% Zn or >8% Al) Piping, piping components, and piping elements exposed to Air with borated water leakage	Loss of material due to boric acid corrosion	Chapter XI.M10, "Boric Acid Corrosion"	No	V.D1.EP-101 V.E.EP-38	V.D2-18(EP-2) V.E-11(EP-38)

Table 3.2-1   Summary of Aging Management Programs for Engineered Safety Features Evaluated in Chapter V of the GALL Report

ID	Type	Component	Aging Effect/Mechanism	Aging Management Programs	Further Evaluation Recommended	Rev2 Item	Rev1 Item
9	PWR	Steel External surfaces, Bolting exposed to Air with borated water leakage	Loss of material due to boric acid corrosion	Chapter XI.M10, "Boric Acid Corrosion"	No	V.A.E-28 V.D1.E-28 V.E.E-28 V.E.E-41	V.A-4(E-28) V.D1-1(E-28) V.E-9(E-28) V.E-2(E-41)
10	BWR/PWR	Cast austenitic stainless steel Piping, piping components, and piping elements exposed to Treated water (borated) >250°C (>482°F), Treated water >250°C (>482°F)	Loss of fracture toughness due to thermal aging embrittlement	Chapter XI.M12, "Thermal Aging Embrittlement of Cast Austenitic Stainless Steel (CASS)"	No	V.D1.E-47 V.D2.E-11	V.D1-16(E-47) V.D2-20(E-11)
11	BWR	Steel Piping, piping components, and piping elements exposed to Steam, Treated water	Wall thinning due to flow-accelerated corrosion	Chapter XI.M17, "Flow-Accelerated Corrosion"	No	V.D2.E-07 V.D2.E-09	V.D2-31(E-07) V.D2-34(E-09)
12	BWR/PWR	Steel, high-strength Closure bolting exposed to Air with steam or water leakage	Cracking due to cyclic loading, stress corrosion cracking	Chapter XI.M18, "Bolting Integrity"	No	V.E.E-03	V.E-3(E-03)
13	BWR/PWR	Steel; stainless steel Bolting, Closure bolting exposed to Air – outdoor (External), Air – indoor, uncontrolled (External)	Loss of material due to general (steel only), pitting, and crevice corrosion	Chapter XI.M18, "Bolting Integrity"	No	V.E.EP-64 V.E.EP-70	V.E-1(EP-1) V.E-4(EP-25)
14	BWR/PWR	Steel Closure bolting exposed to Air with steam or water leakage	Loss of material due to general corrosion	Chapter XI.M18, "Bolting Integrity"	No	V.E.E-02	V.E-6(E-02)

Table 3.2-1 Summary of Aging Management Programs for Engineered Safety Features Evaluated in Chapter V of the GALL Report

ID	Type	Component	Aging Effect/Mechanism	Aging Management Programs	Further Evaluation Recommended	Rev2 Item	Rev1 Item
15	BWR/PWR	Copper alloy, Nickel alloy, Steel; stainless steel, Stainless steel, Steel; stainless steel Bolting, Closure bolting exposed to Any environment, Air – outdoor (External), Raw water, Treated borated water, Fuel oil, Treated water, Air – indoor, uncontrolled (External)	Loss of preload due to thermal effects, gasket creep, and self-loosening	Chapter XI.M18, "Bolting Integrity"	No	V.E.EP-116 V.E.EP-117 V.E.EP-118 V.E.EP-119 V.E.EP-120 V.E.EP-121 V.E.EP-122 V.E.EP-69	N/A N/A N/A N/A N/A N/A V.E-5(EP-24)
16	BWR/PWR	Steel Containment isolation piping and components (Internal surfaces), Piping, piping components, and piping elements exposed to Treated water	Loss of material due to general, pitting, and crevice corrosion	Chapter XI.M2, "Water Chemistry," and Chapter XI.M32, "One-Time Inspection"	No	V.C.EP-62 V.D2.EP-60	V.C-6(E-31) V.D2-33(E-08)
17	BWR	Aluminum, Stainless steel Piping, piping components, and piping elements exposed to Treated water	Loss of material due to pitting and crevice corrosion	Chapter XI.M2, "Water Chemistry," and Chapter XI.M32, "One-Time Inspection"	No	V.D2.EP-71 V.D2.EP-73	V.D2-19(EP-26) V.D2-28(EP-32)
18	BWR/PWR	Stainless steel Containment isolation piping and components (Internal surfaces) exposed to Treated water	Loss of material due to pitting and crevice corrosion	Chapter XI.M2, "Water Chemistry," and Chapter XI.M32, "One-Time Inspection"	No	V.C.EP-63	V.C-4(E-33)
19	BWR/PWR	Stainless steel Heat exchanger tubes exposed to Treated water	Reduction of heat transfer due to fouling	Chapter XI.M2, "Water Chemistry," and Chapter XI.M32, "One-Time Inspection"	No	V.A.EP-74 V.D2.EP-74	V.A-16(EP-34) V.D2-13(EP-34)

Table 3.2-1 Summary of Aging Management Programs for Engineered Safety Features Evaluated in Chapter V of the GALL Report

ID	Type	Component	Aging Effect/Mechanism	Aging Management Programs	Further Evaluation Recommended	Rev2 Item	Rev1 Item
20	PWR	Stainless steel Piping, piping components, and piping elements; tanks exposed to Treated water (borated) >60°C (>140°F)	Cracking due to stress corrosion cracking	Chapter XI.M2, "Water Chemistry"	No	V.A.E-12 V.D1.E-12	V.A-28(E-12) V.D1-31(E-12)
21	PWR	Steel (with stainless steel or nickel-alloy cladding) Safety injection tank (accumulator) exposed to Treated water (borated) >60°C (>140°F)	Cracking due to stress corrosion cracking	Chapter XI.M2, "Water Chemistry"	No	V.D1.E-38	V.D1-33(E-38)
22	PWR	Stainless steel Piping, piping components, and piping elements; tanks exposed to Treated water (borated)	Loss of material due to pitting and crevice corrosion	Chapter XI.M2, "Water Chemistry"	No	V.A.EP-41 V.D1.EP-41	V.A-27(EP-41) V.D1-30(EP-41)
23	BWR/PWR	Steel Heat exchanger components, Containment isolation piping and components (Internal surfaces) exposed to Raw water	Loss of material due to general, pitting, crevice, and microbiologically-influenced corrosion; fouling that leads to corrosion	Chapter XI.M20, "Open-Cycle Cooling Water System"	No	V.A.EP-90 V.C.E-22 V.D1.EP-90 V.D2.EP-90	V.A-10(E-18) V.C-5(E-22) V.D1-7(E-18) V.D2-8(E-18)
24	PWR	Stainless steel Piping, piping components, and piping elements exposed to Raw water	Loss of material due to pitting, crevice, and microbiologically-influenced corrosion	Chapter XI.M20, "Open-Cycle Cooling Water System"	No	V.D1.EP-55	V.D1-25(EP-55)

Table 3.2-1 Summary of Aging Management Programs for Engineered Safety Features Evaluated in Chapter V of the GALL Report

ID	Type	Component	Aging Effect/Mechanism	Aging Management Programs	Further Evaluation Recommended	Rev2 Item	Rev1 Item
25	BWR/PWR	Stainless steel Heat exchanger components, Containment isolation piping and components (Internal surfaces) exposed to Raw water	Loss of material due to pitting, crevice, and microbiologically-influenced corrosion; fouling that leads to corrosion	Chapter XI.M20, "Open-Cycle Cooling Water System"	No	V.A.EP-91 V.C.E-34 V.D1.EP-91 V.D2.EP-91	V.A-8(E-20) V.C-3(E-34) V.D1-5(E-20) V.D2-6(E-20)
26	BWR	Stainless steel Heat exchanger tubes exposed to Raw water	Reduction of heat transfer due to fouling	Chapter XI.M20, "Open-Cycle Cooling Water System"	No	V.D2.E-21	V.D2-12(E-21)
27	BWR/PWR	Stainless steel, Steel Heat exchanger tubes exposed to Raw water	Reduction of heat transfer due to fouling	Chapter XI.M20, "Open-Cycle Cooling Water System"	No	V.A.E-21 V.D1.E-21 V.D2.E-23	V.A-15(E-21) V.D1-11(E-21) V.D2-15(E-23)
28	BWR/PWR	Stainless steel Piping, piping components, and piping elements exposed to Closed-cycle cooling water >60°C (>140°F)	Cracking due to stress corrosion cracking	Chapter XI.M21A, "Closed Treated Water Systems"	No	V.A.EP-98 V.C.EP-98 V.D1.EP-98 V.D2.EP-98	V.A-24(EP-44) V.C-8(EP-44) V.D1-23(EP-44) V.D2-26(EP-44)
29	BWR/PWR	Steel Piping, piping components, and piping elements exposed to Closed-cycle cooling water	Loss of material due to general, pitting, and crevice corrosion	Chapter XI.M21A, "Closed Treated Water Systems"	No	V.C.EP-99	V.C-9(EP-48)
30	BWR/PWR	Steel Heat exchanger components exposed to Closed-cycle cooling water	Loss of material due to general, pitting, crevice, and galvanic corrosion	Chapter XI.M21A, "Closed Treated Water Systems"	No	V.A.EP-92 V.D1.EP-92 V.D2.EP-92	V.A-9(E-17) V.D1-6(E-17) V.D2-7(E-17)

Table 3.2-1   Summary of Aging Management Programs for Engineered Safety Features Evaluated in Chapter V of the GALL Report

ID	Type	Component	Aging Effect/Mechanism	Aging Management Programs	Further Evaluation Recommended	Rev2 Item	Rev1 Item
31	BWR/PWR	Stainless steel Heat exchanger components, Piping, piping components, and piping elements exposed to Closed-cycle cooling water	Loss of material due to pitting and crevice corrosion	Chapter XI.M21A, "Closed Treated Water Systems"	No	V.A.EP-93, V.A.EP-95, V.C.EP-95, V.D1.EP-93, V.D1.EP-95, V.D2.EP-93, V.D2.EP-95	V.A-7(E-19), V.A-23(EP-33), V.C-7(EP-33), V.D1-4(E-19), V.D1-22(EP-33), V.D2-5(E-19), V.D2-25(EP-33)
32	BWR/PWR	Copper alloy Heat exchanger components, Piping, piping components, and piping elements exposed to Closed-cycle cooling water	Loss of material due to pitting, crevice, and galvanic corrosion	Chapter XI.M21A, "Closed Treated Water Systems"	No	V.A.EP-94, V.A.EP-97, V.B.EP-97, V.D1.EP-94, V.D1.EP-97, V.D2.EP-94, V.D2.EP-97	V.A-5(EP-13), V.A-20(EP-36), V.B-6(EP-36), V.D1-2(EP-13), V.D1-17(EP-36), V.D2-3(EP-13), V.D2-21(EP-36)
33	BWR/PWR	Copper alloy, Stainless steel Heat exchanger tubes exposed to Closed-cycle cooling water	Reduction of heat transfer due to fouling	Chapter XI.M21A, "Closed Treated Water Systems"	No	V.A.EP-100, V.A.EP-96, V.D1.EP-96, V.D2.EP-96	V.A-11(EP-39), V.A-13(EP-35), V.D1-19(EP-35), V.D2-10(EP-35)
34	BWR/PWR	Copper alloy (>15% Zn or >8% Al) Piping, piping components, and piping elements, Heat exchanger components exposed to Closed-cycle cooling water	Loss of material due to selective leaching	Chapter XI.M33, "Selective Leaching"	No	V.A.EP-27, V.A.EP-37, V.B.EP-27, V.B.EP-37, V.D1.EP-27, V.D1.EP-37, V.D2.EP-27, V.D2.EP-37	V.A-22(EP-27), V.A-6(EP-37), V.B-7(EP-27), V.B-5(EP-37), V.D1-19(EP-27), V.D1-3(EP-37), V.D2-23(EP-27), V.D2-4(EP-37)
35	PWR	Gray cast iron Motor cooler exposed to Treated water	Loss of material due to selective leaching	Chapter XI.M33, "Selective Leaching"	No	V.A.E-43, V.D1.E-43	V.A-18(E-43), V.D1-13(E-43)
36	PWR	Gray cast iron Piping, piping components, and piping elements exposed to Closed-cycle cooling water	Loss of material due to selective leaching	Chapter XI.M33, "Selective Leaching"	No	V.D1.EP-52	V.D1-20(EP-52)

Table 3.2-1 Summary of Aging Management Programs for Engineered Safety Features Evaluated in Chapter V of the GALL Report

ID	Type	Component	Aging Effect/Mechanism	Aging Management Programs	Further Evaluation Recommended	Rev2 Item	Rev1 Item
37	BWR/PWR	Gray cast iron Piping, piping components, and piping elements exposed to Soil	Loss of material due to selective leaching	Chapter XI.M33, "Selective Leaching"	No	V.B.EP-54 V.D1.EP-54 V.D2.EP-54	V.B-8(EP-54) V.D1-21(EP-54) V.D2-24(EP-54)
38	BWR	Elastomers Elastomer seals and components exposed to Air – indoor, uncontrolled (External)	Hardening and loss of strength due to elastomer degradation	Chapter XI.M36, "External Surfaces Monitoring of Mechanical Components"	No	V.B.EP-59	V.B-4(E-06)
39	BWR/PWR	Steel Containment isolation piping and components (External surfaces) exposed to Condensation (External)	Loss of material due to general corrosion	Chapter XI.M36, "External Surfaces Monitoring of Mechanical Components"	No	V.C.E-30 V.E.E-46	V.C-2(E-30) V.E-10(E-46)
40	BWR/PWR	Steel Ducting, piping, and components (External surfaces), Ducting, closure bolting, Containment isolation piping and components (External surfaces) exposed to Air – indoor, uncontrolled (External)	Loss of material due to general corrosion	Chapter XI.M36, "External Surfaces Monitoring of Mechanical Components"	No	V.A.E-26 V.B.E-26 V.B.E-40 V.C.E-35 V.D2.E-26 V.E.E-44	V.A-1(E-26) V.B-3(E-26) V.B-2(E-40) V.C-1(E-35) V.D2-2(E-26) V.E-7(E-44)
41	BWR/PWR	Steel External surfaces exposed to Air – outdoor (External)	Loss of material due to general corrosion	Chapter XI.M36, "External Surfaces Monitoring of Mechanical Components"	No	V.E.E-45	V.E-8(E-45)
42	BWR/PWR	Aluminum Piping, piping components, and piping elements exposed to Air - outdoor	Loss of material due to pitting and crevice corrosion	Chapter XI.M36, "External Surfaces Monitoring of Mechanical Components"	No	V.E.EP-114	N/A

Table 3.2-1 Summary of Aging Management Programs for Engineered Safety Features Evaluated in Chapter V of the GALL Report

ID	Type	Component	Aging Effect/Mechanism	Aging Management Programs	Further Evaluation Recommended	Rev2 Item	Rev1 Item
43	BWR	Elastomers Elastomer seals and components exposed to Air – indoor, uncontrolled (Internal)	Hardening and loss of strength due to elastomer degradation	Chapter XI.M38, "Inspection of Internal Surfaces in Miscellaneous Piping and Ducting Components"	No	V.B.EP-58	V.B-4(E-06)
44	BWR/PWR	Steel Piping and components (Internal surfaces), Ducting and components (Internal surfaces) exposed to Air – indoor, uncontrolled (Internal)	Loss of material due to general corrosion	Chapter XI.M38, "Inspection of Internal Surfaces in Miscellaneous Piping and Ducting Components"	No	V.A.E-29 V.B.E-25 V.D2.E-29	V.A-19(E-29) V.B-1(E-25) V.D2-16(E-29)
45	PWR	Steel Encapsulation components exposed to Air – indoor, uncontrolled (Internal)	Loss of material due to general, pitting, and crevice corrosion	Chapter XI.M38, "Inspection of Internal Surfaces in Miscellaneous Piping and Ducting Components"	No	V.A.EP-42	V.A-2(EP-42)
46	BWR	Steel Piping and components (Internal surfaces) exposed to Condensation (Internal)	Loss of material due to general, pitting, and crevice corrosion	Chapter XI.M38, "Inspection of Internal Surfaces in Miscellaneous Piping and Ducting Components"	No	V.D2.E-27	V.D2-17(E-27)
47	PWR	Steel Encapsulation components exposed to Air with borated water leakage (Internal)	Loss of material due to general, pitting, crevice, and boric acid corrosion	Chapter XI.M38, "Inspection of Internal Surfaces in Miscellaneous Piping and Ducting Components"	No	V.A.EP-43	V.A-3(EP-43)

ID	Type	Component	Aging Effect/Mechanism	Aging Management Programs	Further Evaluation Recommended	Rev2 Item	Rev1 Item
48	BWR/PWR	Stainless steel Piping, piping components, and piping elements (Internal surfaces); tanks exposed to Condensation (Internal)	Loss of material due to pitting and crevice corrosion	Chapter XI.M38, "Inspection of Internal Surfaces in Miscellaneous Piping and Ducting Components"	No	V.A.EP-81 V.D1.EP-81 V.D2.EP-61	V.A-26(EP-53) V.D1-29(EP-53) V.D2-35(E-14)
49	BWR/PWR	Steel Piping, piping components, and piping elements exposed to Lubricating oil	Loss of material due to general, pitting, and crevice corrosion	Chapter XI.M39, "Lubricating Oil Analysis," and Chapter XI.M32, "One-Time Inspection"	No	V.A.EP-77 V.D1.EP-77 V.D2.EP-77	V.A-25(EP-46) V.D1-28(EP-46) V.D2-30(EP-46)
50	BWR/PWR	Copper alloy, Stainless steel Piping, piping components, and piping elements exposed to Lubricating oil	Loss of material due to pitting and crevice corrosion	Chapter XI.M39, "Lubricating Oil Analysis," and Chapter XI.M32, "One-Time Inspection"	No	V.A.EP-76 V.D1.EP-76 V.D1.EP-80 V.D2.EP-76	V.A-21(EP-45) V.D1-19(EP-45) V.D1-24(EP-51) V.D2-22(EP-45)
51	BWR/PWR	Steel, Copper alloy, Stainless steel Heat exchanger tubes exposed to Lubricating oil	Reduction of heat transfer due to fouling	Chapter XI.M39, "Lubricating Oil Analysis," and Chapter XI.M32, "One-Time Inspection"	No	V.A.EP-75 V.A.EP-78 V.A.EP-79 V.D1.EP-75 V.D1.EP-78 V.D1.EP-79 V.D2.EP-75 V.D2.EP-78 V.D2.EP-79	V.A-17(EP-40) V.A-12(EP-47) V.A-14(EP-50) V.D1-12(EP-40) V.D1-8(EP-47) V.D1-10(EP-50) V.D2-14(EP-40) V.D2-9(EP-47) V.D2-11(EP-50)
52	BWR/PWR	Steel (with coating or wrapping) Piping, piping components, and piping elements exposed to Soil or Concrete	Loss of material due to general, pitting, crevice, and microbiologically-influenced corrosion	Chapter XI.M41, "Buried and Underground Piping and Tanks"	No	V.B.EP-111	V.B-9(E-42)

Table 3.2-1 Summary of Aging Management Programs for Engineered Safety Features Evaluated in Chapter V of the GALL Report

ID	Type	Component	Aging Effect/Mechanism	Aging Management Programs	Further Evaluation Recommended	Rev2 Item	Rev1 Item
53	BWR/PWR	Stainless steel Piping, piping components, and piping elements exposed to Soil or Concrete	Loss of material due to pitting and crevice corrosion	Chapter XI.M41, "Buried and Underground Piping and Tanks"	No	V.D1.EP-72 V.D2.EP-72	V.D1-26(EP-31) V.D2-27(EP-31)
	BWR/PWR	Steel; stainless steel Underground piping, piping components, and piping elements exposed to air-indoor uncontrolled or condensation (external)	Loss of material due to general (steel only), pitting and crevice corrosion	Chapter XI.M41, "Buried and Underground Piping and Tanks"	No	V.E.EP-123	
54	BWR	Stainless steel Piping, piping components, and piping elements exposed to Treated water >60°C (>140°F)	Cracking due to stress corrosion cracking, intergranular stress corrosion cracking	Chapter XI.M7, "BWR Stress Corrosion Cracking," and Chapter XI.M2, "Water Chemistry"	No	V.D2.E-37	V.D2-29(E-37)
55	BWR/PWR	Steel Piping, piping components, and piping elements exposed to Concrete	None	None, provided 1) attributes of the concrete are consistent with ACI 318 or ACI 349 (low water-to-cement ratio, low permeability, and adequate air entrainment) as cited in NUREG-1557, and 2) plant OE indicates no degradation of the concrete	No, if conditions are met.	V.F.EP-112	V.F-17(EP-5)

Table 3.2-1 Summary of Aging Management Programs for Engineered Safety Features Evaluated in Chapter V of the GALL Report

ID	Type	Component	Aging Effect/Mechanism	Aging Management Programs	Further Evaluation Recommended	Rev2 Item	Rev1 Item
56	BWR/PWR	Aluminum Piping, piping components, and piping elements exposed to Air – indoor, uncontrolled (Internal/External)	None	None	NA – No AEM or AMP	V.F.EP-3	V.F-2(EP-3)
57	BWR/PWR	Copper alloy Piping, piping components, and piping elements exposed to Air – indoor, uncontrolled (External), Gas	None	None	NA – No AEM or AMP	V.F.EP-10 V.F.EP-9	V.F-3(EP-10) V.F-4(EP-9)
58	PWR	Copper alloy (≤15% Zn and ≤8% Al) Piping, piping components, and piping elements exposed to Air with borated water leakage	None	None	NA – No AEM or AMP	V.F.EP-12	V.F-5(EP-12)
59	BWR/PWR	Galvanized steel Ducting, piping, and components exposed to Air – indoor, controlled (External)	None	None	NA – No AEM or AMP	V.F.EP-14	V.F-1(EP-14)
60	BWR/PWR	Glass Piping elements exposed to Air – indoor, uncontrolled (External), Lubricating oil, Raw water, Treated water, Treated water (borated), Air with borated water leakage, Condensation (Internal/External), Gas, Closed-cycle cooling water, Air – outdoor	None	None	NA – No AEM or AMP	V.F.EP-15 V.F.EP-16 V.F.EP-28 V.F.EP-29 V.F.EP-30 V.F.EP-65 V.F.EP-66 V.F.EP-67 V.F.EP-68 V.F.EP-87	V.F-6(EP-15) V.F-7(EP-16) V.F-8(EP-28) V.F-10(EP-29) V.F-9(EP-30) N/A N/A N/A N/A
61	BWR/PWR	Nickel alloy Piping, piping components, and piping elements exposed to Air – indoor, uncontrolled (External)	None	None	NA – No AEM or AMP	V.F.EP-17	V.F-11(EP-17)

Table 3.2-1 Summary of Aging Management Programs for Engineered Safety Features Evaluated in Chapter V of the GALL Report

ID	Type	Component	Aging Effect/Mechanism	Aging Management Programs	Further Evaluation Recommended	Rev2 Item	Rev1 Item
62	BWR/PWR	Nickel alloy Piping, piping components, and piping elements exposed to Air with borated water leakage	None	None	NA - No AEM or AMP	V.F.EP-115	N/A
63	BWR/PWR	Stainless steel Piping, piping components, and piping elements exposed to Air – indoor, uncontrolled (External), Air with borated water leakage, Concrete, Gas, Air – indoor, uncontrolled (Internal)	None	None	NA - No AEM or AMP	V.F.EP-18 V.F.EP-19 V.F.EP-20 V.F.EP-22 V.F.EP-82	V.F-12(EP-18) V.F-13(EP-19) V.F-14(EP-20) V.F-15(EP-22) N/A
64	BWR/PWR	Steel Piping, piping components, and piping elements exposed to Air – indoor, controlled (External), Gas	None	None	NA - No AEM or AMP	V.F.EP-4 V.F.EP-7	V.F-16(EP-4) V.F-18(EP-7)

Table 3.2-2 Aging Management Programs Recommended for Engineered Safety Features

GALL Report Chapter/AMP	Program Name
Chapter XI.M2	Water Chemistry
Chapter XI.M7	BWR Stress Corrosion Cracking
Chapter XI.M10	Boric Acid Corrosion
Chapter XI.M12	Thermal Aging Embrittlement of Cast Austenitic Stainless Steel (CASS)
Chapter XI.M17	Flow-Accelerated Corrosion (FAC)
Chapter XI.M18	Bolting Integrity
Chapter XI.M20	Open-Cycle Cooling Water System
Chapter XI.M21A	Closed Treated Water Systems
Chapter XI.M32	One-Time Inspection
Chapter XI.M33	Selective Leaching
Chapter XI.M36	External Surfaces Monitoring of Mechanical Components
Chapter XI.M38	Inspection of Internal Surfaces in Miscellaneous Piping and Ducting Components
Chapter XI.M39	Lubricating Oil Analysis
Chapter XI.M41	Buried and Underground Piping and Tanks
Appendix for GALL	Quality Assurance for Aging Management Programs
SRP-LR Appendix A	Plant-specific AMP

3.3   AGING MANAGEMENT OF AUXILIARY SYSTEMS

Review Responsibilities

Primary - Branch assigned responsibility by PM as described in SRP-LR Section 3.0 of this SRP-LR.

3.3.1   Areas of Review

This section addresses the aging management review (AMR) and the associated aging management program (AMP) of the auxiliary systems for license renewal. For a recent vintage plant, the information related to the auxiliary systems contained in Chapter 9, "Auxiliary Systems," of the plant's FSAR consistent with the "Standard Review Plan for the Review of Safety Analysis Reports for Nuclear Power Plants" (NUREG-0800). The auxiliary systems contained in this review plan section are generally consistent with those contained in NUREG-0800 except for refueling water, chilled water, heat removal, condenser circulating water, and condensate storage system. For older plants, the location of applicable information is plant-specific because an older plant's FSAR may have predated NUREG-0800.

Typical auxiliary systems that are subject to an AMR for license renewal are new fuel storage, spent fuel storage, spent fuel pool cooling and cleanup (BWR/PWR), suppression pool cleanup (BWR), overhead heavy load and light load (related to refueling) handling, open-cycle cooling water, closed-cycle cooling water, ultimate heat sink, compressed air, chemical and volume control (PWR), standby liquid control (BWR), reactor water cleanup (BWR), shutdown cooling (older BWR), control room area ventilation, auxiliary and radwaste area ventilation, primary containment heating and ventilation, diesel generator building ventilation, fire protection, diesel fuel oil, and emergency diesel generator. This review plan section also includes structures and components in nonsafety-related systems that are not connected to safety related SSCs but have a spatial relationship such that their failure could adversely impact the performance of a safety related SSC intended function. Examples of such non-safety related systems may be plant drains, liquid waste processing, potable/sanitary water, water treatment, process sampling, and cooling water systems.

Aging management is reviewed, following the guidance in this SRP-LR Section 3.1, for portions of the chemical and volume control system for PWRs, and for standby liquid control, reactor water cleanup, and shutdown cooling systems extending up to the first isolation valve outside of containment for BWRs (the shutdown cooling systems for older BWRs). The following systems have portions that are classified as Group B quality standard: open-cycle cooling water (service water system), closed-cycle cooling water, compressed air, standby liquid control, shutdown cooling system (older BWR), control room area ventilation and auxiliary and radwaste area ventilation. Aging management for these portions is reviewed following the guidance in Section 3.3. The aging management program for the cooling towers is reviewed following the guidance in Section 3.5 for "Group 6" structures.

The responsible review organization is to review the following license renewal application (LRA) AMR and AMP items assigned to it, per SRP-LR Section 3.0:

AMRs
- AMR results consistent with the GALL Report
- AMR results for which further evaluation is recommended by the GALL Report

- AMR results not consistent with or not addressed in the GALL Report

AMPs
- Consistent with GALL Report AMPs
- Plant-specific AMPs

FSAR Supplement
- The responsible review organization is to review the FSAR Supplement associated with each assigned AMP.

3.3.2 Acceptance Criteria

The acceptance criteria for the areas of review describe methods for determining whether the applicant has met the requirements of the NRC's regulations in 10 CFR 54.21.

3.3.2.1 AMR Results Consistent with the GALL Report

The AMR and the AMPs applicable to the auxiliary system features are described and evaluated in Chapter VII of NUREG-1801 (GALL Report).

The applicant's LRA should provide sufficient information so that the NRC reviewer is able to confirm that the specific LRA AMR item and the associated LRA AMP are consistent with the cited GALL Report AMR item. The reviewer should then confirm that the LRA AMR item is consistent with the GALL Report AMR item to which it is compared.

When the applicant is crediting a different aging management program than recommended in the GALL Report, the reviewer should confirm that the alternate aging management program is valid to use for aging management and will be capable of managing the effects of aging as adequately as the aging management program recommended by the GALL Report.

3.3.2.2 AMR Results for Which Further Evaluation is Recommended by the GALL Report

The basic acceptance criteria, defined in Subsection 3.3.2.1, need to be applied first for all of the AMRs and AMPs reviewed as part of this section. In addition, if the GALL Report AMR item to which the LRA AMR item is compared identifies that "further evaluation is recommended," then additional criteria apply as identified by the GALL Report for each of the following aging effect/aging mechanism combinations. Refer to Table 3.3-1, comparing the "Further Evaluation Recommended" and the "Rev2 Item" columns, for the AMR items that reference the following subsections. The 2005 AMR item counterpart is provided in the "Rev1 Item" column.

3.3.2.2.1 Cumulative Fatigue Damage

Fatigue is a TLAA as defined in 10 CFR 54.3. TLAAs are required to be evaluated in accordance with 10 CFR 54.21(c). This TLAA is addressed separately in Section 4.3, "Metal Fatigue Analysis," or Section 4.7, "Other Plant-Specific Time-Limited Aging Analyses," of this SRP-LR.

3.3.2.2.2   Cracking due to Stress Corrosion Cracking and Cyclic Loading

Cracking due to SCC and cyclic loading could occur in stainless steel PWR non-regenerative heat exchanger components exposed to treated borated water greater than 60°C (>140°F) in the chemical and volume control system. The existing aging management program on monitoring and control of primary water chemistry in PWRs manages the aging effects of cracking due to SCC. However, control of water chemistry does not preclude cracking due to SCC and cyclic loading. Therefore, the effectiveness of the water chemistry control program should be verified to ensure that cracking is not occurring. The GALL Report recommends that a plant-specific aging management program be evaluated to verify the absence of cracking due to SCC and cyclic loading to ensure that these aging effects are managed adequately. An acceptable verification program is to include temperature and radioactivity monitoring of the shell side water, and eddy current testing of tubes.

3.3.2.2.3   Cracking due to Stress Corrosion Cracking

Cracking due to stress corrosion cracking could occur for stainless steel piping, piping components, piping elements and tanks exposed to outdoor air. The possibility of cracking also extends to components exposed to air which has recently been introduced into buildings, i.e., components near intake vents. Cracking is only known to occur in environments containing sufficient halides (primarily chlorides) and in which condensation or deliquescence is possible. Condensation or deliquescence should generally be assumed to be possible. Applicable outdoor air environments (and associated indoor air environments) include, but are not limited to, those within approximately 5 miles of a saltwater coastline, those within 1/2 mile of a highway which is treated with salt in the wintertime, those areas in which the soil contains more than trace chlorides, those plants having cooling towers where the water is treated with chlorine or chlorine compounds, and those areas subject to chloride contamination from other agricultural or industrial sources. This item is applicable for the environments described above.

GALL AMP XI.M36, "External Surfaces Monitoring," is an acceptable method to manage the aging effect. The applicant may demonstrate that this item is not applicable by describing the outdoor air environment present at the plant and demonstrating that external chloride stress corrosion cracking is not expected. The GALL Report recommends further evaluation to determine whether an adequate aging management program is used to manage this aging effect based on the environmental conditions applicable to the plant and ASME Code Section XI requirements applicable to the components.

3.3.2.2.4   Loss of Material due to Cladding Breach

Loss of material due to cladding breach could occur for PWR steel charging pump casings with stainless steel cladding exposed to treated borated water. The GALL Report references NRC Information Notice 94-63, "Boric Acid Corrosion of Charging Pump Casings Caused by Cladding Cracks," and recommends further evaluation of a plant-specific aging management program to ensure that the aging effect is adequately managed. Acceptance criteria are described in Branch Technical Position RLSB-1 (Appendix A.1 of this SRP-LR).

3.3.2.2.5   Loss of Material due to Pitting and Crevice Corrosion

Loss of material due to pitting and crevice corrosion could occur for stainless steel piping, piping components, piping elements, and tanks exposed to outdoor air. The possibility of pitting and crevice corrosion also extends to components exposed to air which has recently been introduced into buildings, i.e., components near intake vents. Pitting and crevice corrosion is only known to occur in environments containing sufficient halides (primarily chlorides) and in which condensation or deliquescence is possible. Condensation or deliquescence should generally be assumed to be possible. Applicable outdoor air environments (and associated indoor air environments) include, but are not limited to, those within approximately 5 miles of a saltwater coastline, those within 1/2 mile of a highway which is treated with salt in the wintertime, those areas in which the soil contains more than trace chlorides, those plants having cooling towers where the water is treated with chlorine or chlorine compounds, and those areas subject to chloride contamination from other agricultural or industrial sources. This item is applicable for the environments described above.

GALL AMP XI.M36, "External Surfaces Monitoring," is an acceptable method to manage the aging effect. The applicant may demonstrate that this item is not applicable by describing the outdoor air environment present at the plant and demonstrating that external pitting or crevice corrosion is not expected. The GALL Report recommends further evaluation to determine whether an adequate aging management program is used to manage this aging effect based on the environmental conditions applicable to the plant and ASME Code Section XI requirements Quality Assurance for Aging Management of Nonsafety-Related Components.

3.3.2.2.6   Quality Assurance for Aging Management of Nonsafety-Related Components

Acceptance criteria are described in Branch Technical Position IQMB-1 (Appendix A.2, of this SRP-LR.)

3.3.2.3   AMR Results Not Consistent with or Not Addressed in the GALL Report

Acceptance criteria are described in Branch Technical Position RLSB-1 (Appendix A.1 of this SRP-LR.)

3.3.2.4   Aging Management Programs

For those AMPs that will be used for aging management and are based on the program elements of an AMP in the GALL Report, the NRC reviewer performs an audit of aging management programs credited in the LRA to confirm consistency with the GALL AMPs identified in the GALL Report, Chapters X and XI.

If the applicant identifies an exception to any of the program elements of the cited GALL Report AMP, the LRA AMP should include a basis demonstrating how the criteria of 10 CFR 54.21(a)(3) would still be met. The NRC reviewer should then confirm that the LRA AMP with all exceptions would satisfy the criteria of 10 CFR 54.21(a)(3). If, while reviewing the LRA AMP, the reviewer identifies a difference between the LRA AMP and the GALL Report AMP that should have been identified as an exception to the GALL Report AMP, the difference should be reviewed and properly dispositioned. The reviewer should document the disposition of all LRA-defined exceptions and staff-identified differences.

The LRA should identify any enhancements that are needed to permit an existing LRA AMP to be declared consistent with the GALL Report AMP to which the LRA AMP is compared. The reviewer is to confirm both that the enhancement, when implemented, would allow the existing LRA AMP to be consistent with the GALL Report AMP and also that the applicant has a commitment in the FSAR Supplement to implement the enhancement prior to the period of extended operation. The reviewer should document the disposition of all enhancements.

If the applicant chooses to use a plant-specific program that is not a GALL AMP, the NRC reviewer should confirm that the plant-specific program satisfies the criteria of Branch Technical Position RLSB-1 (Appendix A.1.2.3 of this SRP-LR).

3.3.2.5 FSAR Supplement

The summary description of the programs and activities for managing the effects of aging for the period of extended operation in the FSAR Supplement should be sufficiently comprehensive, such that later changes can be controlled by 10 CFR 50.59. The description should contain information associated with the bases for determining that aging effects will be managed during the period of extended operation. The description should also contain any future aging management activities, including enhancements and commitments, to be completed before the period of extended operation. Table 3.0-1 of this SRP-LR provides examples of the type of information to be included in the FSAR Supplement. Table 3.3-2 lists the programs that are applicable for this SRP-LR subsection.

3.3.3 Review Procedures

For each area of review, the following review procedures are to be followed.

3.3.3.1 AMR Results Consistent with the GALL Report

The applicant may reference the GALL Report in its LRA, as appropriate, and demonstrate that the AMRs and AMPs at its facility are consistent with those reviewed and approved in the GALL Report. The reviewer should not conduct a re-review of the substance of the matters described in the GALL Report. If the applicant has provided the information necessary to adopt the finding of program acceptability as described and evaluated in the GALL Report, the reviewer should find acceptable the applicant's reference to the GALL Report in its LRA. In making this determination, the reviewer confirms that the applicant has provided a brief description of the system, components, materials, and environment. The reviewer also confirms that the applicant has stated that the applicable aging effects and industry and plant-specific operating experience have been reviewed by the applicant and are evaluated in the GALL Report.

Furthermore, the reviewer should confirm that the applicant has addressed operating experience identified after the issuance of the GALL Report. Performance of this review requires the reviewer to confirm that the applicant has identified those aging effects for the auxiliary system components that are contained in the GALL Report as applicable to its plant.

3.3.3.2 AMR Results Report for Which Further Evaluation is Recommended by the GALL Report

The basic review procedures defined in Subsection 3.3.3.1 need to be applied first for all of the AMRs and AMPs provided in this section. In addition, if the GALL Report AMR item to which the LRA AMR item is compared identifies that "further evaluation is recommended," then additional criteria apply as identified by the GALL Report for each of the following aging effect/aging mechanism combinations. Refer to Table 3.3-1 for the items that reference the following subsections.

3.3.3.2.1   Cumulative Fatigue Damage

Fatigue is a TLAA as defined in 10 CFR 54.3. TLAAs are required to be evaluated in accordance with 10 CFR 54.21(c). The evaluation of this TLAA is addressed separately in Section 4.3 of this SRP-LR.

3.3.3.2.2   Cracking due to Stress Corrosion Cracking and Cyclic Loading

The GALL Report also recommends further evaluation of programs to manage cracking due to SCC and cyclic loading in the stainless steel non-regenerative heat exchangers in the chemical and volume control system (PWR) exposed to treated borated water >60°C (>140°F). The water chemistry program relies on monitoring and control of water chemistry to manage the aging effects of cracking due to SCC and cyclic loading. The GALL Report recommends the effectiveness of the chemistry control program be verified to ensure that cracking is not occurring. The absence of cracking due to SCC and cyclic loading is to be verified. An acceptable verification program is to include temperature and radioactivity monitoring of the shell side water, and eddy current testing of tubes. The reviewer reviews the applicant's proposed program on a case-by-case basis to ensure that an adequate program will be in place for the management of these aging effects.

3.3.3.2.3   Cracking due to Stress Corrosion Cracking

The GALL Report recommends further evaluation to manage cracking due to stress corrosion cracking of stainless steel piping, piping components, piping elements, and tanks exposed to outdoor air environments containing sufficient halides (primarily chlorides) and in which condensation or deliquescence is possible. The possibility of cracking also extends to components exposed to air which has recently been introduced into buildings, i.e., components near intake vents.

The reviewer should determine whether an adequate program is used to manage the aging effect based on the applicable environmental conditions and ASME Code requirements. Cracking is only known to occur in environments containing sufficient halides (primarily chlorides) and in which condensation or deliquescence is possible. Condensation or deliquescence should generally be assumed to be possible. Applicable outdoor air environments (and associated indoor air environments) include, but are not limited to, those within approximately 5 miles of a saltwater coastline, those within 1/2 mile of a highway which is treated with salt in the wintertime, those areas in which the soil contains more than trace chlorides, those plants having cooling towers where the water is treated with chlorine or chlorine compounds, and those areas subject to chloride contamination from other agricultural or industrial sources. This item is applicable for the environments described above. GALL AMP XI.M36, "External Surfaces Monitoring," is an acceptable method to manage the aging effect.

3.3.3.2.4   Loss of Material due to Cladding Breach

The GALL Report recommends further evaluation of programs to manage loss of material due to cladding breach for PWR steel charging pump casings with stainless steel cladding. The GALL Report references NRC Information Notice 94-63, Boric Acid Corrosion of Charging Pump Casings Caused by Cladding Cracks and recommends further evaluation on a plant-specific basis to ensure that the aging effect is adequately managed. The reviewer reviews the applicant's proposed programs on a case-by-case basis to ensure that an adequate program will be in place for the management of general corrosion of these components.

3.3.3.2.5 Loss of Material due to Pitting and Crevice Corrosion

The GALL Report recommends further evaluation to manage loss of material due to pitting and crevice corrosion of stainless steel piping, piping components, piping elements, and tanks exposed to outdoor air environments containing sufficient halides (primarily chlorides) and in which condensation or deliquescence is possible. The possibility of pitting and crevice corrosion also extends to components exposed to air which has recently been introduced into buildings, i.e., components near intake vents. The reviewer should determine whether an adequate program is used to manage the aging effect based on the applicable environmental conditions and ASME Code requirements. Pitting and crevice corrosion is only known to occur in environments containing sufficient halides (primarily chlorides) and in which condensation or deliquescence is possible. Condensation or deliquescence should generally be assumed to be possible. Applicable outdoor air environments (and associated indoor air environments) include, but are not limited to, those within approximately 5 miles of a saltwater coastline, those within 1/2 mile of a highway which is treated with salt in the wintertime, those areas in which the soil contains more than trace chlorides, those plants having cooling towers where the water is treated with chlorine or chlorine compounds, and those areas subject to chloride contamination from other agricultural or industrial sources. This item is applicable for the environments described above. GALL AMP XI.M36, "External Surfaces Monitoring," is an acceptable method to manage the aging effect.

3.3.3.2.6 Quality Assurance for Aging Management of Nonsafety-Related Components

The applicant's aging management programs for license renewal should contain the elements of corrective actions, the confirmation process, and administrative controls. Safety-related components are covered by 10 CFR Part 50, Appendix B, which is adequate to address these program elements. However, Appendix B does not apply to nonsafety-related components that are subject to an AMR for license renewal. Nevertheless, the applicant has the option to expand the scope of its 10 CFR Part 50, Appendix B program to include these components and address the associated program elements. If the applicant chooses this option, the reviewer verifies that the applicant has documented such a commitment in the FSAR Supplement. If the applicant chooses alternative means, the branch responsible for quality assurance should be requested to review the applicant's proposal on a case-by-case basis.

3.3.3.3 AMR Results Not Consistent with or Not Addressed in the GALL Report

The reviewer should confirm that the applicant, in its LRA, has identified applicable aging effects, listed the appropriate combination of materials and environments, and has credited AMPs that will adequately manage the aging effects. The AMP credited by the applicant could be an AMP that is described and evaluated in the GALL Report or a plant-specific program. Review procedures are described in Branch Technical Position RSLB-1 (Appendix A.1 of this SRP-LR).

3.3.3.4 Aging Management Programs

The reviewer confirms that the applicant has identified the appropriate AMPs as described and evaluated in the GALL Report. If the applicant commits to an enhancement to make its LRA AMP consistent with a GALL Report AMP, then the reviewer is to confirm that this enhancement, when implemented, will make the LRA AMP consistent with the GALL Report AMP. If the applicant identifies, in the LRA AMP, an exception to any of the program elements of the GALL Report AMP, the reviewer is to confirm that the LRA AMP with the exception will satisfy the criteria of 10 CFR 54.21(a)(3). If the reviewer identifies a difference, not identified by the LRA, between the LRA AMP and the GALL Report AMP with which the LRA claims to be consistent, the reviewer should confirm that the LRA AMP with this difference satisfies 10 CFR 54.21(a)(3). The reviewer should document the basis for accepting enhancements, exceptions or differences. The AMPs evaluated in the GALL Report pertinent to the auxiliary systems components are summarized in Table 3.3-1 of this SRP-LR. The "Rev 2 Item" (for 2010) and "Rev1 Item" (for 2005 counterpart) columns identify the AMR item numbers in the GALL Report, Chapter VII, presenting detailed information summarized by this row.

3.3.3.5   FSAR Supplement

The reviewer confirms that the applicant has provided in its FSAR supplement information equivalent to that in Table 3.0-1 for aging management of the auxiliary systems. Table 3.3-2 lists the AMPs that are applicable for this SRP-LR subsection. The reviewer also confirms that the applicant has provided information for subsection 3.3.3.3, "AMR Results Not Consistent with or Not Addressed in the GALL Report," equivalent to that in Table 3.0-1.

The staff expects to impose a license condition on any renewed license to require the applicant to update its FSAR to include this FSAR Supplement at the next update required pursuant to 10 CFR 50.71(e)(4). As part of the license condition until the FSAR update is complete, the applicant may make changes to the programs described in its FSAR Supplement without prior NRC approval, provided that the applicant evaluates each such change and finds it acceptable pursuant to the criteria set forth in 10 CFR 50.59. If the applicant updates the FSAR to include the final FSAR supplement before the license is renewed, no condition will be necessary.

As noted in Table 3.0-1, an applicant need not incorporate the implementation schedule into its FSAR. However, the reviewer should confirm that the applicant has identified and committed in the LRA to any future aging management activities, including enhancements and commitments, to be completed before entering the period of extended operation. The staff expects to impose a license condition on any renewed license to ensure that the applicant will complete these activities no later than the committed date.

3.3.4   Evaluation Findings

If the reviewer determines that the applicant has provided information sufficient to satisfy the provisions of this section, then an evaluation finding similar to the following text should be included in the staff's safety evaluation report:

> On the basis of its review, as discussed above, the staff concludes that the applicant has demonstrated that the aging effects associated with the auxiliary systems components will be adequately managed so that the intended functions will be maintained consistent with the CLB for the period of extended operation, as required by 10 CFR 54.21(a)(3).

> The staff also reviewed the applicable FSAR Supplement program summaries and concludes that they adequately describe the AMPs credited for managing aging of the auxiliary systems, as required by 10 CFR 54.21(d).

3.3.5   Implementation

Except in those cases in which the applicant proposes an acceptable alternative method for complying with specified portions of the NRC's regulations, the method described herein will be used by the staff in its evaluation of conformance with NRC regulations.

3.3.6   References

1. NUREG-0800, "Standard Review Plan for the Review of Safety Analysis Reports for Nuclear Power Plants," U.S. Nuclear Regulatory Commission, March 2007.

2. NUREG-1801, "Generic Aging Lessons Learned (GALL) Report," U.S. Nuclear Regulatory Commission, Revision 2, 2010.

3. NEI 95-10, "Industry Guideline for Implementing the Requirements of 10 CFR Part 54 – The License Renewal Rule," Nuclear Energy Institute, Revision 6.

4. ASME Section XI, "Rules for Inservice Inspection of Nuclear Power Plant Components," The ASME Boiler and Pressure Vessel Code, 2004 edition as approved in 10 CFR 50.55a, The American Society of Mechanical Engineers, New York, NY.

5. ASTM D95-83, Standard Test Method for Water in Petroleum Products and Bituminous Materials by Distillation, American Society for Testing and Materials, West Conshohocken, PA, 1990.

Table 3.3-1 Summary of Aging Management Programs for Auxiliary Systems Evaluated in Chapter VII of the GALL Report

ID	Type	Component	Aging Effect/ Mechanism	Aging Management Programs	Further Evaluation Recommended	Rev2 Item	Rev1 Item
1	BWR/ PWR	Steel Cranes: structural girders exposed to Air – indoor, uncontrolled (External)	Cumulative fatigue damage due to fatigue	Fatigue is a time-limited aging analysis (TLAA) to be evaluated for the period of extended operation for structural girders of cranes that fall within the scope of 10 CFR 54 (Standard Review Plan, Section 4.7, "Other Plant-Specific Time-Limited Aging Analyses," for generic guidance for meeting the requirements of 10 CFR 54.21(c)(1))	Yes, TLAA (See subsection 3.3.2.1)	VII.B.A-06	VII.B-2(A-06)
2	BWR/ PWR	Stainless steel, Steel Heat exchanger components and tubes, Piping, piping components, and piping elements exposed to Treated borated water, Air – indoor, uncontrolled, Treated water	Cumulative fatigue damage due to fatigue	Fatigue is a time-limited aging analysis (TLAA) to be evaluated for the period of extended operation. See the SRP, Section 4.3 "Metal Fatigue," for acceptable methods for meeting the requirements of 10 CFR 54.21(c)(1).	Yes, TLAA (See subsection 3.3.2.1)	VII.E1.A-100 VII.E1.A-34 VII.E1.A-57 VII.E3.A-34 VII.E3.A-62 VII.E4.A-62	VII.E1-4(A-100) VII.E1-18(A-34) VII.E1-16(A-57) VII.E3-17(A-34) VII.E3-14(A-62) VII.E4-13(A-62)
3	PWR	Stainless steel Heat exchanger components, non-regenerative exposed to Treated borated water >60°C (>140°F)	Cracking due to stress corrosion cracking; cyclic loading	Chapter XI.M2, "Water Chemistry" The AMP is to be augmented by verifying the absence of cracking due to stress corrosion cracking and cyclic loading. An acceptable verification program is to include temperature and radioactivity monitoring of the shell side water, and eddy current testing of tubes.	Yes, plant-specific (See subsection 3.3.2.2)	VII.E1.A-69	VII.E1-9(A-69)
4	BWR/ PWR	Stainless steel Piping, piping components,	Cracking due to stress	Chapter XI.M36, "External Surfaces Monitoring of	Yes, environmental conditions need to be	VII.C1.AP-209 VII.C2.AP-209 VII.C3.AP-209	N/A N/A N/A

Table 3.3-1 Summary of Aging Management Programs for Auxiliary Systems Evaluated in Chapter VII of the GALL Report

ID	Type	Component	Aging Effect/ Mechanism	Aging Management Programs	Further Evaluation Recommended	Rev2 Item	Rev1 Item
		and piping elements; tanks exposed to Air – outdoor	corrosion cracking	Mechanical Components"	evaluated (See subsection 3.3.2.2.3)	VII.D.AP-209 VII.E1.AP-209 VII.E4.AP-209 VII.F1.AP-209 VII.F2.AP-209 VII.F4.AP-209 VII.G.AP-209 VII.H1.AP-209 VII.H2.AP-209	N/A N/A N/A N/A N/A N/A N/A N/A N/A
5	PWR	Steel (with stainless steel or nickel-alloy cladding) Pump Casings exposed to Treated borated water	Loss of material due to cladding breach	A plant-specific aging management program is to be evaluated. Reference NRC Information Notice 94-63, "Boric Acid Corrosion of Charging Pump Casings Caused by Cladding Cracks."	Yes, verify that plant-specific program addresses clad cracking (See subsection 3.3.2.2.4)	VII.E1.AP-85	VII.E1-21(AP-85)
6	BWR/ PWR	Stainless steel Piping, piping components, and piping elements; tanks exposed to Air – outdoor	Loss of material due to pitting and crevice corrosion	Chapter XI.M36, "External Surfaces Monitoring of Mechanical Components"	Yes, environmental conditions need to be evaluated (See subsection 3.3.2.2.5)	VII.C1.AP-221 VII.C2.AP-221 VII.C3.AP-221 VII.D.AP-221 VII.E1.AP-221 VII.E4.AP-221 VII.F1.AP-221 VII.F2.AP-221 VII.F4.AP-221 VII.G.AP-221 VII.H1.AP-221 VII.H2.AP-221	N/A N/A N/A N/A N/A N/A N/A N/A N/A N/A N/A
7	PWR	Stainless steel High-pressure pump, casing exposed to Treated borated water	Cracking due to cyclic loading	Chapter XI.M1, "ASME Section XI Inservice Inspection, Subsections IWB, IWC, and IWD"	No	VII.E1.AP-115	VII.E1-7(A-76)
8	PWR	Stainless steel Heat exchanger	Cracking due to cyclic	Chapter XI.M1, "ASME Section XI Inservice Inspection, Subsections	No	VII.E1.AP-119	N/A

Table 3.3-1   Summary of Aging Management Programs for Auxiliary Systems Evaluated in Chapter VII of the GALL Report

ID	Type	Component	Aging Effect/ Mechanism	Aging Management Programs	Further Evaluation Recommended	Rev2 Item	Rev1 Item
		components and tubes exposed to Treated borated water >60°C (>140°F)	loading	IWB, IWC, and IWD"			
9	PWR	Steel, Aluminum, Copper alloy (>15% Zn or >8% Al) External surfaces, Piping, piping components, and piping elements, Bolting exposed to Air with borated water leakage	Loss of material due to boric acid corrosion	Chapter XI.M10, "Boric Acid Corrosion"	No	VII.A3.A-79 VII.A3.AP-1 VII.E1.A-79 VII.E1.AP-1 VII.I.A-102 VII.I.A-79 VII.I.AP-66	VII.A3-2(A-79) VII.A3-4(AP-1) VII.E1-1(A-79) VII.E1-10(AP-1) VII.I-2(A-102) VII.I-10(A-79) VII.I-12(AP-66)
10	BWR/ PWR	Steel, high-strength Closure bolting exposed to Air with steam or water leakage	Cracking due to stress corrosion cracking; cyclic loading	Chapter XI.M18, "Bolting Integrity"	No	VII.I.A-04	VII.I-3(A-04)
11	BWR/ PWR	Steel, high-strength High-pressure pump, closure bolting exposed to Air with steam or water leakage	Cracking due to stress corrosion cracking; cyclic loading	Chapter XI.M18, "Bolting Integrity"	No	VII.E1.AP-122	VII.E1-8(A-104)
12	BWR/ PWR	Steel; stainless steel Closure bolting, Bolting exposed to Condensation, Air – indoor, uncontrolled (External), Air – outdoor (External)	Loss of material due to general (steel only), pitting, and crevice corrosion	Chapter XI.M18, "Bolting Integrity"	No	VII.D.AP-121 VII.I.AP-125 VII.I.AP-126	VII.D-1(A-103) VII.I-4(AP-27) VII.I-1(AP-28)

ID	Type	Component	Aging Effect/ Mechanism	Aging Management Programs	Further Evaluation Recommended	Rev2 Item	Rev1 Item
13	BWR/ PWR	Steel Closure bolting exposed to Air with steam or water leakage	Loss of material due to general corrosion	Chapter XI.M18, "Bolting Integrity"	No	VII.I.A-03	VII.I-6(A-03)
14	BWR/ PWR	Steel, Stainless Steel Bolting exposed to Soil	Loss of preload	Chapter XI.M18, "Bolting Integrity"	No	VII.I.AP-242 VII.I.AP-244	N/A N/A
15	BWR/ PWR	Steel; stainless steel, Copper alloy, Nickel alloy, Stainless steel Closure bolting, Bolting exposed to Air – indoor, uncontrolled (External), Any environment, Air – outdoor (External), Raw water, Treated borated water, Fuel oil, Treated water	Loss of preload due to thermal effects, gasket creep, and self-loosening	Chapter XI.M18, "Bolting Integrity"	No	VII.I.AP-124 VII.I.AP-261 VII.I.AP-262 VII.I.AP-263 VII.I.AP-264 VII.I.AP-265 VII.I.AP-266 VII.I.AP-267	VII.I-5(AP-26) N/A N/A N/A N/A N/A N/A
16	BWR	Stainless steel Piping, piping components, and piping elements exposed to Treated water >60°C (>140°F)	Cracking due to stress corrosion cracking, intergranular stress corrosion cracking	Chapter XI.M2, "Water Chemistry," and Chapter XI.M25, "BWR Reactor Water Cleanup System"	No	VII.E3.AP-283	VII.E3-16(A-60)
17	BWR	Stainless steel Heat exchanger tubes exposed to Treated water	Reduction of heat transfer due to fouling	Chapter XI.M2, "Water Chemistry," and Chapter XI.M32, "One-Time Inspection"	No	VII.A4.AP-139	VII.A4-4(AP-62)
18	BWR/ PWR	Stainless steel High-pressure pump,	Cracking due to stress	Chapter XI.M2, "Water Chemistry," and	No	VII.E1.AP-114 VII.E2.AP-181	VII.E1-7(A-76) VII.E2-2(A-59)

Table 3.3-1   Summary of Aging Management Programs for Auxiliary Systems Evaluated in Chapter VII of the GALL Report

ID	Type	Component	Aging Effect/ Mechanism	Aging Management Programs	Further Evaluation Recommended	Rev2 Item	Rev1 Item
		casing, Piping, piping components, and piping elements exposed to Treated borated water >60°C (>140°F), Sodium pentaborate solution >60°C (>140°F)	corrosion cracking	Chapter XI.M32, "One-Time Inspection"			
19	BWR/ PWR	Stainless steel Regenerative heat exchanger components exposed to Treated water >60°C (>140°F)	Cracking due to stress corrosion cracking	Chapter XI.M2, "Water Chemistry," and Chapter XI.M32, "One-Time Inspection"	No	VII.E3.AP-120	VII.E3-19(A-85)
20	BWR/ PWR	Stainless steel, Stainless steel; steel with stainless steel cladding Heat exchanger components exposed to Treated borated water >60°C (>140°F), Treated water >60°C (>140°F)	Cracking due to stress corrosion cracking	Chapter XI.M2, "Water Chemistry," and Chapter XI.M32, "One-Time Inspection"	No	VII.E1.AP-118 VII.E3.AP-112	VII.E1-5(A-84) VII.E3-3(A-71)
21	BWR	Steel Piping, piping components, and piping elements exposed to Treated water	Loss of material due to general, pitting, and crevice corrosion	Chapter XI.M2, "Water Chemistry," and Chapter XI.M32, "One-Time Inspection"	No	VII.E3.AP-106 VII.E4.AP-106	VII.E3-18(A-35) VII.E4-17(A-35)
22	BWR	Copper alloy Piping, piping components, and piping elements exposed to Treated	Loss of material due to general, pitting, crevice, and galvanic	Chapter XI.M2, "Water Chemistry," and Chapter XI.M32, "One-Time Inspection"	No	VII.A4.AP-140 VII.E3.AP-140 VII.E4.AP-140	VII.A4-7(AP-64) VII.E3-9(AP-64) VII.E4-7(AP-64)

ID	Type	Component	Aging Effect/ Mechanism	Aging Management Programs	Further Evaluation Recommended	Rev2 Item	Rev1 Item
		water	corrosion				
23	BWR/ PWR	Aluminum Piping, piping components, and piping elements exposed to Treated water	Loss of material due to pitting and crevice corrosion	Chapter XI.M2, "Water Chemistry," and Chapter XI.M32, "One-Time Inspection"	No	VII.C2.AP-257 VII.H2.AP-258	N/A N/A
24	BWR	Aluminum Piping, piping components, and piping elements exposed to Treated water	Loss of material due to pitting and crevice corrosion	Chapter XI.M2, "Water Chemistry," and Chapter XI.M32, "One-Time Inspection"	No	VII.E4.AP-130	VII.E4-4(AP-38)
25	BWR	Stainless steel, Stainless steel; steel with stainless steel cladding, Aluminum Piping, piping components, and piping elements, Heat exchanger components exposed to Treated water, Sodium pentaborate solution	Loss of material due to pitting and crevice corrosion	Chapter XI.M2, "Water Chemistry," and Chapter XI.M32, "One-Time Inspection"	No	VII.A4.AP-110 VII.A4.AP-111 VII.A4.AP-130 VII.E2.AP-141 VII.E3.AP-110 VII.E3.AP-130 VII.E4.AP-110	VII.A4-11(A-58) VII.A4-2(A-70) VII.A4-5(AP-38) VII.E2-1(AP-73) VII.E3-15(A-58) VII.E3-7(AP-38) VII.E4-14(A-58)
26	BWR/ PWR	Steel (with elastomer lining), Steel (with elastomer lining or stainless steel cladding) Piping, piping components, and piping elements exposed to Treated water	Loss of material due to pitting and crevice corrosion (only for steel after lining/cladding degradation)	Chapter XI.M2, "Water Chemistry," and Chapter XI.M32, "One-Time Inspection"	No	VII.A3.AP-107 VII.A4.AP-108	VII.A3-9(A-39) VII.A4-12(A-40)

Table 3.3-1 Summary of Aging Management Programs for Auxiliary Systems Evaluated in Chapter VII of the GALL Report

ID	Type	Component	Aging Effect/ Mechanism	Aging Management Programs	Further Evaluation Recommended	Rev2 Item	Rev1 Item
27	BWR	Stainless steel Heat exchanger tubes exposed to Treated water	Reduction of heat transfer due to fouling	Chapter XI.M2, "Water Chemistry," and Chapter XI.M32, "One-Time Inspection"	No	VII.E3.AP-139	VII.E3-6(AP-62)
28	BWR/ PWR	Stainless steel, Steel (with stainless steel or nickel-alloy cladding) Spent fuel storage racks (BWR), Spent fuel storage racks (PWR), Piping, piping components, and piping elements, Piping, piping components, and piping elements; tanks exposed Treated water >60°C (>140°F), Treated borated water >60°C (>140°F)	Cracking due to stress corrosion cracking	Chapter XI.M2, "Water Chemistry"	No	VII.A2.A-96 VII.A2.A-97 VII.A3.A-56 VII.E1.AP-82	VII.A2-6(A-96) VII.A2-7(A-97) VII.A3-10(A-56) VII.E1-20(AP-82)
29	BWR/ PWR	Steel (with stainless steel cladding); stainless steel Piping, piping components, and piping elements exposed to Treated borated water	Loss of material due to pitting and crevice corrosion	Chapter XI.M2, "Water Chemistry"	No	VII.A2.AP-79 VII.A3.AP-79 VII.E1.AP-79	VII.A2-1(AP-79) VII.A3-8(AP-79) VII.E1-17(AP-79)
30	BWR/ PWR	Concrete; cementitious material Piping, piping components, and piping elements exposed to Raw	Changes in material properties due to aggressive chemical attack	Chapter XI.M20, "Open-Cycle Cooling Water System"	No	VII.C1.AP-250	N/A

Table 3.3-1 Summary of Aging Management Programs for Auxiliary Systems Evaluated in Chapter VII of the GALL Report

ID	Type	Component	Aging Effect/ Mechanism	Aging Management Programs	Further Evaluation Recommended	Rev2 Item	Rev1 Item
		Water					
	BWR/ PWR	Fiberglass, HDPE Piping, piping components, and piping elements exposed to Raw water (internal)	Cracking, blistering, change in color due to water absorption	Chapter XI.M20, "Open-Cycle Cooling Water System"	No	VII.C1.AP-238 VII.C1.AP-239	N/A N/A
31	BWR/ PWR	Concrete; cementitious material Piping, piping components, and piping elements exposed to Raw Water	Cracking due to settling	Chapter XI.M20, "Open-Cycle Cooling Water System"	No	VII.C1.AP-248	N/A
32	BWR/ PWR	Reinforced concrete, asbestos cement Piping, piping components, and piping elements exposed to Raw water	Cracking due to aggressive chemical attack and leaching; Changes in material properties due to aggressive chemical attack	Chapter XI.M20, "Open-Cycle Cooling Water System"	No	VII.C1.AP-155	N/A
	BWR/ PWR	Elastomer seals and components exposed to raw water	Hardening and loss of strength due to elastomer degradation; loss of material due to erosion	Chapter XI.M20, "Open-Cycle Cooling Water System"	No	VII.C1.AP-75 VII.C1.AP-76	VII.C1-1(AP-75) VII.C1-2(AP-76)
33	BWR/ PWR	Concrete; cementitious material Piping, piping	Loss of material due to abrasion, cavitation,	Chapter XI.M20, "Open-Cycle Cooling Water System"	No	VII.C1.AP-249	N/A

Table 3.3-1   Summary of Aging Management Programs for Auxiliary Systems Evaluated in Chapter VII of the GALL Report

ID	Type	Component	Aging Effect/ Mechanism	Aging Management Programs	Further Evaluation Recommended	Rev2 Item	Rev1 Item
		components, and piping elements exposed to Raw Water	aggressive chemical attack, and leaching				
34	BWR/ PWR	Nickel alloy, Copper alloy Piping, piping components, and piping elements exposed to Raw water	Loss of material due to general, pitting, and crevice corrosion	Chapter XI.M20, "Open-Cycle Cooling Water System"	No	VII.C1.AP-206 VII.C3.AP-195 VII.C3.AP-206	VII.C1-13(AP-53) VII.C3-2(A-43) VII.C3-6(AP-53)
35	BWR/ PWR	Copper alloy Piping, piping components, and piping elements exposed to Raw water	Loss of material due to general, pitting, crevice, and microbiologically-influenced corrosion	Chapter XI.M20, "Open-Cycle Cooling Water System"	No	VII.H2.AP-193	VII.H2-11(AP-45)
36	BWR/ PWR	Copper alloy Piping, piping components, and piping elements exposed to Raw water	Loss of material due to general, pitting, crevice, and microbiologically-influenced corrosion; fouling that leads to corrosion	Chapter XI.M20, "Open-Cycle Cooling Water System"	No	VII.C1.AP-196	VII.C1-9(A-44)
37	BWR/ PWR	Steel (with coating or lining) Piping, piping components, and piping elements exposed to Raw water	Loss of material due to general, pitting, crevice, and microbiologically-influenced corrosion; fouling that leads to	Chapter XI.M20, "Open-Cycle Cooling Water System"	No	VII.C1.AP-194 VII.C3.AP-194 VII.H2.AP-194	VII.C1-19(A-38) VII.C3-10(A-38) VII.H2-22(A-38)

Table 3.3-1 Summary of Aging Management Programs for Auxiliary Systems Evaluated in Chapter VII of the GALL Report

ID	Type	Component	Aging Effect/Mechanism	Aging Management Programs	Further Evaluation Recommended	Rev2 Item	Rev1 Item
			corrosion; lining/coating degradation				
38	BWR/PWR	Copper alloy, Steel Heat exchanger components exposed to Raw water	Loss of material due to general, pitting, crevice, galvanic, and microbiologically-influenced corrosion; fouling that leads to corrosion	Chapter XI.M20, "Open-Cycle Cooling Water System"	No	VII.C1.AP-179 VII.C1.AP-183	VII.C1-3(A-65) VII.C1-5(A-64)
39	BWR/PWR	Stainless steel Piping, piping components, and piping elements exposed to Raw water	Loss of material due to pitting and crevice corrosion	Chapter XI.M20, "Open-Cycle Cooling Water System"	No	VII.C3.A-53	VII.C3-7(A-53)
40	BWR/PWR	Stainless steel Piping, piping components, and piping elements exposed to Raw water	Loss of material due to pitting and crevice corrosion; fouling that leads to corrosion	Chapter XI.M20, "Open-Cycle Cooling Water System"	No	VII.C1.A-54	VII.C1-15(A-54)
41	BWR/PWR	Stainless steel Piping, piping components, and piping elements exposed to Raw water	Loss of material due to pitting, crevice, and microbiologically-influenced corrosion	Chapter XI.M20, "Open-Cycle Cooling Water System"	No	VII.H2.AP-55	VII.H2-18(AP-55)
42	BWR/PWR	Copper alloy, Titanium, Stainless steel Heat exchanger tubes exposed to Raw water	Reduction of heat transfer due to fouling	Chapter XI.M20, "Open-Cycle Cooling Water System"	No	VII.C1.A-72 VII.C1.AP-153 VII.C1.AP-187 VII.C3.AP-187 VII.G.AP-187 VII.H2.AP-187	VII.C1-6(A-72) N/A VII.C1-7(AP-61) VII.C3-1(AP-61) VII.G-7(AP-61) VII.H2-6(AP-61)

Table 3.3-1		Summary of Aging Management Programs for Auxiliary Systems Evaluated in Chapter VII of the GALL Report				
ID	Type	Component	Aging Effect/ Mechanism	Aging Management Programs	Further Evaluation Recommended	Rev2 Item	Rev1 Item
43	BWR/ PWR	Stainless steel Piping, piping components, and piping elements exposed to Closed-cycle cooling water >60°C (>140°F)	Cracking due to stress corrosion cracking	Chapter XI.M21A, "Closed Treated Water Systems"	No	VII.C2.AP-186 VII.E3.AP-186 VII.E4.AP-186	VII.C2-11(AP-60) VII.E3-13(AP-60) VII.E4-11(AP-60)
44	BWR/ PWR	Stainless steel; steel with stainless steel cladding Heat exchanger components exposed to Closed-cycle cooling water >60°C (>140°F)	Cracking due to stress corrosion cracking	Chapter XI.M21A, "Closed Treated Water Systems"	No	VII.E3.AP-192	VII.E3-2(A-68)
45	BWR/ PWR	Steel Piping, piping components, and piping elements; tanks exposed to Closed-cycle cooling water	Loss of material due to general, pitting, and crevice corrosion	Chapter XI.M21A, "Closed Treated Water Systems"	No	VII.C2.AP-202 VII.F1.AP-202 VII.F2.AP-202 VII.F3.AP-202 VII.F4.AP-202 VII.H2.AP-202	VII.C2-14(A-25) VII.F1-20(A-25) VII.F2-18(A-25) VII.F3-20(A-25) VII.F4-16(A-25) VII.H2-23(A-25)
46	BWR/ PWR	Steel, Copper alloy Heat exchanger components, Piping, piping components, and piping elements exposed to Closed-cycle cooling water	Loss of material due to general, pitting, crevice, and galvanic corrosion	Chapter XI.M21A, "Closed Treated Water Systems"	No	VII.A3.AP-189 VII.A3.AP-199 VII.A4.AP-189 VII.A4.AP-199 VII.C2.AP-189 VII.C2.AP-199 VII.E1.AP-189 VII.E1.AP-199 VII.E1.AP-203 VII.E3.AP-189 VII.E3.AP-199 VII.E4.AP-189 VII.E4.AP-199 VII.F1.AP-189 VII.F1.AP-199 VII.F1.AP-203 VII.F2.AP-189	VII.A3-3(A-63) VII.A3-5(AP-12) VII.A4-3(A-63) VII.A4-6(AP-12) VII.C2-1(A-63) VII.C2-4(AP-12) VII.E1-6(A-63) VII.E1-11(AP-12) VII.E1-2(AP-34) VII.E3-4(A-63) VII.E3-8(AP-12) VII.E4-2(A-63) VII.E4-5(AP-12) VII.F1-11(A-63) VII.F1-15(AP-12)

Table 3.3-1   Summary of Aging Management Programs for Auxiliary Systems Evaluated in Chapter VII of the GALL Report

ID	Type	Component	Aging Effect/ Mechanism	Aging Management Programs	Further Evaluation Recommended	Rev2 Item	Rev1 Item
						VII.F2.AP-199 VII.F3.AP-189 VII.F3.AP-199 VII.F3.AP-203 VII.F4.AP-189 VII.F4.AP-199 VII.H1.AP-199 VII.H2.AP-199	VII.F1-8(AP-34) VII.F2-9(A-63) VII.F2-13(AP-12) VII.F3-11(A-63) VII.F3-15(AP-12) VII.F3-8(AP-34) VII.F4-8(A-63) VII.F4-11(AP-12) VII.H1-2(AP-12) VII.H2-8(AP-12)
47	BWR	Stainless steel; steel with stainless steel cladding Heat exchanger components exposed to Closed-cycle cooling water	Loss of material due to microbiologically-influenced corrosion	Chapter XI.M21A, "Closed Treated Water Systems"	No	VII.E3.AP-191 VII.E4.AP-191	VII.E3-1(A-67) VII.E4-1(A-67)
48	BWR/ PWR	Aluminum Piping, piping components, and piping elements exposed to Closed-cycle cooling water	Loss of material due to pitting and crevice corrosion	Chapter XI.M21A, "Closed Treated Water Systems"	No	VII.C2.AP-254 VII.H2.AP-255	N/A N/A
49	BWR/ PWR	Stainless steel Piping, piping components, and piping elements exposed to Closed-cycle cooling water	Loss of material due to pitting and crevice corrosion	Chapter XI.M21A, "Closed Treated Water Systems"	No	VII.C2.A-52	VII.C2-10(A-52)
50	BWR/ PWR	Stainless steel, Copper Alloy, Steel Heat exchanger tubes exposed to Closed-cycle cooling water	Reduction of heat transfer due to fouling	Chapter XI.M21A, "Closed Treated Water Systems"	No	VII.C2.AP-188 VII.C2.AP-205 VII.E3.AP-188 VII.E4.AP-188 VII.F1.AP-204 VII.F1.AP-205	VII.C2-3(AP-63) VII.C2-2(AP-80) VII.E3-5(AP-63) VII.E4-3(AP-63) VII.F1-13(AP-77)

ID	Type	Component	Aging Effect/ Mechanism	Aging Management Programs	Further Evaluation Recommended	Rev2 Item	Rev1 Item
						VII.F2.AP-204 VII.F2.AP-205 VII.F3.AP-204 VII.F3.AP-205 VII.F4.AP-204	VII.F1-12(AP-80) VII.F2-11(AP-77) VII.F2-10(AP-80) VII.F3-13(AP-77) VII.F3-12(AP-80) VII.F4-9(AP-77)
51	BWR/ PWR	Boraflex Spent fuel storage racks: neutron-absorbing sheets (PWR), Spent fuel storage racks: neutron-absorbing sheets (BWR) exposed to Treated borated water, Treated water	Reduction of neutron-absorbing capacity due to boraflex degradation	Chapter XI.M22, "Boraflex Monitoring"	No	VII.A2.A-86 VII.A2.A-87	VII.A2-4(A-86) VII.A2-2(A-87)
52	BWR/ PWR	Steel Cranes: rails and structural girders exposed to Air – indoor, uncontrolled (External)	Loss of material due to general corrosion	Chapter XI.M23, "Inspection of Overhead Heavy Load and Light Load (Related to Refueling) Handling Systems"	No	VII.B.A-07	VII.B-3(A-07)
53	BWR/ PWR	Steel Cranes - rails exposed to Air – indoor, uncontrolled (External)	Loss of material due to wear	Chapter XI.M23, "Inspection of Overhead Heavy Load and Light Load (Related to Refueling) Handling Systems"	No	VII.B.A-05	VII.B-1(A-05)
54	BWR/ PWR	Copper alloy Piping, piping components, and piping elements exposed to	Loss of material due to general, pitting, and crevice corrosion	Chapter XI.M24, "Compressed Air Monitoring"	No	VII.D.AP-240	N/A

Table 3.3-1 Summary of Aging Management Programs for Auxiliary Systems Evaluated in Chapter VII of the GALL Report

ID	Type	Component	Aging Effect/ Mechanism	Aging Management Programs	Further Evaluation Recommended	Rev2 Item	Rev1 Item
		Condensation					
55	BWR/ PWR	Steel Piping, piping components, and piping elements: compressed air system exposed to Condensation (Internal)	Loss of material due to general and pitting corrosion	Chapter XI.M24, "Compressed Air Monitoring"	No	VII.D.A-26	VII.D-2(A-26)
56	BWR/ PWR	Stainless steel Piping, piping components, and piping elements exposed to Condensation (Internal)	Loss of material due to pitting and crevice corrosion	Chapter XI.M24, "Compressed Air Monitoring"	No	VII.D.AP-81	VII.D-4(AP-81)
57	BWR/ PWR	Elastomers Fire barrier penetration seals exposed to Air - indoor, uncontrolled, Air – outdoor	Increased hardness; shrinkage; loss of strength due to weathering	Chapter XI.M26, "Fire Protection"	No	VII.G.A-19 VII.G.A-20	VII.G-1(A-19) VII.G-2(A-20)
58	BWR/ PWR	Steel Halon/carbon dioxide fire suppression system piping, piping components, and piping elements exposed to Air – indoor, uncontrolled (External)	Loss of material due to general, pitting, and crevice corrosion	Chapter XI.M26, "Fire Protection"	No	VII.G.AP-150	N/A
59	BWR/ PWR	Steel Fire rated doors exposed to Air - indoor, uncontrolled,	Loss of material due to wear	Chapter XI.M26, "Fire Protection"	No	VII.G.A-21 VII.G.A-22	VII.G-3(A-21) VII.G-4(A-22)

Table 3.3-1   Summary of Aging Management Programs for Auxiliary Systems Evaluated in Chapter VII of the GALL Report

ID	Type	Component	Aging Effect/ Mechanism	Aging Management Programs	Further Evaluation Recommended	Rev2 Item	Rev1 Item
		Air – outdoor					
60	BWR/ PWR	Reinforced concrete Structural fire barriers: walls, ceilings and floors exposed to Air - indoor, uncontrolled	Concrete cracking and spalling due to aggressive chemical attack, and reaction with aggregates	Chapter XI.M26, "Fire Protection," and Chapter XI.S6, "Structures Monitoring"	No	VII.G.A-90	VII.G-28(A-90)
61	BWR/ PWR	Reinforced concrete Structural fire barriers: walls, ceilings and floors exposed to Air – outdoor	Cracking, loss of material due to freeze-thaw, aggressive chemical attack, and reaction with aggregates	Chapter XI.M26, "Fire Protection," and Chapter XI.S6, "Structures Monitoring"	No	VII.G.A-92	VII.G-30(A-92)
62	BWR/ PWR	Reinforced concrete Structural fire barriers: walls, ceilings and floors exposed to Air - indoor, uncontrolled, Air – outdoor	Loss of material due to corrosion of embedded steel	Chapter XI.M26, "Fire Protection," and Chapter XI.S6, "Structures Monitoring"	No	VII.G.A-91 VII.G.A-93	VII.G-29(A-91) VII.G-31(A-93)
63	BWR/ PWR	Steel Fire Hydrants exposed to Air – outdoor	Loss of material due to general, pitting, and crevice corrosion	Chapter XI.M27, "Fire Water System"	No	VII.G.AP-149	N/A
64	BWR/ PWR	Steel, Copper alloy Piping, piping components, and piping elements exposed to Raw water	Loss of material due to general, pitting, crevice, and microbiologically-influenced corrosion; fouling	Chapter XI.M27, "Fire Water System"	No	VII.G.A-33 VII.G.AP-197	VII.G-24(A-33) VII.G-12(A-45)

ID	Type	Component	Aging Effect/ Mechanism	Aging Management Programs	Further Evaluation Recommended	Rev2 Item	Rev1 Item
			that leads to corrosion				
65	BWR/ PWR	Aluminum Piping, piping components, and piping elements exposed to Raw water	Loss of material due to pitting and crevice corrosion	Chapter XI.M27, "Fire Water System"	No	VII.G.AP-180	VII.G-8(AP-83)
66	BWR/ PWR	Stainless steel Piping, piping components, and piping elements exposed to Raw water	Loss of material due to pitting and crevice corrosion; fouling that leads to corrosion	Chapter XI.M27, "Fire Water System"	No	VII.G.A-55	VII.G-19(A-55)
67	BWR/ PWR	Steel Tanks exposed to Air – outdoor (External)	Loss of material due to general, pitting, and crevice corrosion	Chapter XI.M29, "Aboveground Metallic Tanks"	No	VII.H1.A-95	VII.H1-11(A-95)
68	BWR/ PWR	Steel Piping, piping components, and piping elements exposed to Fuel oil	Loss of material due to general, pitting, and crevice corrosion	Chapter XI.M30, "Fuel Oil Chemistry", and Chapter XI.M32, "One-Time Inspection"	No	VII.G.AP-234	VII.G-21(A-28)
69	BWR/ PWR	Copper alloy Piping, piping components, and piping elements exposed to Fuel oil	Loss of material due to general, pitting, crevice, and microbiologically-influenced corrosion	Chapter XI.M30, "Fuel Oil Chemistry," and Chapter XI.M32, "One-Time Inspection"	No	VII.G.AP-132 VII.H1.AP-132 VII.H2.AP-132	VII.G-10(AP-44) VII.H1-3(AP-44) VII.H2-9(AP-44)
70	BWR/ PWR	Steel Piping, piping components, and piping elements; tanks exposed to Fuel oil	Loss of material due to general, pitting, crevice, and microbiologically-influenced	Chapter XI.M30, "Fuel Oil Chemistry," and Chapter XI.M32, "One-Time Inspection"	No	VII.H1.AP-105 VII.H2.AP-105	VII.H1-10(A-30) VII.H2-24(A-30)

Table 3.3-1   Summary of Aging Management Programs for Auxiliary Systems Evaluated in Chapter VII of the GALL Report

ID	Type	Component	Aging Effect/ Mechanism	Aging Management Programs	Further Evaluation Recommended	Rev2 Item	Rev1 Item
			corrosion; fouling that leads to corrosion				
71	BWR/ PWR	Stainless steel, Aluminum Piping, piping components, and piping elements exposed to Fuel oil	Loss of material due to pitting, crevice, and microbiologically-influenced corrosion	Chapter XI.M30, "Fuel Oil Chemistry," and Chapter XI.M32, "One-Time Inspection"	No	VII.G.AP-136 VII.H1.AP-129 VII.H1.AP-136 VII.H2.AP-129 VII.H2.AP-136	VII.G-17(AP-54) VII.H1-1(AP-35) VII.H1-6(AP-54) VII.H2-7(AP-35) VII.H2-16(AP-54)
72	BWR/ PWR	Gray cast iron, Copper alloy (>15% Zn or >8% Al) Piping, piping components, and piping elements, Heat exchanger components exposed to Treated water, Closed-cycle cooling water, Soil, Raw water	Loss of material due to selective leaching	Chapter XI.M33, "Selective Leaching"	No	VII.A3.AP-31 VII.A3.AP-43 VII.A4.AP-31 VII.A4.AP-32 VII.A4.AP-43 VII.C1.A-02 VII.C1.A-47 VII.C1.A-51 VII.C1.A-66 VII.C2.A-50 VII.C2.AP-31 VII.C2.AP-32 VII.C2.AP-43 VII.C3.A-02 VII.C3.A-47 VII.C3.A-51 VII.E1.AP-31 VII.E1.AP-43 VII.E1.AP-65 VII.E3.AP-31 VII.E3.AP-32 VII.E3.AP-43 VII.E4.AP-31 VII.E4.AP-43 VII.F1.AP-31 VII.F1.AP-43 VII.F1.AP-65 VII.F2.AP-31 VII.F2.AP-43	VII.A3-7(AP-31) VII.A3-6(AP-43) VII.A4-10(AP-31) VII.A4-9(AP-32) VII.A4-8(AP-43) VII.C1-12(A-02) VII.C1-10(A-47) VII.C1-11(A-51) VII.C1-4(A-66) VII.C2-8(A-50) VII.C2-9(AP-31) VII.C2-7(AP-32) VII.C2-6(AP-43) VII.C3-5(A-02) VII.C3-3(A-47) VII.C3-4(A-51) VII.E1-14(AP-31) VII.E1-13(AP-43) VII.E1-3(AP-65) VII.E3-12(AP-31) VII.E3-11(AP-32) VII.E3-10(AP-43) VII.E4-10(AP-31)

Table 3.3-1 Summary of Aging Management Programs for Auxiliary Systems Evaluated in Chapter VII of the GALL Report

ID	Type	Component	Aging Effect/ Mechanism	Aging Management Programs	Further Evaluation Recommended	Rev2 Item	Rev1 Item
						VII.F3.A-50 VII.F3.AP-43 VII.F3.AP-65 VII.F4.AP-31 VII.F4.AP-43 VII.G.A-02 VII.G.A-47 VII.G.A-51 VII.G.AP-31 VII.H1.A-02 VII.H1.AP-43 VII.H2.A-02 VII.H2.A-47 VII.H2.A-51 VII.H2.AP-43	VII.E4-9(AP-32) VII.E4-8(AP-43) VII.F1-18(AP-31) VII.F1-17(AP-43) VII.F1-9(AP-65) VII.F2-16(AP-31) VII.F2-15(AP-43) VII.F3-18(A-50) VII.F3-17(AP-43) VII.F3-9(AP-65) VII.F4-14(AP-31) VII.F4-13(AP-43) VII.G-15(A-02) VII.G-13(A-47) VII.G-14(A-51) VII.G-16(AP-31) VII.H1-5(A-02) VII.H1-4(AP-43) VII.H2-15(A-02) VII.H2-13(A-47) VII.H2-14(A-51) VII.H2-12(AP-43)
73	BWR/ PWR	Concrete; cementitious material Piping, piping components, and piping elements exposed to Air – outdoor	Changes in material properties due to aggressive chemical attack	Chapter XI.M36, "External Surfaces Monitoring of Mechanical Components"	No	VII.C1.AP-253	N/A

Table 3.3-1   Summary of Aging Management Programs for Auxiliary Systems Evaluated in Chapter VII of the GALL Report

ID	Type	Component	Aging Effect/ Mechanism	Aging Management Programs	Further Evaluation Recommended	Rev2 Item	Rev1 Item
74	BWR/ PWR	Concrete; cementitious material Piping, piping components, and piping elements exposed to Air - outdoor	Cracking due to settling	Chapter XI.M36, "External Surfaces Monitoring of Mechanical Components"	No	VII.C1.AP-251	N/A
75	BWR/ PWR	Reinforced concrete, asbestos cement Piping, piping components, and piping elements exposed to Air – outdoor	Cracking due to aggressive chemical attack and leaching; Changes in material properties due to aggressive chemical attack	Chapter XI.M36, "External Surfaces Monitoring of Mechanical Components"	No	VII.C1.AP-156	N/A
76	BWR/ PWR	Elastomers Elastomer: seals and components exposed to Air – indoor, uncontrolled (Internal/External)	Hardening and loss of strength due to elastomer degradation	Chapter XI.M36, "External Surfaces Monitoring of Mechanical Components"	No	VII.F1.AP-102 VII.F2.AP-102 VII.F3.AP-102 VII.F4.AP-102	VII.F1-7(A-17) VII.F2-7(A-17) VII.F3-7(A-17) VII.F4-6(A-17)
77	BWR/ PWR	Concrete; cementitious material Piping, piping components, and piping elements exposed to Air - outdoor	Loss of material due to abrasion, cavitation, aggressive chemical attack, and leaching	Chapter XI.M36, "External Surfaces Monitoring of Mechanical Components"	No	VII.C1.AP-252	N/A
78	BWR/ PWR	Steel Piping and components (External surfaces), Ducting and components (External surfaces), Ducting; closure bolting	Loss of material due to general corrosion	Chapter XI.M36, "External Surfaces Monitoring of Mechanical Components"	No	VII.D.A-80 VII.F1.A-10 VII.F1.A-105 VII.F2.A-10 VII.F2.A-105 VII.F3.A-10 VII.F3.A-105	VII.D-3(A-80) VII.F1-2(A-10) VII.F1-4(A-105) VII.F2-2(A-10) VII.F2-4(A-105) VII.F3-2(A-10) VII.F3-4(A-105)

Table 3.3-1 Summary of Aging Management Programs for Auxiliary Systems Evaluated in Chapter VII of the GALL Report

ID	Type	Component	Aging Effect/ Mechanism	Aging Management Programs	Further Evaluation Recommended	Rev2 Item	Rev1 Item
		exposed to Air – indoor, uncontrolled (External), Air – indoor, uncontrolled (External), Air – outdoor (External), Condensation (External)				VII.F4.A-10 VII.F4.A-105 VII.I.A-105 VII.I.A-77 VII.I.A-78 VII.I.A-81	VII.F4-1(A-10) VII.F4-3(A-105) VII.I-7(A-105) VII.I-8(A-77) VII.I-9(A-78) VII.I-11(A-81)
79	BWR/ PWR	Copper alloy Piping, piping components, and piping elements exposed to Condensation (External)	Loss of material due to general, pitting, and crevice corrosion	Chapter XI.M36, "External Surfaces Monitoring of Mechanical Components"	No	VII.F1.AP-109 VII.F2.AP-109 VII.F3.AP-109 VII.F4.AP-109	VII.F1-16(A-46) VII.F2-14(A-46) VII.F3-16(A-46) VII.F4-12(A-46)
80	BWR/ PWR	Steel Heat exchanger components, Piping, piping components, and piping elements exposed to Air – indoor, uncontrolled (External), Air – outdoor (External)	Loss of material due to general, pitting, and crevice corrosion	Chapter XI.M36, "External Surfaces Monitoring of Mechanical Components"	No	VII.F1.AP-41 VII.F2.AP-41 VII.F3.AP-41 VII.F4.AP-41 VII.G.AP-40 VII.G.AP-41 VII.H1.A-24 VII.H2.AP-40 VII.H2.AP-41	VII.F1-10(AP-41) VII.F2-8(AP-41) VII.F3-10(AP-41) VII.F4-7(AP-41) VII.G-6(AP-40) VII.G-5(AP-41) VII.H1-8(A-24) VII.H2-4(AP-40) VII.H2-3(AP-41)
81	BWR/ PWR	Copper alloy, Aluminum Piping, piping components, and piping elements exposed to Air – outdoor (External), Air - outdoor	Loss of material due to pitting and crevice corrosion	Chapter XI.M36, "External Surfaces Monitoring of Mechanical Components"	No	VII.I.AP-159 VII.I.AP-256	N/A N/A
82	BWR/ PWR	Elastomers Elastomer: seals and components	Loss of material due to wear	Chapter XI.M36, "External Surfaces Monitoring of	No	VII.F1.AP-113 VII.F2.AP-113 VII.F3.AP-113	VII.F1-5(A-73) VII.F2-5(A-73) VII.F3-5(A-73)

Table 3.3-1   Summary of Aging Management Programs for Auxiliary Systems Evaluated in Chapter VII of the GALL Report

ID	Type	Component	Aging Effect/ Mechanism	Aging Management Programs	Further Evaluation Recommended	Rev2 Item	Rev1 Item
		exposed to Air – indoor, uncontrolled (External)		Mechanical Components"		VII.F4.AP-113	VII.F4-4(A-73)
83	BWR/ PWR	Stainless steel Diesel engine exhaust piping, piping components, and piping elements exposed to Diesel exhaust	Cracking due to stress corrosion cracking	Chapter XI.M38, "Inspection of Internal Surfaces in Miscellaneous Piping and Ducting Components"	No	VII.H2.AP-128	VII.H2-1(AP-33)
85	BWR/ PWR	Elastomers Elastomer seals and components exposed to Closed-cycle cooling water	Hardening and loss of strength due to elastomer degradation	Chapter XI.M38, "Inspection of Internal Surfaces in Miscellaneous Piping and Ducting Components"	No	VII.C2.AP-259	N/A
86	BWR/ PWR	Elastomers Elastomers, linings, Elastomer: seals and components exposed to Treated borated water, Treated water, Raw water	Hardening and loss of strength due to elastomer degradation	Chapter XI.M38, "Inspection of Internal Surfaces in Miscellaneous Piping and Ducting Components"	No	VII.A3.AP-100 VII.A4.AP-101	VII.A3-1(A-15) VII.A4-1(A-16)
88	BWR/ PWR	Steel; stainless steel Piping, piping components, and piping elements, Piping, piping components, and piping elements, diesel engine exhaust exposed to Raw water (potable), Diesel exhaust	Loss of material due to general (steel only), pitting, and crevice corrosion	Chapter XI.M38, "Inspection of Internal Surfaces in Miscellaneous Piping and Ducting Components"	No	VII.E5.AP-270 VII.H2.AP-104	N/A VII.H2-2(A-27)

Table 3.3-1   Summary of Aging Management Programs for Auxiliary Systems Evaluated in Chapter VII of the GALL Report

ID	Type	Component	Aging Effect/ Mechanism	Aging Management Programs	Further Evaluation Recommended	Rev2 Item	Rev1 Item
89	BWR/ PWR	Steel, Copper alloy Piping, piping components, and piping elements exposed to Moist air or condensation (Internal)	Loss of material due to general, pitting, and crevice corrosion	Chapter XI.M38, "Inspection of Internal Surfaces in Miscellaneous Piping and Ducting Components"	No	VII.G.A-23 VII.G.AP-143 VII.H2.A-23	VII.G-23(A-23) VII.G-9(AP-78) VII.H2-21(A-23)
90	BWR/ PWR	Steel Ducting and components (Internal surfaces) exposed to Condensation (Internal)	Loss of material due to general, pitting, crevice, and (for drip pans and drain lines) microbiologically-influenced corrosion	Chapter XI.M38, "Inspection of Internal Surfaces in Miscellaneous Piping and Ducting Components"	No	VII.F1.A-08 VII.F2.A-08 VII.F3.A-08 VII.F4.A-08	VII.F1-3(A-08) VII.F2-3(A-08) VII.F3-3(A-08) VII.F4-2(A-08)
91	BWR/ PWR	Steel Piping, piping components, and piping elements; tanks exposed to Waste Water	Loss of material due to general, pitting, crevice, and microbiologically-influenced corrosion	Chapter XI.M38, "Inspection of Internal Surfaces in Miscellaneous Piping and Ducting Components"	No	VII.E5.AP-281	N/A
92	BWR/ PWR	Aluminum Piping, piping components, and piping elements exposed to Condensation (Internal)	Loss of material due to pitting and crevice corrosion	Chapter XI.M38, "Inspection of Internal Surfaces in Miscellaneous Piping and Ducting Components"	No	VII.F1.AP-142 VII.F2.AP-142 VII.F3.AP-142 VII.F4.AP-142	VII.F1-14(AP-74) VII.F2-12(AP-74) VII.F3-14(AP-74) VII.F4-10(AP-74)
93	BWR/ PWR	Copper alloy Piping, piping components, and piping elements exposed to Raw water	Loss of material due to pitting and crevice corrosion	Chapter XI.M38, "Inspection of Internal Surfaces in Miscellaneous Piping and Ducting Components"	No	VII.E5.AP-271	N/A

Table 3.3-1 Summary of Aging Management Programs for Auxiliary Systems Evaluated in Chapter VII of the GALL Report

ID	Type	Component	Aging Effect/ Mechanism	Aging Management Programs	Further Evaluation Recommended	Rev2 Item	Rev1 Item
		(potable)					
94	BWR/ PWR	Stainless steel Ducting and components exposed to Condensation	Loss of material due to pitting and crevice corrosion	Chapter XI.M38, "Inspection of Internal Surfaces in Miscellaneous Piping and Ducting Components"	No	VII.F1.AP-99 VII.F2.AP-99 VII.F3.AP-99	VII.F1-1(A-09) VII.F2-1(A-09) VII.F3-1(A-09)
95	BWR/ PWR	Copper alloy, Stainless steel, Nickel alloy, Steel Piping, piping components, and piping elements, Heat exchanger components, Piping, piping components, and piping elements; tanks exposed to Waste water, Condensation (Internal)	Loss of material due to pitting, crevice, and microbiologically-influenced corrosion	Chapter XI.M38, "Inspection of Internal Surfaces in Miscellaneous Piping and Ducting Components"	No	VII.E5.AP-272 VII.E5.AP-273 VII.E5.AP-274 VII.E5.AP-275 VII.E5.AP-276 VII.E5.AP-278 VII.E5.AP-279 VII.E5.AP-280	N/A N/A N/A N/A N/A N/A N/A
96	BWR/ PWR	Elastomers Elastomer: seals and components exposed to Air – indoor, uncontrolled (Internal)	Loss of material due to wear	Chapter XI.M38, "Inspection of Internal Surfaces in Miscellaneous Piping and Ducting Components"	No	VII.F1.AP-103 VII.F2.AP-103 VII.F3.AP-103 VII.F4.AP-103	VII.F1-6(A-18) VII.F2-6(A-18) VII.F3-6(A-18) VII.F4-5(A-18)
97	BWR/ PWR	Steel Piping, piping components, and piping elements, Reactor coolant pump oil collection system: tanks, Reactor coolant pump oil collection system: piping, tubing, valve bodies exposed	Loss of material due to general, pitting, and crevice corrosion	Chapter XI.M39, "Lubricating Oil Analysis," and Chapter XI.M32, "One-Time Inspection"	No	VII.C1.AP-127 VII.C2.AP-127 VII.E1.AP-127 VII.E4.AP-127 VII.F1.AP-127 VII.F2.AP-127 VII.F3.AP-127 VII.F4.AP-127 VII.G.AP-116 VII.G.AP-117 VII.G.AP-127	VII.C1-17(AP-30) VII.C2-13(AP-30) VII.E1-19(AP-30) VII.E4-16(AP-30) VII.F1-19(AP-30) VII.F2-17(AP-

Table 3.3-1   Summary of Aging Management Programs for Auxiliary Systems Evaluated in Chapter VII of the GALL Report

ID	Type	Component	Aging Effect/ Mechanism	Aging Management Programs	Further Evaluation Recommended	Rev2 Item	Rev1 Item
		to Lubricating oil				VII.H2.AP-127	30) VII.F3-19(AP-30) VII.F4-15(AP-30) VII.G-27(A-82) VII.G-26(A-83) VII.G-22(AP-30) VII.H2-20(AP-30)
98	BWR/ PWR	Steel Heat exchanger components exposed to Lubricating oil	Loss of material due to general, pitting, crevice, and microbiologically-influenced corrosion; fouling that leads to corrosion	Chapter XI.M39, "Lubricating Oil Analysis," and Chapter XI.M32, "One-Time Inspection"	No	VII.H2.AP-131	VII.H2-5(AP-39)
99	BWR/ PWR	Copper alloy, Aluminum Piping, piping components, and piping elements exposed to Lubricating oil	Loss of material due to pitting and crevice corrosion	Chapter XI.M39, "Lubricating Oil Analysis," and Chapter XI.M32, "One-Time Inspection"	No	VII.C1.AP-133 VII.C2.AP-133 VII.E1.AP-133 VII.E4.AP-133 VII.G.AP-133 VII.H2.AP-133 VII.H2.AP-162	VII.C1-8(AP-47) VII.C2-5(AP-47) VII.E1-12(AP-47) VII.E4-6(AP-47) VII.G-11(AP-47) VII.H2-10(AP-47) N/A
100	BWR/ PWR	Stainless steel Piping, piping components, and piping elements exposed to Lubricating oil	Loss of material due to pitting, crevice, and microbiologically-influenced corrosion	Chapter XI.M39, "Lubricating Oil Analysis," and Chapter XI.M32, "One-Time Inspection"	No	VII.C1.AP-138 VII.C2.AP-138 VII.E1.AP-138 VII.E4.AP-138 VII.G.AP-138 VII.H2.AP-138	VII.C1-14(AP-59) VII.C2-12(AP-59) VII.E1-15(AP-59) VII.E4-12(AP-

Table 3.3-1   Summary of Aging Management Programs for Auxiliary Systems Evaluated in Chapter VII of the GALL Report

ID	Type	Component	Aging Effect/ Mechanism	Aging Management Programs	Further Evaluation Recommended	Rev2 Item	Rev1 Item
							59) VII.G-18(AP-59) VII.H2-17(AP-59)
101	BWR/ PWR	Aluminum Heat exchanger tubes exposed to Lubricating oil	Reduction of heat transfer due to fouling	Chapter XI.M39, "Lubricating Oil Analysis," and Chapter XI.M32, "One-Time Inspection"	No	VII.H2.AP-154	N/A
102	BWR/ PWR	Boral®, boron steel, and other materials (excluding Boraflex) Spent fuel storage racks: neutron-absorbing sheets (PWR), Spent fuel storage racks: neutron-absorbing sheets (BWR) exposed to Treated borated water, Treated water	Reduction of neutron-absorbing capacity; change in dimensions and loss of material due to effects of SFP environment	Chapter XI.M40, "Monitoring of Neutron-Absorbing Materials other than Boraflex"	No	VII.A2.AP-235 VII.A2.AP-236	VII.A2-5(A-88) VII.A2-3(A-89)
103	BWR/ PWR	Reinforced concrete, asbestos cement Piping, piping components, and piping elements exposed to Soil or concrete	Cracking due to aggressive chemical attack and leaching; Changes in material properties due to aggressive chemical attack	Chapter XI.M41, "Buried and Underground Piping and Tanks"	No	VII.C1.AP-157	N/A
104	BWR/ PWR	HDPE, Fiberglass Piping, piping components, and	Cracking, blistering, change in color	Chapter XI.M41, "Buried and Underground Piping and Tanks"	No	VII.C1.AP-175 VII.C1.AP-176	N/A N/A

Table 3.3-1 Summary of Aging Management Programs for Auxiliary Systems Evaluated in Chapter VII of the GALL Report

ID	Type	Component	Aging Effect/ Mechanism	Aging Management Programs	Further Evaluation Recommended	Rev2 Item	Rev1 Item
		piping elements exposed to Soil or concrete	due to water absorption				
105	BWR/ PWR	Concrete cylinder piping, Asbestos cement pipe Piping, piping components, and piping elements exposed to Soil or concrete	Cracking, spalling, corrosion of rebar due to exposure of rebar	Chapter XI.M41, "Buried and Underground Piping and Tanks"	No	VII.C1.AP-177 VII.C1.AP-178 VII.C1.AP-237	N/A N/A N/A
106	BWR/ PWR	Steel (with coating or wrapping) Piping, piping components, and piping elements exposed to Soil or concrete	Loss of material due to general, pitting, crevice, and microbiologically-influenced corrosion	Chapter XI.M41, "Buried and Underground Piping and Tanks"	No	VII.C1.AP-198 VII.C3.AP-198 VII.G.AP-198 VII.H1.AP-198	VII.C1-18(A-01) VII.C3-9(A-01) VII.G-25(A-01) VII.H1-9(A-01)
107	BWR/ PWR	Stainless steel Piping, piping components, and piping elements exposed to Soil or concrete	Loss of material due to pitting and crevice corrosion	Chapter XI.M41, "Buried and Underground Piping and Tanks"	No	VII.C1.AP-137 VII.C3.AP-137 VII.G.AP-137 VII.H1.AP-137 VII.H2.AP-137	VII.C1-16(AP-56) VII.C3-8(AP-56) VII.G-20(AP-56) VII.H1-7(AP-56) VII.H2-19(AP-56)
108	BWR/ PWR	Titanium, Super austenitic, Aluminum, Copper Alloy, Stainless Steel Piping, piping components, and piping elements, Bolting exposed to Soil or concrete	Loss of material due to pitting and crevice corrosion	Chapter XI.M41, "Buried and Underground Piping and Tanks"	No	VII.C1.AP-171 VII.C1.AP-172 VII.C1.AP-173 VII.C1.AP-174 VII.I.AP-243	N/A N/A N/A N/A N/A

Table 3.3-1   Summary of Aging Management Programs for Auxiliary Systems Evaluated in Chapter VII of the GALL Report

ID	Type	Component	Aging Effect/ Mechanism	Aging Management Programs	Further Evaluation Recommended	Rev2 Item	Rev1 Item
109	BWR/ PWR	Steel Bolting exposed to Soil or concrete	Loss of material due to general, pitting and crevice corrosion	Chapter XI.M41, "Buried and Underground Piping and Tanks"	No	VII.I.AP-241	N/A
	BWR/ PWR	Underground Aluminum, Copper Alloy, Stainless Steel and Steel Piping, piping components, and piping elements	Loss of material due to general (steel only), pitting and crevice corrosion	Chapter XI.M41, "Buried and Underground Piping and Tanks"	No	VII.I.AP-284	N/A
110	BWR	Stainless steel Piping, piping components, and piping elements exposed to Treated water >60°C (>140°F)	Cracking due to stress corrosion cracking	Chapter XI.M7, "BWR Stress Corrosion Cracking," and Chapter XI.M2, "Water Chemistry"	No	VII.E4.A-61	VII.E4-15(A-61)
111	BWR/ PWR	Steel Structural steel exposed to Air – indoor, uncontrolled (External)	Loss of material due to general, pitting, and crevice corrosion	Chapter XI.S6, "Structures Monitoring"	No	VII.A1.A-94	VII.A1-1(A-94)
112	BWR/ PWR	Steel Piping, piping components, and piping elements exposed to Concrete	None	None, provided 1) attributes of the concrete are consistent with ACI 318 or ACI 349 (low water-to-cement ratio, low permeability, and adequate air entrainment) as cited in NUREG-1557, and 2) plant OE indicates no degradation of the concrete	No, if conditions are met.	VII.J.AP-282	VII.J-21(AP-3)
113	BWR/ PWR	Aluminum Piping, piping components, and piping elements exposed to Air – dry	None	None	NA - No AEM or AMP	VII.J.AP-134 VII.J.AP-135 VII.J.AP-36 VII.J.AP-37	N/A N/A VII.J-1(AP-36) VII.J-2(AP-37)

Table 3.3-1   Summary of Aging Management Programs for Auxiliary Systems Evaluated in Chapter VII of the GALL Report

ID	Type	Component	Aging Effect/ Mechanism	Aging Management Programs	Further Evaluation Recommended	Rev2 Item	Rev1 Item
		(Internal/External), Air – indoor, uncontrolled (Internal/External), Air – indoor, controlled (External), Gas					
114	BWR/ PWR	Copper alloy Piping, piping components, and piping elements exposed to Air – indoor, uncontrolled (Internal/External), Air – dry, Gas	None	None	NA - No AEM or AMP	VII.J.AP-144 VII.J.AP-8 VII.J.AP-9	N/A VII.J-3(AP-8) VII.J-4(AP-9)
115	PWR	Copper alloy (≤15% Zn and ≤8% Al) Piping, piping components, and piping elements exposed to Air with borated water leakage	None	None	NA - No AEM or AMP	VII.J.AP-11	VII.J-5(AP-11)
116	BWR/ PWR	Galvanized steel Piping, piping components, and piping elements exposed to Air – indoor, uncontrolled	None	None	NA - No AEM or AMP	VII.J.AP-13	VII.J-6(AP-13)
117	BWR/ PWR	Glass Piping elements exposed to Air – indoor, uncontrolled (External), Lubricating oil, Closed-cycle cooling water, Air – outdoor, Fuel oil, Raw water, Treated water,	None	None	NA - No AEM or AMP	VII.J.AP-14 VII.J.AP-15 VII.J.AP-166 VII.J.AP-167 VII.J.AP-48 VII.J.AP-49 VII.J.AP-50 VII.J.AP-51 VII.J.AP-52 VII.J.AP-96	VII.J-8(AP-14) VII.J-10(AP-15) N/A N/A VII.J-7(AP-48) VII.J-9(AP-49) VII.J-11(AP-50) VII.J-13(AP-51) VII.J-12(AP-52) N/A

Table 3.3-1   Summary of Aging Management Programs for Auxiliary Systems Evaluated in Chapter VII of the GALL Report

ID	Type	Component	Aging Effect/ Mechanism	Aging Management Programs	Further Evaluation Recommended	Rev2 Item	Rev1 Item
		Treated borated water, Air with borated water leakage, Condensation (Internal/External) Gas				VII.J.AP-97 VII.J.AP-98	N/A N/A
118	BWR/ PWR	Nickel alloy Piping, piping components, and piping elements exposed to Air – indoor, uncontrolled (External)	None	None	NA - No AEM or AMP	VII.J.AP-16	VII.J-14(AP-16)
119	BWR/ PWR	Nickel alloy, PVC, Glass Piping, piping components, and piping elements exposed to Air with borated water leakage, Air – indoor, uncontrolled, Condensation (Internal), Waste Water	None	None	NA - No AEM or AMP	VII.J.AP-260 VII.J.AP-268 VII.J.AP-269 VII.J.AP-277	N/A N/A N/A N/A
120	BWR/ PWR	Stainless steel Piping, piping components, and piping elements exposed to Air – indoor, uncontrolled (Internal/External), Air – indoor, uncontrolled (External), Air with borated water leakage, Concrete, Air – dry, Gas	None	None	NA - No AEM or AMP	VII.J.AP-123 VII.J.AP-17 VII.J.AP-18 VII.J.AP-19 VII.J.AP-20 VII.J.AP-22	N/A VII.J-15(AP-17) VII.J-16(AP-18) VII.J-17(AP-19) VII.J-18(AP-20) VII.J-19(AP-22)

Table 3.3-1 Summary of Aging Management Programs for Auxiliary Systems Evaluated in Chapter VII of the GALL Report

ID	Type	Component	Aging Effect/ Mechanism	Aging Management Programs	Further Evaluation Recommended	Rev2 Item	Rev1 Item
121	BWR/ PWR	Steel Piping, piping components, and piping elements exposed to Air – indoor, controlled (External), Air – dry, Gas	None	None	NA - No AEM or AMP	VII.J.AP-2 VII.J.AP-4 VII.J.AP-6	VII.J-20(AP-2) VII.J-22(AP-4) VII.J-23(AP-6)
122	BWR/ PWR	Titanium Heat exchanger components, Piping, piping components, and piping elements exposed to Air – indoor, uncontrolled or Air – outdoor	None	None	NA - No AEM or AMP	VII.J.AP-151 VII.J.AP-160	N/A N/A
123	BWR/ PWR	Titanium (ASTM Grades 1,2, 7, 11, or 12 that contains > 5% aluminum or more than 0.20% oxygen or any amount of tin) Heat exchanger components other than tubes, Piping, piping components, and piping elements exposed to Raw water	None	None	NA - No AEM or AMP	VII.C1.AP-152 VII.C1.AP-161	N/A N/A

Table 3.3-2 Aging Management Programs Recommended for Aging Management of Auxiliary Systems

GALL Report Chapter/AMP	Program Name
Chapter XI.M1	ASME Section XI Inservice Inspection, Subsections IWB, IWC, and IWD
Chapter XI.M2	Water Chemistry
Chapter XI.M7	BWR Stress Corrosion Cracking
Chapter XI.M10	Boric Acid Corrosion
Chapter XI.M18	Bolting Integrity
Chapter XI.M20	Open-Cycle Cooling Water System
Chapter XI.M21A	Closed Treated Water Systems
Chapter XI.M22	Boraflex Monitoring
Chapter XI.M23	Inspection of Overhead Heavy and Light Loads (Related to Refueling) Handling Systems
Chapter XI.M24	Compressed Air Monitoring
Chapter XI.M25	BWR Reactor Cleanup System
Chapter XI.M26	Fire Protection
Chapter XI.M27	Fire Water System
Chapter XI.M29	Aboveground Metallic Tanks
Chapter XI.M30	Fuel Oil Chemistry
Chapter XI.M32	One-Time Inspection
Chapter XI.M33	Selective Leaching
Chapter XI.M36	External Surfaces Monitoring of Mechanical Components
Chapter XI.M38	Inspection of Internal Surfaces in Miscellaneous Piping and Ducting Components
Chapter XI.M39	Lubricating Oil Analysis
Chapter XI.M40	Monitoring of Neutron-Absorbing Materials Other than Boraflex
Chapter XI.M41	Buried and Underground Piping and Tanks
Chapter XI.S6	Structures Monitoring
Appendix for GALL	Quality Assurance for Aging Management Programs
SRP-LR Appendix A	Plant-specific AMP

3.4   AGING MANAGEMENT OF STEAM AND POWER CONVERSION SYSTEM

Review Responsibilities

Primary - Branch assigned responsibility by PM as described in Section 3.0 of this SRP-LR.

3.4.1   Areas of Review

This section addresses the aging management review (AMR) and the associated aging management program (AMP) of the steam and power conversion system. For a recent vintage plant, the information related to the steam and power conversion system is contained in Chapter 10, "Steam and Power Conversion System," of the plant's FSAR, consistent with the "Standard Review Plan for the Review of Safety Analysis Reports for Nuclear Power Plants" (NUREG-0800). The steam and power conversion systems contained in this review plan section are generally consistent with those contained in NUREG-0800 except for the condenser circulating water and the condensate storage systems. For older plants, the location of applicable information is plant-specific because an older plant's FSAR may have predated NUREG-0800.

Typical steam and power conversion systems that are subject to an AMR for license renewal are steam turbine, main steam, extraction steam, feedwater, condensate, steam generator blowdown, and auxiliary feedwater. This review plan section also includes structures and components in nonsafety-related systems that are not connected to safety-related SSCs but have a spatial relationship such that their failure could adversely impact the performance of a safety-related SSC-intended function. Examples of such nonsafety-related systems may be extraction steam, plant heating steam/auxiliary boilers and hot water heating systems.

The aging management for the steam generator is reviewed following the guidance in Section 3.1 of this SRP-LR. The aging management for portions of the BWR main steam and main feedwater systems, extending from the reactor vessel to the outermost containment isolation valve, is reviewed separately following the guidance in Section 3.1 of this SRP-LR.

The responsible review organization is to review the following LRA AMR and AMP items assigned to it, per SRP-LR Section 3.0:

AMRs
- AMR results consistent with the GALL Report
- AMR results for which further evaluation is recommended by the GALL Report
- AMR results not consistent with or not addressed in the GALL Report

AMPs
- Consistent with GALL Report AMPs
- Plant-specific AMPs

FSAR Supplement
- The responsible review organization is to review the FSAR Supplement associated with each assigned AMP.

3.4.2 Acceptance Criteria

The acceptance criteria for the areas of review describe methods for determining whether the applicant has met the requirements of the NRC's regulations in 10 CFR 54.21.

3.4.2.1 AMR Results Consistent with the GALL Report

The AMR and the AMPs applicable to the steam and power conversion system are described and evaluated in Chapter VIII of NUREG-1801 (GALL Report).

The applicant's LRA should provide sufficient information so that the NRC reviewer is able to confirm that the specific LRA AMR item and the associated LRA AMP are consistent with the cited GALL Report AMR item. The reviewer should then confirm that the LRA AMR item is consistent with the GALL Report AMR item to which it is compared.

When the applicant is crediting a different aging management program than recommended in the GALL Report, the reviewer should confirm that the alternate aging management program is valid to use for aging management and will be capable of managing the effects of aging as adequately as the aging management program recommended by the GALL Report.

3.4.2.2 AMR Results for Which Further Evaluation is Recommended by the GALL Report

The basic acceptance criteria, defined in Subsection 3.4.2.1, need to be applied first for all of the AMRs and AMPs reviewed as part of this section. In addition, if the GALL Report AMR item to which the LRA AMR item is compared identifies that "further evaluation is recommended," then additional criteria apply as identified by the GALL Report for each of the following aging effect/aging mechanism combinations. Refer to Table 3.4-1, comparing the "Further Evaluation Recommended" and the "Rev2 Item" columns, for the AMR items that reference the following subsections. The 2005 AMR item counterpart is provided in the "Rev1 Item" column.

3.4.2.2.1 Cumulative Fatigue Damage

Fatigue is a TLAA as defined in 10 CFR 54.3. TLAAs are required to be evaluated in accordance with 10 CFR 54.21(c). This TLAA is addressed separately in Section 4.3, "Metal Fatigue Analysis," of this SRP-LR. The related GALL Report items invoked by the subsection are VIII.D1.S-11, VIII.D2.S-11, VIII.G.S-11, VIII.B1.S-08, VIII.B2.S-08.

3.4.2.2.2 Cracking due to Stress Corrosion Cracking (SCC)

Cracking due to stress corrosion cracking could occur for stainless steel piping, piping components, piping elements, and tanks exposed to outdoor air. The possibility of cracking also extends to components exposed to air which has recently been introduced into buildings, i.e., components near intake vents. Cracking is only known to occur in environments containing sufficient halides (primarily chlorides) and in which condensation or deliquescence is possible. Condensation or deliquescence should generally be assumed to be possible. Applicable outdoor air environments (and associated indoor air environments) include, but are not limited to, those within approximately 5 miles of a saltwater coastline, those within 1/2 mile of a highway which is treated with salt in the wintertime, those areas in which the soil contains more than trace chlorides, those plants having cooling towers where the water is treated with chlorine or chlorine compounds, and those areas subject to chloride contamination from other agricultural or industrial sources. This item is applicable for the environments described above.

GALL AMP XI.M36, "External Surfaces Monitoring," is an acceptable method to manage the aging effect. The applicant may demonstrate that this item is not applicable by describing the outdoor air environment present at the plant and demonstrating that external chloride stress corrosion cracking is not expected. The GALL Report recommends further evaluation to determine whether an adequate aging management program is used to manage this aging effect based on the environmental conditions applicable to the plant and ASME Code Section XI requirements applicable to the components.

3.4.2.2.3   Loss of Material due to Pitting and Crevice Corrosion

Loss of material due to pitting and crevice corrosion could occur for stainless steel piping, piping components, piping elements, and tanks exposed to outdoor air. The possibility of pitting and crevice corrosion also extends to components exposed to air which has recently been introduced into buildings, i.e., components near intake vents. Pitting and crevice corrosion is only known to occur in environments containing sufficient halides (primarily chlorides) and in which condensation or deliquescence is possible. Condensation or deliquescence should generally be assumed to be possible. Applicable outdoor air environments (and associated indoor air environments) include, but are not limited to, those within approximately 5 miles of a saltwater coastline, those within 1/2 mile of a highway which is treated with salt in the wintertime, those areas in which the soil contains more than trace chlorides, those plants having cooling towers where the water is treated with chlorine or chlorine compounds, and those areas subject to chloride contamination from other agricultural or industrial sources. This item is applicable for the environments described above.

GALL AMP XI.M36, "External Surfaces Monitoring," is an acceptable method to manage the aging effect. The applicant may demonstrate that this item is not applicable by describing the outdoor air environment present at the plant and demonstrating that external pitting or crevice corrosion is not expected. The GALL Report recommends further evaluation to determine whether an adequate aging management program is used to manage this aging effect based on the environmental conditions applicable to the plant and ASME Code Section XI requirements Quality Assurance for Aging Management of Nonsafety-Related Components.

3.4.2.2.4 Quality Assurance for Aging Management of Nonsafety-Related Components

Acceptance criteria are described in Branch Technical Position IQMB-1 (Appendix A.2, of this SRP-LR).

3.4.2.3   AMR Results Not Consistent with or Not Addressed in the GALL Report

Acceptance criteria are described in Branch Technical Position RLSB-1 (Appendix A.1 of this SRP-LR).

3.4.2.4   Aging Management Programs

For those AMPs that will be used for aging management and are based on the program elements of an AMP in the GALL Report, the NRC reviewer performs an audit of aging management programs credited in the LRA to confirm consistency with the GALL AMPs identified in the GALL Report, Chapters X and XI.

If the applicant identifies an exception to any of the program elements of the cited GALL Report AMP, the LRA AMP should include a basis demonstrating how the criteria of 10 CFR 54.21(a)(3) would still be met. The NRC reviewer should then confirm that the LRA AMP with all exceptions would satisfy the criteria of 10 CFR 54.21(a)(3). If, while reviewing the LRA AMP, the reviewer identifies a difference between the LRA AMP and the GALL Report AMP that should have been identified as an exception to the GALL Report AMP, the difference should be reviewed and properly dispositioned. The reviewer should document the disposition of all LRA-defined exceptions and staff-identified differences.

The LRA should identify any enhancements that are needed to permit an existing LRA AMP to be declared consistent with the GALL Report AMP to which the LRA AMP is compared. The reviewer is to confirm both that the enhancement, when implemented, would allow the existing LRA AMP to be consistent with the GALL Report AMP and also that the applicant has a commitment in the FSAR Supplement to implement the enhancement prior to the period of extended operation. The reviewer should document the disposition of all enhancements.

If the applicant chooses to use a plant-specific program that is not a GALL AMP, the NRC reviewer should confirm that the plant-specific program satisfies the criteria of Branch Technical Position RLSB-1 (Appendix A.1.2.3 of this SRP-LR).

3.4.2.5   FSAR Supplement

The summary description of the programs and activities for managing the effects of aging for the period of extended operation in the FSAR Supplement should be sufficiently comprehensive that later changes can be controlled by 10 CFR 50.59. The description should contain information associated with the bases for determining that aging effects will be managed during the period of extended operation. The description should also contain any future aging management activities, including enhancements and commitments, to be completed before the period of extended operation. Table 3.0-1 of this SRP-LR provides examples of the type of information to be included in the FSAR Supplement. Table 3.4-2 lists the programs that are applicable for this SRP-LR subsection.

3.4.3   Review Procedures

For each area of review, the following review procedures discussed below are to be followed.

3.4.3.1   AMR Results Consistent with the GALL Report

The applicant may reference the GALL Report in its LRA, as appropriate, and demonstrate that the AMRs and AMPs at its facility are consistent with those reviewed and approved in the GALL Report. The reviewer should not conduct a re-review of the substance of the matters described in the GALL Report. If the applicant has provided the information necessary to adopt the finding of program acceptability as described and evaluated in the GALL Report, the reviewer should find acceptable the applicant's reference to the GALL Report in its LRA. In making this determination, the reviewer confirms that the applicant has provided a brief description of the system, components, materials, and environment. The reviewer also confirms that the applicant has stated that the applicable aging effects and industry and plant-specific operating experience have been reviewed by the applicant and are evaluated in the GALL Report.

Furthermore, the reviewer should confirm that the applicant has addressed operating experience identified after the issuance of the GALL Report. Performance of this review requires the reviewer to confirm that the applicant has identified those aging effects for the steam and power conversion system components that are contained in the GALL Report as applicable to its plant.

3.4.3.2 AMR Results for Which Further Evaluation is Recommended by the GALL Report

The basic review procedures defined in Subsection 3.4.3.1 need to be applied first for all of the AMRs and AMPs provided in this section. In addition, if the GALL Report AMR item to which the LRA AMR item is compared identifies that "further evaluation is recommended," then additional criteria apply as identified by the GALL Report for each of the following aging effect/aging mechanism combinations. Refer to Table 3.4-1 for the Rev 2 item references for the following subsections.

3.4.3.2.1 Cumulative Fatigue Damage

Fatigue is a TLAA as defined in 10 CFR 54.3. TLAAs are required to be evaluated in accordance with 10 CFR 54.21(c). The reviewer reviews the evaluation of this TLAA separately following the guidance in Section 4.3 of this SRP-LR.

3.4.3.2.2 Cracking due to Stress Corrosion Cracking

The GALL Report recommends further evaluation to manage cracking due to stress corrosion cracking of stainless steel piping, piping components, piping elements, and tanks exposed to outdoor air environments containing sufficient halides (primarily chlorides) and in which condensation or deliquescence is possible. The possibility of cracking also extends to components exposed to air which has recently been introduced into buildings, i.e., components near intake vents.

The reviewer should determine whether an adequate program is used to manage the aging effect based on the applicable environmental conditions and ASME Code requirements. Cracking is only known to occur in environments containing sufficient halides (primarily chlorides) and in which condensation or deliquescence is possible. Condensation or deliquescence should generally be assumed to be possible. Applicable outdoor air environments (and associated indoor air environments) include, but are not limited to, those within approximately 5 miles of a saltwater coastline, those within 1/2 mile of a highway which is treated with salt in the wintertime, those areas in which the soil contains more than trace chlorides, those plants having cooling towers where the water is treated with chlorine or chlorine compounds, and those areas subject to chloride contamination from other agricultural or industrial sources. This item is applicable for the environments described above. GALL AMP XI.M36, "External Surfaces Monitoring," is an acceptable method to manage the aging effect.

3.4.3.2.3 Loss of Material due to Pitting and Crevice Corrosion

The GALL Report recommends further evaluation to manage loss of material due to pitting and crevice corrosion of stainless steel piping, piping components, piping elements, and tanks exposed to outdoor air environments containing sufficient halides (primarily chlorides) and in which condensation or deliquescence is possible. The possibility of pitting and crevice corrosion also extends to components exposed to air which has recently been introduced into buildings, i.e., components near intake vents.

The reviewer should determine whether an adequate program is used to manage the aging effect based on the applicable environmental conditions and ASME Code requirements. Pitting and crevice corrosion is only known to occur in environments containing sufficient halides (primarily chlorides) and in which condensation or deliquescence is possible. Condensation or deliquescence should generally be assumed to be possible. Applicable outdoor air environments (and associated indoor air environments) include, but are not limited to, those within approximately 5 miles of a saltwater coastline, those within 1/2 mile of a highway which is treated with salt in the wintertime, those areas in which the soil contains more than trace chlorides, those plants having cooling towers where the water is treated with chlorine or chlorine compounds, and those areas subject to chloride contamination from other agricultural or industrial sources. This item is applicable for the environments described above. GALL AMP XI.M36, "External Surfaces Monitoring," is an acceptable method to manage the aging effect.

3.4.3.2.4 Quality Assurance for Aging Management of Nonsafety-Related Components

The applicant's aging management programs for license renewal should contain the elements of corrective actions, the confirmation process, and administrative controls. Safety-related components are covered by 10 CFR Part 50, Appendix B, which is adequate to address these program elements. However, Appendix B does not apply to nonsafety-related components that are subject to an aging management review for license renewal. Nevertheless, the applicant has the option to expand the scope of its 10 CFR Part 50, Appendix B program to include these components and address these program elements. If the applicant chooses this option, the reviewer verifies that the applicant has documented such a commitment in the FSAR Supplement. If the applicant chooses alternative means, the branch responsible for quality assurance should be requested to review the applicant's proposal on a case-by-case basis.

3.4.3.3 AMR Results Not Consistent with or Not Addressed in the GALL Report

The reviewer should confirm that the applicant, in its LRA, has identified applicable aging effects, listed the appropriate combination of materials and environments, and has credited AMPs that will adequately manage the aging effects. The AMP credited by the applicant could be an AMP that is described and evaluated in the GALL Report or a plant-specific program. Review procedures are described in Branch Technical Position RSLB-1 (Appendix A.1 of this SRP-LR).

3.4.3.4 Aging Management Programs

The reviewer confirms that the applicant has identified the appropriate AMPs as described and evaluated in the GALL Report. If the applicant commits to an enhancement to make its LRA AMP consistent with a GALL Report AMP, then the reviewer is to confirm that this enhancement, when implemented, will make the LRA AMP consistent with the GALL Report AMP. If the applicant identifies, in the LRA AMP, an exception to any of the program elements of the GALL Report AMP, the reviewer is to confirm that the LRA AMP with the exception will satisfy the criteria of 10 CFR 54.21(a)(3). If the reviewer identifies a difference, not identified by the LRA, between the LRA AMP and the GALL Report AMP with which the LRA claims to be consistent, the reviewer should confirm that the LRA AMP with this difference satisfies 10 CFR 54.21(a)(3). The reviewer should document the basis for accepting enhancements, exceptions, or differences. The AMPs evaluated in the GALL Report pertinent to the steam and power conversion system are summarized in Table 3.4-1 of this SRP-LR. The "Rev 2 Item" (for 2010) and "Rev1 Item" (for 2005 counterpart) columns identify the AMR item numbers in the GALL Report, Chapter VIII, presenting detailed information summarized by this row.

Table 3.4-1 of this SRP-LR may identify a plant-specific aging management program. If the applicant chooses to use a plant-specific program that is not a GALL AMP, the NRC reviewer should confirm that the plant-specific program satisfies the criteria of Branch Technical Position RLSB-1 (Appendix A.1.2.3 of this SRP-LR).

3.4.3.5   FSAR Supplement

The reviewer confirms that the applicant has provided in the FSAR supplement information equivalent to that in Table 3.0-1 for aging management of the steam and power conversion systems. Table 3.4-2 lists the AMPs that are applicable for this SRP-LR subsection. The reviewer also confirms that the applicant has provided information for Subsection 3.4.3.3, "AMR Results Not Consistent with or Not Addressed in the GALL Report," equivalent to that in Table 3.0-1.

The staff expects to impose a license condition on any renewed license to require the applicant to update its FSAR to include this FSAR Supplement at the next update required pursuant to 10 CFR 50.71(e)(4). As part of the license condition until the FSAR update is complete, the applicant may make changes to the programs described in its FSAR Supplement without prior NRC approval, provided that the applicant evaluates each such change and finds it acceptable pursuant to the criteria set forth in 10 CFR 50.59. If the applicant updates the FSAR to include the final FSAR supplement before the license is renewed, no condition will be necessary.

As noted in Table 3.0-1, the applicant need not incorporate the implementation schedule into its FSAR. However, the reviewer should confirm that the applicant has identified and committed in the LRA to any future aging management activities, including enhancements and commitments, to be completed before entering the period of extended operation. The staff expects to impose a license condition on any renewed license to ensure that the applicant will complete these activities no later than the committed date.

3.4.4   Evaluation Findings

If the reviewer determines that the applicant has provided information sufficient to satisfy the provisions of this section, then an evaluation finding similar to the following text should be included in the staff's safety evaluation report:

> On the basis of its review, as discussed above, the staff concludes that the applicant has demonstrated that the aging effects associated with the steam and power conversion system components will be adequately managed so that the intended functions will be maintained consistent with the CLB for the period of extended operation, as required by 10 CFR 54.21(a)(3).

> The staff also reviewed the applicable FSAR Supplement program summaries and concludes that they adequately describe the AMPs credited for managing aging of the steam and power conversion system, as required by 10 CFR 54.21(d).

3.4.5   Implementation

Except in those cases in which the applicant proposes an acceptable alternative method for complying with specified portions of the NRC's regulations, the method described herein will be used by the staff in its evaluation of conformance with NRC regulations.

3.4.6 References

1. NUREG-0800, "Standard Review Plan for the Review of Safety Analysis Reports for Nuclear Power Plants," U.S. Nuclear Regulatory Commission, March 2007.

2. NUREG-1801, "Generic Aging Lessons Learned (GALL)," U.S. Nuclear Regulatory Commission, Revision 2, 2010.

3. NEI 95-10, "Industry Guideline for Implementing the Requirements of 10 CFR Part 54 – The License Renewal Rule," Nuclear Energy Institute, Revision 6.

ID	Type	Component	Aging Effect/Mechanism	Aging Management Programs	Further Evaluation Recommended	Rev2 Item	Rev1 Item
1	BWR/PWR	Steel Piping, piping components, and piping elements exposed to Steam or Treated water	Cumulative fatigue damage due to fatigue	Fatigue is a time-limited aging analysis (TLAA) to be evaluated for the period of extended operation. See the SRP, Section 4.3 "Metal Fatigue," for acceptable methods for meeting the requirements of 10 CFR 54.21(c)(1).	Yes, TLAA (See subsection 3.4.2.1)	VIII.B1.S-08 VIII.B2.S-08 VIII.D1.S-11 VIII.D2.S-11 VIII.G.S-11	VIII.B1-10(S-08) VIII.B2-5(S-08) VIII.D1-7(S-11) VIII.D2-6(S-11) VIII.G-37(S-11)
2	BWR/PWR	Stainless steel Piping, piping components, and piping elements; tanks exposed to Air – outdoor	Cracking due to stress corrosion cracking	Chapter XI.M36, "External Surfaces Monitoring of Mechanical Components"	Yes, environmental conditions need to be evaluated (See subsection 3.4.2.2)	VIII.A.SP-118 VIII.B1.SP-118 VIII.B2.SP-118 VIII.C.SP-118 VIII.D1.SP-118 VIII.D2.SP-118 VIII.E.SP-118 VIII.F.SP-118 VIII.G.SP-118	N/A N/A N/A N/A N/A N/A N/A N/A
3	BWR/PWR	Stainless steel Piping, piping components, and piping elements; tanks exposed to Air – outdoor	Loss of material due to pitting and crevice corrosion	Chapter XI.M36, "External Surfaces Monitoring of Mechanical Components"	Yes, environmental conditions need to be evaluated (See subsection 3.4.2.3)	VIII.A.SP-127 VIII.B1.SP-127 VIII.B2.SP-127 VIII.C.SP-127 VIII.D1.SP-127 VIII.D2.SP-127 VIII.E.SP-127 VIII.F.SP-127 VIII.G.SP-127	N/A N/A N/A N/A N/A N/A N/A N/A
4	PWR	Steel External surfaces, Bolting exposed to Air with borated water leakage	Loss of material due to boric acid corrosion	Chapter XI.M10, "Boric Acid Corrosion"	No	VIII.H.S-30 VIII.H.S-40	VIII.H-9(S-30) VIII.H-2(S-40)

Table 3.4-1 Summary of Aging Management Programs for Steam and Power Conversion System Evaluated in Chapter VIII of the GALL Report

ID	Type	Component	Aging Effect/Mechanism	Aging Management Programs	Further Evaluation Recommended	Rev2 Item	Rev1 Item
5	BWR/PWR	Steel Piping, piping components, and piping elements exposed to Steam, Treated water	Wall thinning due to flow-accelerated corrosion	Chapter XI.M17, "Flow-Accelerated Corrosion"	No	VIII.A.S-15 VIII.B1.S-15 VIII.B2.S-15 VIII.C.S-15 VIII.D1.S-16 VIII.D2.S-16 VIII.E.S-16 VIII.F.S-16 VIII.G.S-16	VIII.A-17(S-15) VIII.B1-9(S-15) VIII.B2-4(S-15) VIII.C-5(S-15) VIII.D1-9(S-16) VIII.D2-8(S-16) VIII.E-35(S-16) VIII.F-26(S-16) VIII.G-39(S-16)
6	BWR/PWR	Steel, Stainless Steel Bolting exposed to Soil	Loss of preload	Chapter XI.M18, "Bolting Integrity"	No	VIII.H.SP-142 VIII.H.SP-144	N/A N/A
7	BWR/PWR	High-strength steel Closure bolting exposed to Air with steam or water leakage	Cracking due to cyclic loading, stress corrosion cracking	Chapter XI.M18, "Bolting Integrity"	No	VIII.H.S-03	VIII.H-3(S-03)
8	BWR/PWR	Steel; stainless steel Bolting, Closure bolting exposed to Air – outdoor (External), Air – indoor, uncontrolled (External)	Loss of material due to general (steel only), pitting, and crevice corrosion	Chapter XI.M18, "Bolting Integrity"	No	VIII.H.SP-82 VIII.H.SP-84	VIII.H-1(S-32) VIII.H-4(S-34)
9	BWR/PWR	Steel Closure bolting exposed to Air with steam or water leakage	Loss of material due to general corrosion	Chapter XI.M18, "Bolting Integrity"	No	VIII.H.S-02	VIII.H-6(S-02)

ID	Type	Component	Aging Effect/Mechanism	Aging Management Programs	Further Evaluation Recommended	Rev2 Item	Rev1 Item
10	BWR/PWR	Copper alloy, Nickel alloy, Steel; stainless steel, Steel; stainless steel Bolting, Closure bolting exposed to Any environment, Air – outdoor (External), Air – indoor, uncontrolled (External)	Loss of preload due to thermal effects, gasket creep, and self-loosening	Chapter XI.M18, "Bolting Integrity"	No	VIII.H.SP-149 VIII.H.SP-150 VIII.H.SP-151 VIII.H.SP-83	N/A N/A N/A VIII.H-5(S-33)
11	BWR/PWR	Stainless steel Piping, piping components, and piping elements, Tanks, Heat exchanger components exposed to Steam, Treated water >60°C (>140°F)	Cracking due to stress corrosion cracking	Chapter XI.M2, "Water Chemistry," and Chapter XI.M32, "One-Time Inspection"	No	VIII.A.SP-98 VIII.B1.SP-88 VIII.B1.SP-98 VIII.B2.SP-98 VIII.C.SP-88 VIII.D1.SP-88 VIII.E.SP-88 VIII.E.SP-97 VIII.F.SP-85 VIII.F.SP-88 VIII.G.SP-88	VIII.A-11(SP-45) VIII.B1-5(SP-17) VIII.B1-2(SP-44) VIII.B2-1(SP-45) VIII.C-2(SP-17) VIII.D1-5(SP-17) VIII.E-30(SP-17) VIII.E-38(SP-42) VIII.F-3(S-39) VIII.F-24(SP-17) VIII.G-33(SP-17)
12	BWR/PWR	Steel; stainless steel Tanks exposed to Treated water	Loss of material due to general (steel only), pitting, and crevice corrosion	Chapter XI.M2, "Water Chemistry," and Chapter XI.M32, "One-Time Inspection"	No	VIII.E.SP-75 VIII.G.SP-75	VIII.E-40(S-13) VIII.G-41(S-13)
13	PWR	Steel Piping, piping components, and piping elements exposed to Treated water	Loss of material due to general, pitting, and crevice corrosion	Chapter XI.M2, "Water Chemistry," and Chapter XI.M32, "One-Time Inspection"	No	VIII.B1.SP-74 VIII.D1.SP-74 VIII.F.SP-74 VIII.G.SP-74	VIII.B1-11(S-10) VIII.D1-8(S-10) VIII.F-25(S-10) VIII.G-38(S-10)

Table 3.4-1 Summary of Aging Management Programs for Steam and Power Conversion System Evaluated in Chapter VIII of the GALL Report

ID	Type	Component	Aging Effect/Mechanism	Aging Management Programs	Further Evaluation Recommended	Rev2 Item	Rev1 Item
14	BWR/PWR	Steel Piping, piping components, and piping elements, PWR heat exchanger components exposed to Steam, Treated water	Loss of material due to general, pitting, and crevice corrosion	Chapter XI.M2, "Water Chemistry," and Chapter XI.M32, "One-Time Inspection"	No	VIII.A.SP-71 VIII.B1.SP-71 VIII.B2.SP-160 VIII.B2.SP-73 VIII.C.SP-71 VIII.C.SP-73 VIII.D2.SP-73 VIII.E.SP-73 VIII.E.SP-78 VIII.F.SP-78	VIII.A-15(S-04) VIII.B1-8(S-07) VIII.B2-3(S-05) VIII.B2-6(S-09) VIII.C-3(S-04) VIII.C-6(S-09) VIII.D2-7(S-09) VIII.E-33(S-09) VIII.E-37(S-19) VIII.F-28(S-19)
15	BWR/PWR	Steel Heat exchanger components exposed to Treated water	Loss of material due to general, pitting, crevice, and galvanic corrosion	Chapter XI.M2, "Water Chemistry," and Chapter XI.M32, "One-Time Inspection"	No	VIII.E.SP-77	VIII.E-7(S-18)
16	BWR/PWR	Copper alloy, Stainless steel, Nickel alloy, Aluminum Piping, piping components, and piping elements, Heat exchanger components and tubes, PWR heat exchanger components exposed to Treated water, Steam	Loss of material due to pitting and crevice corrosion	Chapter XI.M2, "Water Chemistry," and Chapter XI.M32, "One-Time Inspection"	No	VIII.A.SP-101 VIII.A.SP-155 VIII.B1.SP-155 VIII.B1.SP-157 VIII.B1.SP-87 VIII.B2.SP-155 VIII.C.SP-87 VIII.D1.SP-87 VIII.D1.SP-90 VIII.D2.SP-87 VIII.D2.SP-90 VIII.E.SP-80 VIII.E.SP-81 VIII.E.SP-87 VIII.E.SP-90 VIII.F.SP-101 VIII.F.SP-81 VIII.F.SP-87 VIII.F.SP-90 VIII.G.SP-87 VIII.G.SP-90	VIII.A-5(SP-61) VIII.A-12(SP-43) VIII.B1-3(SP-43) VIII.B1-1(SP-18) VIII.B1-4(SP-16) VIII.B2-2(SP43) VIII.C-1(SP-16) VIII.D1-4(SP-16) VIII.D1-1(SP-24) VIII.D2-4(SP-16) VIII.D2-1(SP-24) VIII.E-4(S-21) VIII.E-36(S-22) VIII.E-29(SP-16) VIII.E-15(SP-24) VIII.F-15(SP-61) VIII.F-27(S-22) VIII.F-23(SP-16) VIII.F-12(SP-24) VIII.G-32(SP-16) VIII.G-17(SP-24)

Table 3.4-1 Summary of Aging Management Programs for Steam and Power Conversion System Evaluated in Chapter VIII of the GALL Report

ID	Type	Component	Aging Effect/Mechanism	Aging Management Programs	Further Evaluation Recommended	Rev2 Item	Rev1 Item
17	PWR	Copper alloy Heat exchanger tubes exposed to Treated water	Reduction of heat transfer due to fouling	Chapter XI.M2, "Water Chemistry," and Chapter XI.M32, "One-Time Inspection"	No	VIII.F.SP-100	VIII.F-7(SP-58)
18	BWR/PWR	Copper alloy, Stainless steel Heat exchanger tubes exposed to Treated water	Reduction of heat transfer due to fouling	Chapter XI.M2, "Water Chemistry," and Chapter XI.M32, "One-Time Inspection"	No	VIII.E.SP-100 VIII.E.SP-96 VIII.F.SP-96 VIII.G.SP-100	VIII.E-10(SP-58) VIII.E-13(SP-40) VIII.F-10(SP-40) VIII.G-10(SP-58)
19	BWR/PWR	Stainless steel, Steel Heat exchanger components exposed to Raw water	Loss of material due to general, pitting, crevice, galvanic, and microbiologically-influenced corrosion; fouling that leads to corrosion	Chapter XI.M20, "Open-Cycle Cooling Water System"	No	VIII.E.SP-117 VIII.E.SP-146 VIII.F.SP-146 VIII.G.SP-117 VIII.G.SP-146	VIII.E-3(S-26) VIII.E-6(S-24) VIII.F-5(S-24) VIII.G-4(S-26) VIII.G-7(S-24)
20	BWR/PWR	Copper alloy, Stainless steel Piping, piping components, and piping elements exposed to Raw water	Loss of material due to pitting, crevice, and microbiologically-influenced corrosion	Chapter XI.M20, "Open-Cycle Cooling Water System"	No	VIII.A.SP-31 VIII.E.SP-31 VIII.E.SP-36 VIII.F.SP-31 VIII.F.SP-36 VIII.G.SP-31 VIII.G.SP-36	VIII.A-4(SP-31) VIII.E-18(SP-31) VIII.E-27(SP-36) VIII.F-14(SP-31) VIII.F-22(SP-36) VIII.G-20(SP-31) VIII.G-30(SP-36)
21	PWR	Stainless steel Heat exchanger components exposed to Raw water	Loss of material due to pitting, crevice, and microbiologically-influenced corrosion; fouling that leads to corrosion	Chapter XI.M20, "Open-Cycle Cooling Water System"	No	VIII.F.SP-117	VIII.F-2(S-26)
22	BWR/PWR	Stainless steel, Copper alloy, Steel Heat exchanger tubes, Heat exchanger components exposed to Raw water	Reduction of heat transfer due to fouling	Chapter XI.M20, "Open-Cycle Cooling Water System"	No	VIII.E.S-28 VIII.E.SP-56 VIII.F.S-28 VIII.F.SP-56 VIII.G.S-27 VIII.G.S-28 VIII.G.SP-56	VIII.E-12(S-28) VIII.E-9(SP-56) VIII.F-9(S-28) VIII.F-6(SP-56) VIII.G-16(S-27) VIII.G-13(S-28) VIII.G-9(SP-56)

Table 3.4-1 Summary of Aging Management Programs for Steam and Power Conversion System Evaluated in Chapter VIII of the GALL Report

ID	Type	Component	Aging Effect/Mechanism	Aging Management Programs	Further Evaluation Recommended	Rev2 Item	Rev1 Item
23	BWR/PWR	Stainless steel Piping, piping components, and piping elements exposed to Closed-cycle cooling water >60°C (>140°F)	Cracking due to stress corrosion cracking	Chapter XI.M21A, "Closed Treated Water Systems"	No	VIII.E.SP-54 VIII.F.SP-54 VIII.G.SP-54	VIII.E-25(SP-54) VIII.F-21(SP-54) VIII.G-28(SP-54)
24	BWR/PWR	Steel Heat exchanger components exposed to Closed-cycle cooling water	Loss of material due to general, pitting, crevice, and galvanic corrosion	Chapter XI.M21A, "Closed Treated Water Systems"	No	VIII.A.S-23	VIII.A-1(S-23)
25	BWR/PWR	Steel Heat exchanger components exposed to Closed-cycle cooling water	Loss of material due to general, pitting, crevice, and galvanic corrosion	Chapter XI.M21A, "Closed Treated Water Systems"	No	VIII.E.S-23 VIII.F.S-23 VIII.G.S-23	VIII.E-5(S-23) VIII.F-4(S-23) VIII.G-5(S-23)
26	BWR/PWR	Stainless steel Heat exchanger components, Piping, piping components, and piping elements exposed to Closed-cycle cooling water	Loss of material due to pitting and crevice corrosion	Chapter XI.M21A, "Closed Treated Water Systems"	No	VIII.E.S-25 VIII.E.SP-39 VIII.F.S-25 VIII.F.SP-39 VIII.G.S-25 VIII.G.SP-39	VIII.E-2(S-25) VIII.E-24(SP-39) VIII.F-1(S-25) VIII.F-20(SP-39) VIII.G-2(S-25) VIII.G-27(SP-39)
27	BWR/PWR	Copper alloy Piping, piping components, and piping elements exposed to Closed-cycle cooling water	Loss of material due to pitting, crevice, and galvanic corrosion	Chapter XI.M21A, "Closed Treated Water Systems"	No	VIII.E.SP-8 VIII.F.SP-8 VIII.G.SP-8	VIII.E-16(SP-8) VIII.F-13(SP-8) VIII.G-18(SP-8)

Table 3.4-1 Summary of Aging Management Programs for Steam and Power Conversion System Evaluated in Chapter VIII of the GALL Report

ID	Type	Component	Aging Effect/Mechanism	Aging Management Programs	Further Evaluation Recommended	Rev2 Item	Rev1 Item
28	BWR/PWR	Steel, Stainless steel, Copper alloy Heat exchanger components and tubes, Heat exchanger tubes exposed to Closed-cycle cooling water	Reduction of heat transfer due to fouling	Chapter XI.M21A, "Closed Treated Water Systems"	No	VIII.A.SP-64 VIII.E.SP-41 VIII.E.SP-57 VIII.E.SP-64 VIII.F.SP-41 VIII.F.SP-64 VIII.G.SP-41 VIII.G.SP-64	VIII.A-2(SP-64) VIII.E-11(SP-41) VIII.E-8(SP-57) VIII.E-14(SP-64) VIII.F-8(SP-41) VIII.F-11(SP-64) VIII.G-11(SP-41) VIII.G-14(SP-64)
29	BWR/PWR	Steel Tanks exposed to Air – outdoor (External)	Loss of material due to general, pitting, and crevice corrosion	Chapter XI.M29, "Aboveground Metallic Tanks"	No	VIII.E.S-31 VIII.G.S-31	VIII.E-39(S-31) VIII.G-40(S-31)
30	BWR/PWR	Steel, Stainless Steel, Aluminum Tanks exposed to Soil or Concrete, Air – outdoor (External)	Loss of material due to general, pitting, and crevice corrosion	Chapter XI.M29, "Aboveground Metallic Tanks"	No	VIII.E.SP-115 VIII.E.SP-138 VIII.E.SP-140 VIII.G.SP-116	N/A N/A N/A N/A
31	BWR/PWR	Stainless steel, Aluminum Tanks exposed to Soil or Concrete	Loss of material due to pitting, and crevice corrosion	Chapter XI.M29, "Aboveground Metallic Tanks"	No	VIII.E.SP-137 VIII.E.SP-139	N/A N/A
32	BWR/PWR	Gray cast iron Piping, piping components, and piping elements exposed to Soil	Loss of material due to selective leaching	Chapter XI.M33, "Selective Leaching"	No	VIII.E.SP-26 VIII.G.SP-26	VIII.E-22(SP-26) VIII.G-25(SP-26)

Table 3.4-1 Summary of Aging Management Programs for Steam and Power Conversion System Evaluated in Chapter VIII of the GALL Report

ID	Type	Component	Aging Effect/Mechanism	Aging Management Programs	Further Evaluation Recommended	Rev2 Item	Rev1 Item
33	BWR/PWR	Gray cast iron, Copper alloy (>15% Zn or >8% Al) Piping, piping components, and piping elements exposed to Treated water, Raw water, Closed-cycle cooling water	Loss of material due to selective leaching	Chapter XI.M33, "Selective Leaching"	No	VIII.A.SP-27 VIII.A.SP-28 VIII.A.SP-30 VIII.E.SP-27 VIII.E.SP-29 VIII.E.SP-30 VIII.E.SP-55 VIII.F.SP-27 VIII.F.SP-29 VIII.F.SP-55 VIII.G.SP-27 VIII.G.SP-28 VIII.G.SP-29 VIII.G.SP-30 VIII.G.SP-55	VIII.A-8(SP-27) VIII.A-7(SP-28) VIII.A-6(SP-30) VIII.E-23(SP-27) VIII.E-19(SP-29) VIII.E-20(SP-30) VIII.E-21(SP-55) VIII.F-19(SP-27) VIII.F-16(SP-29) VIII.F-17(SP-30) VIII.F-18(SP-55) VIII.G-26(SP-27) VIII.G-24(SP-28) VIII.G-21(SP-29) VIII.G-22(SP-30) VIII.G-23(SP-55)
34	BWR/PWR	Steel External surfaces exposed to Air – indoor, uncontrolled (External), Air – outdoor (External), Condensation (External)	Loss of material due to general corrosion	Chapter XI.M36, "External Surfaces Monitoring of Mechanical Components"	No	VIII.H.S-29 VIII.H.S-41 VIII.H.S-42	VIII.H-7(S-29) VIII.H-8(S-41) VIII.H-10(S-42)
35	BWR/PWR	Aluminum Piping, piping components, and piping elements exposed to Air - outdoor	Loss of material due to pitting and crevice corrosion	Chapter XI.M36, "External Surfaces Monitoring of Mechanical Components"	No	VIII.H.SP-147	N/A
36	PWR	Steel Piping, piping components, and piping elements exposed to Air – outdoor (Internal)	Loss of material due to general, pitting, and crevice corrosion	Chapter XI.M38 "Inspection of Internal Surfaces in Miscellaneous Piping and Ducting Components"	No	VIII.B1.SP-59	VIII.B1-6(SP-59)

Table 3.4-1 Summary of Aging Management Programs for Steam and Power Conversion System Evaluated in Chapter VIII of the GALL Report

ID	Type	Component	Aging Effect/Mechanism	Aging Management Programs	Further Evaluation Recommended	Rev2 Item	Rev1 Item
37	PWR	Steel Piping, piping components, and piping elements exposed to Condensation (Internal)	Loss of material due to general, pitting, and crevice corrosion	Chapter XI.M38, "Inspection of Internal Surfaces in Miscellaneous Piping and Ducting Components"	No	VIII.B1.SP-60 VIII.G.SP-60	VIII.B1-7(SP-60) VIII.G-34(SP-60)
38	PWR	Steel Piping, piping components, and piping elements exposed to Raw water	Loss of material due to general, pitting, crevice, galvanic, and microbiologically-influenced corrosion; fouling that leads to corrosion	Chapter XI.M38, "Inspection of Internal Surfaces in Miscellaneous Piping and Ducting Components"	No	VIII.G.SP-136	VIII.G-36(S-12)
39	BWR/PWR	Stainless steel Piping, piping components, and piping elements exposed to Condensation (Internal)	Loss of material due to pitting and crevice corrosion	Chapter XI.M38, "Inspection of Internal Surfaces in Miscellaneous Piping and Ducting Components"	No	VIII.B1.SP-110 VIII.B2.SP-110	N/A N/A
40	BWR/PWR	Steel Piping, piping components, and piping elements exposed to Lubricating oil	Loss of material due to general, pitting, and crevice corrosion	Chapter XI.M39, "Lubricating Oil Analysis," and Chapter XI.M32, "One-Time Inspection"	No	VIII.A.SP-91 VIII.D1.SP-91 VIII.D2.SP-91 VIII.E.SP-91 VIII.G.SP-91	VIII.A-14(SP-25) VIII.D1-6(SP-25) VIII.D2-5(SP-25) VIII.E-32(SP-25) VIII.G-35(SP-25)
41	PWR	Steel Heat exchanger components exposed to Lubricating oil	Loss of material due to general, pitting, crevice, and microbiologically-influenced corrosion	Chapter XI.M39, "Lubricating Oil Analysis," and Chapter XI.M32, "One-Time Inspection"	No	VIII.G.SP-76	VIII.G-6(S-17)

Table 3.4-1   Summary of Aging Management Programs for Steam and Power Conversion System Evaluated in Chapter VIII of the GALL Report

ID	Type	Component	Aging Effect/Mechanism	Aging Management Programs	Further Evaluation Recommended	Rev2 Item	Rev1 Item
42	PWR	Aluminum Piping, piping components, and piping elements exposed to Lubricating oil	Loss of material due to pitting and crevice corrosion	Chapter XI.M39, "Lubricating Oil Analysis," and Chapter XI.M32, "One-Time Inspection"	No	VIII.G.SP-114	N/A
43	BWR/PWR	Copper alloy Piping, piping components, and piping elements exposed to Lubricating oil	Loss of material due to pitting and crevice corrosion	Chapter XI.M39, "Lubricating Oil Analysis," and Chapter XI.M32, "One-Time Inspection"	No	VIII.A.SP-92 VIII.D1.SP-92 VIII.D2.SP-92 VIII.E.SP-92 VIII.G.SP-92	VIII.A-3(SP-32) VIII.D1-2(SP-32) VIII.D2-2(SP-32) VIII.E-17(SP-32) VIII.G-19(SP-32)
44	BWR/PWR	Stainless steel Piping, piping components, and piping elements, Heat exchanger components exposed to Lubricating oil	Loss of material due to pitting, crevice, and microbiologically-influenced corrosion	Chapter XI.M39, "Lubricating Oil Analysis," and Chapter XI.M32, "One-Time Inspection"	No	VIII.A.SP-95 VIII.D1.SP-95 VIII.D2.SP-95 VIII.E.SP-95 VIII.G.SP-79 VIII.G.SP-95	VIII.A-9(SP-38) VIII.D1-3(SP-38) VIII.D2-3(SP-38) VIII.E-26(SP-38) VIII.G-3(S-20) VIII.G-29(SP-38)
45	PWR	Aluminum Heat exchanger components and tubes exposed to Lubricating oil	Reduction of heat transfer due to fouling	Chapter XI.M39, "Lubricating Oil Analysis," and Chapter XI.M32, "One-Time Inspection"	No	VIII.G.SP-113	N/A
46	PWR	Stainless steel, Steel, Copper alloy Heat exchanger tubes exposed to Lubricating oil	Reduction of heat transfer due to fouling	Chapter XI.M39, "Lubricating Oil Analysis," and Chapter XI.M32, "One-Time Inspection"	No	VIII.G.SP-102 VIII.G.SP-103 VIII.G.SP-99	VIII.G-12(SP-62) VIII.G-15(SP-63) VIII.G-8(SP-53)
47	BWR/PWR	Steel (with coating or wrapping) Piping, piping components, and piping elements; tanks exposed to Soil or Concrete	Loss of material due to general, pitting, crevice, and microbiologically-influenced corrosion	Chapter XI.M41, "Buried and Underground Piping and Tanks"	No	VIII.E.SP-145 VIII.G.SP-145	VIII.E-1(S-01) VIII.G-1(S-01)

Table 3.4-1 Summary of Aging Management Programs for Steam and Power Conversion System Evaluated in Chapter VIII of the GALL Report

ID	Type	Component	Aging Effect/Mechanism	Aging Management Programs	Further Evaluation Recommended	Rev2 Item	Rev1 Item
48	BWR/PWR	Stainless Steel Bolting exposed to Soil	Loss of material due to pitting and crevice corrosion	Chapter XI.M41, "Buried and Underground Piping and Tanks"	No	VIII.H.SP-143	N/A
49	BWR/PWR	Stainless steel Piping, piping components, and piping elements exposed to Soil or Concrete	Loss of material due to pitting and crevice corrosion	Chapter XI.M41, "Buried and Underground Piping and Tanks"	No	VIII.E.SP-94 VIII.G.SP-94	VIII.E-28(SP-37) VIII.G-31(SP-37)
50	BWR/PWR	Steel Bolting exposed to Soil	Loss of material due to general, pitting and crevice corrosion	Chapter XI.M41, "Buried and Underground Piping and Tanks"	No	VIII.H.SP-141	N/A
	BWR/PWR	Underground Stainless Steel and Steel Piping, piping components, and piping elements	Loss of material due to general (steel only), pitting and crevice corrosion	Chapter XI.M41, "Buried and Underground Piping and Tanks"	No	VIII.H.SP-161	N/A
51	BWR/PWR	Steel Piping, piping components, and piping elements exposed to Concrete	None	None, provided 1) attributes of the concrete are consistent with ACI 318 or ACI 349 (low water-to-cement ratio, low permeability, and adequate air entrainment) as cited in NUREG-1557, and 2) plant OE indicates no degradation of the concrete	No, if conditions are met.	VIII.I.SP-154	VIII.I-14(SP-2)

Table 3.4-1 Summary of Aging Management Programs for Steam and Power Conversion System Evaluated in Chapter VIII of the GALL Report

ID	Type	Component	Aging Effect/Mechanism	Aging Management Programs	Further Evaluation Recommended	Rev2 Item	Rev1 Item
52	BWR/PWR	Aluminum Piping, piping components, and piping elements exposed to Gas, Air – indoor, uncontrolled (Internal/External)	None	None	NA - No AEM or AMP	VIII.I.SP-23 VIII.I.SP-93	VIII.I-1-(SP-23) N/A
53	PWR	Copper alloy (≤15% Zn and ≤8% Al) Piping, piping components, and piping elements exposed to Air with borated water leakage	None	None	NA - No AEM or AMP	VIII.I.SP-104	N/A
54	BWR/PWR	Copper alloy Piping, piping components, and piping elements exposed to Gas, Air – indoor, uncontrolled (External)	None	None	NA - No AEM or AMP	VIII.I.SP-5 VIII.I.SP-6	VIII.I-3(SP-5) VIII.I-2(SP-6)
55	BWR/PWR	Glass Piping elements exposed to Lubricating oil, Air – outdoor, Condensation (Internal/External), Raw water, Treated water, Air with borated water leakage, Gas, Closed-cycle cooling water, Air – indoor, uncontrolled (External)	None	None	NA - No AEM or AMP	VIII.I.SP-10 VIII.I.SP-108 VIII.I.SP-111 VIII.I.SP-33 VIII.I.SP-34 VIII.I.SP-35 VIII.I.SP-67 VIII.I.SP-68 VIII.I.SP-69 VIII.I.SP-70 VIII.I.SP-9	VIII.I-6(SP-10) N/A N/A VIII.I-4(SP-33) VIII.I-7(SP-34) VIII.I-8(SP-35) N/A N/A N/A N/A VIII.I-5(SP-9)

Table 3.4-1 Summary of Aging Management Programs for Steam and Power Conversion System Evaluated in Chapter VIII of the GALL Report

ID	Type	Component	Aging Effect/Mechanism	Aging Management Programs	Further Evaluation Recommended	Rev2 Item	Rev1 Item
56	BWR/PWR	Nickel alloy Piping, piping components, and piping elements exposed to Air – indoor, uncontrolled (External)	None	None	NA - No AEM or AMP	VIII.I.SP-11	VIII.I-9(SP-11)
57	BWR/PWR	Nickel alloy, PVC Piping, piping components, and piping elements exposed to Air with borated water leakage, Air – indoor, uncontrolled, Condensation (Internal)	None	None	NA - No AEM or AMP	VIII.I.SP-148 VIII.I.SP-152 VIII.I.SP-153	N/A N/A N/A
58	BWR/PWR	Stainless steel Piping, piping components, and piping elements exposed to Air – indoor, uncontrolled (External), Concrete, Gas, Air – indoor, uncontrolled (Internal)	None	None	NA - No AEM or AMP	VIII.I.SP-12 VIII.I.SP-13 VIII.I.SP-15 VIII.I.SP-86	VIII.I-10(SP-12) VIII.I-11(SP-13) VIII.I-12(SP-15) N/A
59	BWR/PWR	Steel Piping, piping components, and piping elements exposed to Air – indoor controlled (External), Gas	None	None	NA - No AEM or AMP	VIII.I.SP-1 VIII.I.SP-4	VIII.I-13(SP-1) VIII.I-15(SP-4)

Table 3.4-2 Aging Management Programs Recommended for Aging Management of Steam and Power Conversion System

GALL Report Chapter/AMP	Program Name
Chapter XI.M2	Water Chemistry
Chapter XI.M10	Boric Acid Corrosion
Chapter XI.M17	Flow-Accelerated Corrosion
Chapter XI.M18	Bolting Integrity
Chapter XI.M20	Open-Cycle Cooling Water System
Chapter XI.M21A	Closed Treated Water Systems
Chapter XI.M29	Aboveground Metallic Tanks
Chapter XI.M32	One-Time Inspection
Chapter XI.M33	Selective Leaching
Chapter XI.M36	External Surfaces Monitoring of Mechanical Components
Chapter XI.M38	Inspection of Internal Surfaces in Miscellaneous Piping and Ducting Components
Chapter XI.M39	Lubricating Oil Analysis
Chapter XI.M41	Buried and Underground Piping and Tanks
Appendix for GALL	Quality Assurance for Aging Management Programs
SRP-LR Appendix A	Plant-specific AMP

3.5 AGING MANAGEMENT OF CONTAINMENTS, STRUCTURES, AND COMPONENT SUPPORTS

Review Responsibilities

Primary - Branch assigned responsibility by PM as described in SRP-LR Section 3.0.

3.5.1 Areas of Review

This section addresses the aging management review (AMR) and the associated aging management program (AMP) for containments, structures, and component supports. For a recent vintage plant, the information related to containments, structures, and component supports is contained in Chapter 3, "Design of Structures, Components, Equipment, and Systems," of the plant's FSAR, consistent with the "Standard Review Plan for the Review of Safety Analysis Reports for Nuclear Power Plants" (NUREG-0800). For older vintage plants, the location of applicable information is plant-specific because an older plant's FSAR may have predated NUREG-0800. The scope of this section is PWR and BWR containment structures, safety-related and other structures, and component supports.

The PWR containment structures consist of concrete (reinforced or prestressed) and steel containments. The BWR containment structures consist of Mark I, Mark II, and Mark III steel and concrete (reinforced or pre-stressed) containments.

The safety-related structures (other than containments) are organized into nine groups: Group 1: BWR reactor building, PWR shield building, control room/building; Group 2: BWR reactor building with steel superstructure; Group 3: auxiliary building, diesel generator building, radwaste building, turbine building, switchgear room, yard structures (auxiliary feedwater pump house, utility/piping tunnels, security lighting poles, manholes, duct banks), SBO structures (transmission towers, startup transformer circuit breaker foundation, electrical enclosure); Group 4: containment internal structures, excluding refueling canal; Group 5: fuel storage facility, refueling canal; Group 6: water-control structures (e.g., intake structure, cooling tower, and spray pond); Group 7: concrete tanks and missile barriers; Group 8: steel tank foundations and missile barriers; and Group 9: BWR unit vent stack.

The component supports are organized into seven groups: Group B1.1: supports for ASME Class 1 piping and components; Group B1.2: supports for ASME Class 2 and 3 piping and components; Group B1.3: supports for ASME Class MC components; Group B2: supports for cable tray, conduit, HVAC ducts, TubeTrack®, instrument tubing, non-ASME piping and components; Group B3: anchorage of racks, panels, cabinets, and enclosures for electrical equipment and instrumentation; Group B4: supports for miscellaneous equipment (e.g., EDG, HVAC components); and Group B5: supports for miscellaneous structures (e.g., platforms, pipe whip restraints, jet impingement shields, masonry walls).

The responsible review organization is to review the following license renewal application (LRA) AMR and AMP items assigned to it, per SRP-LR section 3.0, for review:

AMRs

- AMR results consistent with the GALL Report
- AMR results for which further evaluation is recommended by the GALL Report
- AMR results that are not consistent with or not addressed in the GALL Report

AMPs

- Consistent with GALL AMPs
- Plant-specific AMPs

FSAR Supplement

- The responsible review organization is to review the FSAR Supplement associated with each assigned AMP.

3.5.2 Acceptance Criteria

The acceptance criteria for the areas of review describe methods for determining whether the applicant has met the requirements of the NRC's regulations in 10 CFR 54.21.

3.5.2.1 AMR Results Consistent with the GALL Report

The AMRs and the AMPs applicable to structures and component supports are described and evaluated in Chapters II and III of the NUREG-1801, (GALL Report).

The applicant's LRA should provide sufficient information so that the reviewer is able to confirm that the specific LRA AMR item and the associated LRA AMP are consistent with the cited GALL Report AMR item. The staff reviewer should then confirm that the LRA AMR item is consistent with the GALL Report AMR item to which it is compared.

When the applicant is crediting a different aging management program than recommended in the GALL Report, the reviewer should confirm that the alternate aging management program is valid to use for aging management and will be capable of managing the effects of aging as adequately as the aging management program recommended by the GALL Report.

3.5.2.2 AMR Results for Which Further Evaluation is Recommended by the GALL Report

The basic acceptance criteria defined in Section 3.5.2.1 need to be applied first for all of the AMRs and AMPs as part of this section. In addition, if the GALL Report AMR item to which the LRA AMR item is compared identifies that "further evaluation is recommended," then additional criteria apply as identified by the GALL Report for each of the following aging effect/aging mechanism combinations. Refer to Table 3.5-1, comparing the "Further Evaluation Recommended" column and the "Rev2 Item" column, for the AMR items that reference the following subsections. The 2005 AMR item counterpart is provided in the "Rev1 Item" column.

3.5.2.2.1 PWR and BWR Containments

3.5.2.2.1.1 Cracking and Distortion due to Increased Stress Levels from Settlement; Reduction of Foundation Strength, and Cracking due to Differential Settlement and Erosion of Porous Concrete Subfoundations

Cracking and distortion due to increased stress levels from settlement could occur in PWR and BWR concrete and steel containments. The existing program relies on ASME Section XI, Subsection IWL to manage these aging effects. Also, reduction of foundation strength and cracking, due to differential settlement and erosion of porous concrete subfoundations could occur in all types of PWR and BWR containments. The existing program relies on the structures monitoring program to manage these aging effects. However, some plants may rely on a de-watering system to lower the site ground water level. If the plant's current licensing basis (CLB) credits a de-watering system to control settlement, the GALL Report recommends further evaluation to verify the continued functionality of the de-watering system during the period of extended operation.

3.5.2.2.1.2   Reduction of Strength and Modulus due to Elevated Temperature

Reduction of strength and modulus of concrete due to elevated temperatures could occur in PWR and BWR concrete and steel containments. The implementation of 10 CFR 50.55a and ASME Section XI, Subsection IWL would not be able to identify the reduction of strength and modulus of concrete due to elevated temperature. Subsection CC-3440 of ASME Section III, Division 2, specifies the concrete temperature limits for normal operation or any other long-term period. The GALL Report recommends further evaluation of a plant-specific aging management program if any portion of the concrete containment components exceeds specified temperature limits, i.e., general area temperature greater than 66°C (150°F) and local area temperature greater than 93°C (200°F). Higher temperatures may be allowed if tests and/or calculations are provided to evaluate the reduction in strength and modulus of elasticity and these reductions are applied to the design calculations. Acceptance criteria are described in Branch Technical Position RLSB-1 (Appendix A.1 of this SRP-LR).

3.5.2.2.1.3   Loss of Material due to General, Pitting and Crevice Corrosion

1. Loss of material due to general, pitting, and crevice corrosion could occur in steel elements of inaccessible areas for all types of PWR and BWR containments. The existing program relies on ASME Section XI, Subsection IWE, and 10 CFR Part 50, Appendix J, to manage this aging effect. The GALL Report recommends further evaluation of plant-specific programs to manage this aging effect if corrosion is indicated from the IWE examinations. Acceptance criteria are described in Branch Technical Position RLSB-1 (Appendix A.1 of this SRP-LR).

2. Loss of material due to general, pitting, and crevice corrosion could occur in steel torus shell of Mark I containments. The existing program relies on ASME Section XI, Subsection IWE, and 10 CFR Part 50, Appendix J, to manage this aging effect. The GALL Report recommends further evaluation of plant-specific programs to manage this aging effect if corrosion is significant. Acceptance criteria are described in Branch Technical Position RLSB-1 (Appendix A.1 of this SRP-LR).

3. Loss of material due to general, pitting, and crevice corrosion could occur in steel torus ring girders and downcomers of Mark I containments, downcomers of Mark II containments, and interior surface of suppression chamber shell of Mark III containments. The existing program relies on ASME Section XI, Subsection IWE to manage this aging effect. The GALL Report recommends further evaluation of plant-specific programs to manage this aging effect if corrosion is significant. Acceptance criteria are described in Branch Technical Position RLSB-1 (Appendix A.1 of this SRP-LR).

3.5.2.2.1.4   Loss of Prestress due to Relaxation, Shrinkage, Creep, and Elevated Temperature

Loss of prestress forces due to relaxation, shrinkage, creep, and elevated temperature for PWR prestressed concrete containments and BWR Mark II prestressed concrete containments is a Time-Limited Aging Analysis (TLAA) as defined in 10 CFR 54.3. TLAAs are required to be evaluated in accordance with 10 CFR 54.21(c). The evaluation of this TLAA is addressed separately in Section 4.5, "Concrete Containment Tendon Prestress Analysis," of this SRP-LR.

3.5.2.2.1.5   Cumulative Fatigue Damage

If included in the current licensing basis, fatigue analyses of suppression pool steel shells (including welded joints) and penetrations (including penetration sleeves, dissimilar metal welds, and penetration bellows) for all types of PWR and BWR containments and BWR vent header, vent line bellows, and downcomers are TLAAs as defined in 10 CFR 54.3. TLAAs are required to be evaluated in accordance with 10 CFR 54.21(c). The evaluation of this TLAA is addressed separately in Section 4.6, "Containment Liner Plates, Metal Containments, and Penetrations Fatigue Analysis," of this SRP-LR.

3.5.2.2.1.6   Cracking due to Stress Corrosion Cracking

Cracking due to stress corrosion cracking of stainless steel penetration bellows and dissimilar metal welds could occur in all types of PWR and BWR containments. The existing program relies on ASME Section XI, Subsection IWE and10 CFR Part 50, Appendix J, to manage this aging effect. The GALL Report recommends further evaluation of additional appropriate examinations/evaluations implemented to detect these aging effects for stainless steel penetration bellows and dissimilar metal welds.

3.5.2.2.1.7   Loss of Material (Scaling, Spalling) and Cracking due to Freeze-Thaw

Loss of material (scaling, spalling) and cracking due to freeze-thaw could occur in inaccessible areas of PWR and BWR concrete containments. The GALL Report recommends further evaluation of this aging effect for plants located in moderate to severe weathering conditions.

3.5.2.2.1.8   Cracking due to Expansion from Reaction with Aggregates

Cracking due to expansion from reaction with aggregates could occur in inaccessible areas of concrete elements of PWR and BWR concrete and steel containments. The GALL Report recommends further evaluation to determine if a plant-specific aging management program is required to manage this aging effect. Acceptance criteria are described in Branch Technical Position RLSB-1 (Appendix A.1 of this SRP-LR).

3.5.2.2.1.9   Increase in Porosity and Permeability due to Leaching of Calcium Hydroxide and Carbonation

Increase in porosity and permeability due to leaching of calcium hydroxide and carbonation could occur in inaccessible areas of concrete elements of PWR and BWR concrete and steel containments. The GALL Report recommends further evaluation if leaching is observed in accessible areas that impact intended functions. Acceptance criteria are described in Branch Technical Position RLSB-1 (Appendix A.1 of this SRP-LR).

3.5.2.2.2 Safety-Related and Other Structures and Component Supports

3.5.2.2.2.1 Aging Management of Inaccessible Areas

1. Loss of material (spalling, scaling) and cracking due to freeze-thaw could occur in below-grade inaccessible concrete areas of Groups 1-3, 5 and 7-9 structures. The GALL Report recommends further evaluation of this aging effect for inaccessible areas of these Groups of structures for plants located in moderate to severe weathering conditions.

2. Cracking due to expansion and reaction with aggregates could occur in below-grade inaccessible concrete areas for Groups 1-5 and 7-9 structures. The GALL Report recommends further evaluation of inaccessible areas of these Groups of structures if concrete was not constructed in accordance with the recommendations in the GALL Report.

3. Cracking and distortion due to increased stress levels from settlement could occur in below-grade inaccessible concrete areas of structures for all Groups, and reduction in foundation strength, and cracking due to differential settlement and erosion of porous concrete subfoundations could occur in below-grade inaccessible concrete areas of Groups 1-3, 5 -9 structures. The existing program relies on structure monitoring programs to manage these aging effects. Some plants may rely on a de-watering system to lower the site ground water level. If the plant's CLB credits a de-watering system, the GALL Report recommends verification of the continued functionality of the de-watering system during the period of extended operation. The GALL Report recommends no further evaluation if this activity is included in the scope of the applicant's structures monitoring program.

4. Increase in porosity and permeability, and loss of strength due to leaching of calcium hydroxide and carbonation could occur in below-grade inaccessible concrete areas of Groups 1-5 and 7-9 structures. The GALL Report recommends further evaluation if leaching is observed in accessible areas that impact intended functions.

3.5.2.2.2.2 Reduction of Strength and Modulus due to Elevated Temperature

Reduction of strength and modulus of concrete due to elevated temperatures could occur in PWR and BWR Group 1-5 concrete structures. For any concrete elements that exceed specified temperature limits, further evaluations are recommended. Appendix A of ACI 349-85 specifies the concrete temperature limits for normal operation or any other long-term period. The temperatures shall not exceed 66°C (150°F) except for local areas, which are allowed to have increased temperatures not to exceed 93°C (200°F). The GALL Report recommends further evaluation of a plant-specific program if any portion of the safety-related and other concrete structures exceeds specified temperature limits, i.e., general area temperature greater than 66°C (150°F) and local area temperature greater than 93°C (200°F). Higher temperatures may be allowed if tests and/or calculations are provided to evaluate the reduction in strength and modulus of elasticity and these reductions are applied to the design calculations. The acceptance criteria are described in Branch Technical Position RLSB-1 (Appendix A.1 of this SRP-LR).

3.5.2.2.2.3   Aging Management of Inaccessible Areas for Group 6 Structures

The GALL Report recommends further evaluation for inaccessible areas of certain Group 6 structure/aging effect combinations as identified below, whether or not they are covered by inspections in accordance with the GALL Report, Chapter XI.S7, "Regulatory Guide 1.127, Inspection of Water-Control Structures Associated with Nuclear Power Plants," or FERC/US Army Corp of Engineers dam inspection and maintenance procedures.

1. Loss of material (spalling, scaling) and cracking due to freeze-thaw could occur in below-grade inaccessible concrete areas of Group 6 structures. The GALL Report recommends further evaluation of this aging effect for inaccessible areas for plants located in moderate to severe weathering conditions.

2. Cracking due to expansion and reaction with aggregates could occur in below-grade inaccessible reinforced concrete areas of Group 6 structures. The GALL Report recommends further evaluation to determine if a plant-specific aging management program is required to manage this aging effect. Acceptance criteria are described in Branch Technical Position RLSB-1 (Appendix A.1 of this SRP-LR).

3. Increase in porosity and permeability and loss of strength due to leaching of calcium hydroxide and carbonation could occur in inaccessible areas of concrete elements of Group 6 structures. The GALL Report recommends further evaluation if leaching is observed in accessible areas that impact intended functions. Acceptance criteria are described in Branch Technical Position RLSB-1 (Appendix A.1 of this SRP-LR).

3.5.2.2.2.4   Cracking due to Stress Corrosion Cracking, and Loss of Material due to Pitting and Crevice Corrosion

Cracking due to stress corrosion cracking and loss of material due to pitting and crevice corrosion could occur for Group 7 and 8 stainless steel tank liners exposed to standing water. The GALL Report recommends further evaluation of plant-specific programs to manage these aging effects. The acceptance criteria are described in Branch Technical Position RLSB-1 (Appendix A.1 of this SRP-LR).

3.5.2.2.2.5   Cumulative Fatigue Damage due to Fatigue

Fatigue of component support members, anchor bolts, and welds for Groups B1.1, B1.2, and B1.3 component supports is a TLAA as defined in 10 CFR 54.3 only if a CLB fatigue analysis exists. TLAAs are required to be evaluated in accordance with 10 CFR 54.21(c). The evaluation of this TLAA is addressed separately in Section 4.3, "Metal Fatigue Analysis," of this SRP-LR.

3.5.2.2.3   Quality Assurance for Aging Management of Nonsafety-Related Components

Acceptance criteria are described in Branch Technical Position IQMB-1 (Appendix A.2 of this SRP-LR).

3.5.2.3   AMR Results Not Consistent with or Not Addressed in the GALL Report

Acceptance criteria are described in Branch Technical Position RLSB-1 (Appendix A.1 of this SRP-LR).

3.5.2.4 Aging Management Programs

For those AMPs that will be used for aging management and are based on the program elements of an AMP in the GALL Report, the NRC reviewer performs an audit of aging management programs credited in the LRA to confirm consistency with the GALL AMPs identified in the GALL Report, Chapters X and XI.

If the applicant identifies an exception to any of the program elements of the cited GALL Report AMP, the LRA AMP should include a basis demonstrating how the criteria of 10 CFR 54.21(a)(3) would still be met. The NRC reviewer should then confirm that the LRA AMP with all exceptions would satisfy the criteria of 10 CFR 54.21(a)(3). If, while reviewing the LRA AMP, the reviewer identifies a difference from the GALL Report AMP that should have been identified as an exception to the GALL Report AMP, this difference should be reviewed and properly dispositioned. The reviewer should document the disposition of all LRA-defined exceptions and staff-identified differences.

The LRA should identify any enhancements that are needed to permit an existing LRA AMP to be declared consistent with the GALL AMP to which the LRA AMP is compared. The reviewer is to confirm both that the enhancement, when implemented, would allow the existing LRA AMP to be consistent with the GALL AMP and that the applicant has a commitment in the FSAR supplement to implement the enhancement prior to the period of extended operation. The reviewer should document the disposition of all enhancements.

If the applicant chooses to use a plant-specific program that is not a GALL AMP, the NRC reviewer should confirm that the plant-specific program satisfies the criteria of Branch Technical Position RLSB-1 (Appendix A.1.2.3 of this SRP-LR).

3.5.2.5 FSAR Supplement

The summary description of the programs and activities for managing the effects of aging for the period of extended operation in the FSAR supplement should be appropriate, such that later changes can be controlled by 10 CFR 50.59. The description should contain information associated with the bases for determining that aging effects are managed during the period of extended operation. The description should also contain any future aging management activities, including enhancements and commitments, to be completed before the period of extended operation. Table 3.0-1 of this SRP-LR provides examples of the type of information to be included in the FSAR Supplement. Table 3.5-2 lists the programs that are applicable for this SRP-LR subsection.

3.5.3 Review Procedures

For each area of review, the review procedures below are to be followed.

3.5.3.1 AMR Results Consistent with the GALL Report

The applicant may reference the GALL Report in its LRA, as appropriate, and demonstrate that the AMRs and AMPs at its facility are consistent with those reviewed and approved in the GALL Report. The reviewer should not conduct a re-review of the substance of the matters described in the GALL Report. If the applicant has provided the information necessary to adopt the finding of program acceptability as described and evaluated in the GALL Report, the reviewer should find acceptable the applicant's reference to GALL in its LRA. In making this determination, the reviewer confirms that the applicant has provided a brief description of the system, components, materials, and environment. The reviewer also confirms that the applicant has stated that the applicable aging effects and industry and plant-specific operating experience have been reviewed by the applicant and are evaluated in the GALL Report.

Furthermore, the reviewer should confirm that the applicant has addressed operating experience identified after the issuance of the GALL Report. Performance of this review requires the reviewer to confirm that the applicant has identified those aging effects for the structures and component supports that are contained in the GALL Report as applicable to its plant.

3.5.3.2 AMR Results for Which Further Evaluation is Recommended by the GALL Report

The basic review procedures defined in Section 3.5.3.1 need to be applied first for all of the AMRs and AMPs provided in this section. In addition, if the GALL AMR item to which the LRA AMR item is compared identifies that further evaluation is recommended, then additional criteria apply as identified by the GALL Report for each of the following aging effect/aging mechanism combinations.

3.5.3.2.1 PWR and BWR Containments

3.5.3.2.1.1 Cracking and Distortion due to Increased Stress Levels from Settlement; Reduction of Foundation Strength and Cracking due to Differential Settlement and Erosion of Porous Concrete Subfoundations

The GALL Report recommends further evaluation of aging management of (1) cracking and distortion due to increases in component stress level from settlement for PWR and BWR concrete and steel containments and (2) reduction of foundation strength and cracking due to differential settlement and erosion of porous concrete subfoundations for all types of PWR and BWR containments if a de-watering system is relied upon to control settlement. The reviewer reviews and confirms that, if the applicant credits a de-watering system in its CLB, the applicant has committed to monitor the functionality of the de-watering system under the applicant's ASME Code Section XI, Subsection IWL or the structures monitoring program. If not, the reviewer evaluates the plant-specific program for monitoring the de-watering system during the period of extended operation.

3.5.3.2.1.2 Reduction of Strength and Modulus due to Elevated Temperature

The GALL Report recommends further evaluation of programs to manage reduction of strength and modulus of concrete due to elevated temperature for PWR and BWR concrete and steel containments. The GALL Report notes that the implementation of ASME Section XI, Subsection IWL examinations and 10 CFR 50.55a would not be able to detect the reduction of concrete strength and modulus due to elevated temperature and also notes that no mandated aging management exists for managing this aging effect.

The GALL Report recommends that a plant-specific evaluation be performed if any portion of the concrete containment components exceeds specified temperature limits, i.e., general temperature greater than 66°C (150°F) and local area temperature greater than 93°C (200°F). The GALL Report also states that higher temperatures may be allowed if tests and/or calculations are provided to evaluate the reduction in strength and modulus of elasticity and these reductions are applied to the design calculations. The reviewer reviews and confirms that the applicant's discussion in the renewal application indicates that the affected PWR and BWR containment components are not exposed to a temperature that exceeds the temperature limits. If the limits are exceeded the reviewer reviews technical basis (i.e., tests and/or calculations) provided by the applicant to justify the higher temperature. Otherwise, the reviewer reviews the applicant's proposed programs to ensure that the effects of elevated temperature will be adequately managed during the period of extended operation.

3.5.3.2.1.3   Loss of Material due to General, Pitting, and Crevice Corrosion

1. The GALL Report identifies programs to manage loss of material due to general, pitting, and crevice corrosion in inaccessible areas of the steel elements in drywell and torus or the steel liner and integral attachments for all types of PWR and BWR containments. The aging management program consists of ASME Section XI, Subsection IWE, and 10 CFR Part 50, Appendix J, leak tests. Subsection IWE exempts from examination portions of the containments that are inaccessible, such as embedded or inaccessible portions of steel liners and steel elements in drywell and torus, and integral attachments.

   To cover the inaccessible areas, 10 CFR 50.55a(b)(2)(ix) requires that the applicant shall evaluate the acceptability of inaccessible areas when conditions exist in accessible areas that could indicate the presence of, or result in, degradation to such inaccessible areas. In addition, the GALL Report recommends further evaluation of plant-specific programs to manage the aging effects for inaccessible areas if specific recommendations defined in the GALL Report cannot be satisfied. The reviewer reviews the applicant's proposed aging management program to confirm that, where appropriate, an effective inspection program has been developed and implemented to ensure that the aging effects in inaccessible areas are adequately managed.

2. The GALL Report identifies programs to manage loss of material due to general, pitting, and crevice corrosion in steel torus shell of Mark I containments. The aging management program consists of ASME Section XI, Subsection IWE, and 10 CFR Part 50, Appendix J, leak tests. In addition, the GALL Report recommends further evaluation of plant-specific programs to manage the aging effects if corrosion is significant. Further evaluation of torus shell corrosion is warranted as a result of industry-wide operating experience that identified a number of incidences of torus corrosion. The reviewer reviews the applicant's proposed aging management program to confirm that, where appropriate, an effective inspection program has been developed and implemented to ensure that the aging effects are adequately managed. A plant-specific program may include the re-coating of the torus, if necessary.

3. The GALL Report identifies programs to manage loss of material due to general, pitting, and crevice corrosion in steel torus ring girders and downcomers of Mark I containments, suppression chambers and downcomers of Mark II containments, and interior surface of suppression chamber shell of Mark III containments. The GALL Report recommends GLAL AMP XI.S1, "ASME Section XI, Subsection IWE," for aging management. In addition, the GALL Report recommends further evaluation of plant-specific programs to manage the aging effects if plant operating experience identified significant corrosion of the torus ring girders, downcomers and suppression chambers.

3.5.3.2.1.4   Loss of Prestress due to Relaxation, Shrinkage, Creep, and Elevated Temperature

Loss of prestress is a TLAA as defined in 10 CFR 54.3. TLAAs are required to be evaluated in accordance with 10 CFR 54.21(c). The evaluation of this TLAA is addressed separately in Section 4.5 of this SRP-LR.

3.5.3.2.1.5   Cumulative Fatigue Damage

Fatigue analyses included in current licensing basis for the containment liner plate, penetrations (including penetration sleeves, dissimilar metal welds, and penetration bellows) for all types of PWR and BWR containments and BWR suppression pool steel shells, vent header, vent line bellows, and downcomers are TLAAs as defined in 10 CFR 54.3. TLAAs are required to be evaluated in accordance with 10 CFR 54.21(c). The evaluation of this TLAA is addressed separately in Section 4.6 of this SRP-LR.

3.5.3.2.1.6   Cracking due to Stress Corrosion Cracking

The GALL Report recommends further evaluation of programs to manage cracking due to SCC for stainless steel penetration sleeves, dissimilar metal welds, and penetration bellows in all types of PWR and BWR containments. Transgranular stress corrosion cracking (TGSCC) is a concern for dissimilar metal welds. In the case of bellows assemblies, SCC may cause aging effects particularly if the material is not shielded from a corrosive environment. Containment ISI IWE and leak rate testing may not be sufficient to detect cracks, especially for dissimilar metal welds. Additional appropriate examinations to detect SCC in bellows assemblies and dissimilar metal welds are recommended to address this issue. The reviewer reviews and evaluates the applicant's proposed programs to confirm that adequate inspection methods will be implemented to ensure that cracks are detected.

3.5.3.2.1.7   Loss of Material (Scaling, Spalling) and Cracking due to Freeze-Thaw

The GALL Report recommends further evaluation of programs to manage loss of material (scaling, spalling) and cracking due to freeze-thaw for concrete elements of PWR and BWR containments. Containment ISI Subsection IWL may not be sufficient for plants located in moderate to severe weathering conditions. Evaluation is needed for plants that are located in moderate to severe weathering conditions (weathering index >100 day-inch/yr) (NUREG-1557). The weathering index for the continental United States is shown in ASTM C33-90, Fig. 1. A plant-specific program is not required if documented evidence confirms that where the existing concrete had air content of 3% to 8% (including tolerance), and subsequent inspection of accessible areas did not exhibit degradation related to freeze-thaw. Such inspections are considered a part of the evaluation. The reviewer reviews and confirms that the applicant has satisfied the recommendations for inaccessible concrete as identified in the GALL Report. Otherwise, the reviewer reviews the applicant's proposed aging management program to verify that, where appropriate, an effective inspection program has been developed and implemented to ensure that these aging effects in inaccessible areas for plants located in moderate to severe weathering conditions are adequately managed.

3.5.3.2.1.8   Cracking due to Expansion from Reaction with Aggregates

The GALL Report recommends further evaluation of programs to manage cracking due to expansion and reaction with aggregates in inaccessible areas of concrete elements of PWR and

BWR concrete and steel containments. A plant-specific aging management program is not necessary if (1) investigations, tests, and petrographic examinations of aggregates performed in accordance with ASTM C295 and other ASTM reactivity tests, as required, can demonstrate that those aggregates do not adversely react within concrete or (2) for potentially reactive aggregates, aggregate concrete reaction is not significant if the structure was constructed in accordance with ACI 318. The reviewer confirms that the applicant has satisfied these conditions for inaccessible concrete as identified in the GALL Report. Otherwise, the reviewer reviews the applicant's proposed aging management program to verify that, where appropriate, an effective inspection program has been developed and implemented to ensure that this aging effect in inaccessible areas is adequately managed.

3.5.3.2.1.9 Increase in Porosity and Permeability due to Leaching of Calcium Hydroxide and Carbonation

The GALL Report recommends further evaluation of programs to manage increase in porosity and permeability due to leaching of calcium hydroxide and carbonation in inaccessible areas of PWR and BWR concrete and steel containments. A plant-specific aging management program is not required, even if reinforced concrete is exposed to flowing water if (1) there is evidence in the accessible areas that the flowing water has not caused leaching and carbonation, or (2) evaluation determined that the observed leaching of calcium hydroxide and carbonation in accessible areas has no impact on the intended function of the concrete structure. The reviewer confirms that the applicant has satisfied these conditions as identified in the GALL Report. Otherwise, the reviewer reviews the applicant's proposed aging management program to verify that, where appropriate, an effective inspection program has been developed and implemented to ensure that this aging effect in inaccessible areas is adequately managed.

3.5.3.2.2 Safety-Related and Other Structures, and Component Supports

3.5.3.2.2.1 Aging Management of Inaccessible Areas

1. The GALL Report recommends further evaluation of programs to manage loss of material (spalling, scaling) and cracking due to freeze-thaw in below-grade inaccessible concrete areas of Groups 1-3, 5, and 7-9 structures. Structure monitoring programs may not be sufficient for plants located in moderate to severe weathering conditions. Further evaluation is needed for plants that are located in moderate to severe weathering conditions (weathering index >100 day-inch/yr) (NUREG-1557). The weathering index for the continental United States is shown in ASTM C33-90, Fig. 1. A plant-specific program is not required if documented evidence confirms that where the existing concrete had air content of 3% to 8% and subsequent inspection did not exhibit degradation related to freeze-thaw. Such inspections should be considered a part of the evaluation. The reviewer confirms that the applicant has satisfied these conditions as identified in the GALL Report. Otherwise, the reviewer reviews the applicant's proposed aging management program to verify that, where appropriate, an effective inspection program has been developed and implemented to ensure that this aging effect in inaccessible areas for plants located in moderate to severe weathering conditions is adequately managed.

2. The GALL Report recommends further evaluation to determine if a plant-specific program is required to manage cracking due to expansion from reaction with aggregates in below-grade inaccessible concrete areas of Groups 1-5 and 7-9 structures. A plant-specific program is not required if (1) investigations, tests, and petrographic

examinations of aggregates performed in accordance with ASTM C295 and other ASTM reactivity tests, as required, can demonstrate that those aggregates do not adversely react within reinforced concrete, or (2) for potentially reactive aggregates, aggregate-reinforced concrete reaction is not significant if the structure was constructed in accordance with ACI 318.The reviewer confirms that the applicant has satisfied these conditions as identified in the GALL Report. Otherwise, the reviewer reviews the applicant's proposed aging management program to verify that, where appropriate, an effective inspection program has been developed and implemented to ensure that the aging effect is adequately managed.

3. The GALL Report recommends further evaluation of aging management of (a) cracking and distortion due to increased stress levels from settlement for inaccessible concrete areas of structures for all Groups and (b) reduction of foundation strength, and cracking due to differential settlement and erosion of porous concrete subfoundations for inaccessible concrete areas of Groups 1-3, and 5-9 structures if a de-watering system is relied upon to manage the aging effect. The reviewer confirms that, if the applicant's plant credits a de-watering system in its CLB, the applicant has committed to monitor the functionality of the de-watering system under the applicant's structures monitoring program. If not, the reviewer reviews and evaluates the plant-specific program for monitoring the de-watering system during the period of extended operation.

4. The GALL Report recommends further evaluation of programs to manage increase in porosity and permeability due to leaching of calcium hydroxide and carbonation in below-grade inaccessible concrete areas of Groups 1-5, and 7-9 structures. A plant-specific aging management program is not required for the reinforced concrete exposed to flowing water if (1) there is evidence in the accessible areas that the flowing water has not caused leaching of calcium hydroxide and carbonation or (2) evaluation determined that the observed leaching of calcium hydroxide and carbonation in accessible areas has no impact on the intended function of the concrete structure. The reviewer confirms that the applicant has satisfied these conditions as identified in the GALL Report. Otherwise, the reviewer reviews the applicant's proposed aging management program to verify that, where appropriate, an effective inspection program has been developed and implemented to ensure that this aging effect in inaccessible areas is adequately managed.

3.5.3.2.2.2   Reduction of Strength and Modulus due to Elevated Temperature

The GALL Report recommends further evaluation of programs to manage reduction of strength and modulus of concrete structures due to elevated temperature for PWR and BWR safety-related and other structures.

The GALL Report recommends that a plant-specific evaluation be performed if any portion of the concrete Groups 1-5 structures exceeds specified temperature limits, i.e., general temperature greater than 66°C (150°F) and local area temperature greater than 93°C (200°F). The GALL Report also states that higher temperatures may be allowed if tests and/or calculations are provided to evaluate the reduction in strength and modulus of elasticity and these reductions are applied to the design calculations. The reviewer reviews and confirms that the applicant's discussion in the renewal application indicates that the affected Groups 1-5 structures are not exposed to temperature that exceeds the temperature limits. If the limits are exceeded the reviewer reviews the technical basis (i.e., tests and/or calculations) provided by the applicant to justify the higher temperature. Otherwise the reviewer reviews the applicant's proposed programs on a case-by-case basis to ensure that the effects of elevated temperature will be adequately managed during the period of extended operation.

3.5.3.2.2.3   Aging Management of Inaccessible Areas for Group 6 Structures

The GALL Report recommends further evaluation for inaccessible areas of certain Group 6 structure/aging effect combinations as identified below, whether or not they are covered by inspections in accordance with the GALL Report, Chapter XI.S7, "Regulatory Guide 1.127, Inspection of Water-Control Structures Associated with Nuclear Power Plants," or FERC/US Army Corp of Engineers dam inspection and maintenance procedures.

1. Loss of material (spalling, scaling) and cracking due to freeze-thaw could occur in below-grade inaccessible concrete areas of Group 6 structures. Further evaluation is needed for plants that are located in moderate to severe weathering conditions (weathering index >100 day-inch/yr) (NUREG-1557, Ref. 7). The weathering index for the continental US is shown in ASTM C33-90, Fig. 1. A plant-specific program is not required if documented evidence confirms that where the existing concrete had air content of 3% to 8% and subsequent inspection of accessible areas did not exhibit degradation related to freeze-thaw. Such inspections should be considered a part of the evaluation. The reviewer reviews and confirms that the applicant has satisfied these conditions as identified in the GALL Report. Otherwise, the reviewer reviews the applicant's proposed aging management program to determine that, where appropriate, an effective inspection program has been developed and implemented to ensure that this aging effect in inaccessible areas for plants located in moderate to severe weathering conditions will be adequately managed.

2. Cracking due to expansion from reaction with aggregates could occur in below-grade inaccessible concrete areas of Group 6 structures. The GALL Report recommends further evaluation to determine if a plant-specific program is required to manage the aging effect. A plant specific program is not required if (1) investigations, tests, and petrographic examinations of aggregates performed in accordance with ASTM C295 and other ASTM reactivity tests, as required, can demonstrate that those aggregates do not adversely react within reinforced concrete or (2) for potentially reactive aggregates, aggregate-reinforced concrete reaction is not significant if the structure was constructed in accordance with ACI 318. The reviewer confirms that the applicant has satisfied these conditions for inaccessible concrete as identified in the GALL Report. Otherwise, the reviewer reviews the applicant's proposed aging management program to verify that, where appropriate, an effective inspection program has been developed and implemented to ensure that the aging effect will be adequately managed.

3. Increase in porosity and permeability due to leaching of calcium hydroxide and carbonation could occur in below-grade inaccessible concrete areas of Group 6 structures. The GALL Report recommends further evaluation to determine if a plant-specific program is required to manage the aging effect. A plant-specific program is not required for the reinforced exposed to flowing water if (1) there is evidence in the accessible areas that the flowing water has not caused leaching and carbonation, or (2) evaluation determined that the observed leaching of calcium hydroxide and carbonation in accessible areas has no impact on the intended function of the concrete structure. The reviewer confirms that the applicant has satisfied these conditions as identified in the GALL Report. Otherwise, the reviewer reviews the applicant's proposed aging management program to verify that, where appropriate, an effective inspection program

has been developed and implemented to ensure that this aging effect in inaccessible areas will be adequately managed.

3.5.3.2.2.4 Cracking due to Stress Corrosion Cracking and Loss of Material due to Pitting and Crevice Corrosion

The GALL Report recommends further evaluation of plant-specific programs to manage cracking due to SCC and loss of material due to pitting and crevice corrosion for stainless steel tank liners exposed to standing water. The reviewer reviews the applicant's proposed aging management program on a case-by-case basis to ensure that the intended functions will be maintained during the period of the extended operation.

3.5.3.2.2.5 Cumulative Fatigue Damage

Fatigue of support members, anchor bolts, and welds for Groups B1.1, B1.2, and B1.3 component supports is a TLAA as defined in 10 CFR 54.3 only if a CLB fatigue analysis exists. TLAAs are required to be evaluated in accordance with 10 CFR 54.21(c). The evaluation of this TLAA is addressed separately in Section 4.3 of this SRP-LR.

3.5.3.2.3 Quality Assurance for Aging Management of Nonsafety-Related Components

The applicant's aging management programs for license renewal should contain the elements of corrective actions, the confirmation process, and administrative controls. Safety-related components are covered by 10 CFR Part 50 Appendix B, which is adequate to address these program elements. However, Appendix B does not apply to nonsafety-related components that are subject to an AMR for license renewal. Nevertheless, an applicant has the option to expand the scope of its 10 CFR Part 50 Appendix B program to include these components and address these program elements. If the applicant chooses this option, the reviewer verifies that the applicant has documented such a commitment in the FSAR supplement. If the applicant chooses alternative means, the branch responsible for quality assurance should be requested to review the applicant's proposal on a case-by-case basis.

3.5.3.3 AMR Results Not Consistent with or Not Addressed in the GALL Report

The reviewer should confirm that the applicant, in their LRA, has identified applicable aging effects, listed the appropriate combination of materials and environments, and credited AMPs that will adequately manage the aging effects. The AMP credited by the applicant could be an AMP that is described and evaluated in the GALL Report or a plant-specific program. Review procedures are described in Branch Technical Position RLSB-1 (Appendix A.1 of this SRP-LR).

3.5.3.4 Aging Management Programs

The reviewer confirms that the applicant has identified the appropriate AMPs as described and evaluated in the GALL Report. If the applicant commits to an enhancement to make its LRA AMP consistent with a GALL Report AMP, then the reviewer is to confirm that this enhancement, when implemented, will make the LRA AMP consistent with the GALL Report AMP. If the applicant identifies, in the LRA AMP, an exception to any of the program elements of the GALL Report AMP, the reviewer is to confirm that the LRA AMP with the exception will satisfy the criteria of 10 CFR 54.21(a)(3). If the reviewer identifies a difference, not identified by the LRA, between the LRA AMP and the GALL Report AMP, with which the LRA claims to be consistent, the reviewer should confirm that the LRA AMP with this difference satisfies 10 CFR

54.21(a)(3). The reviewer should document the basis for accepting enhancements, exceptions, or differences. The AMPs evaluated in the GALL Report pertinent to the containments, structures, and component supports are summarized in Table 3.5-1 of this SRP-LR. The "Rev 2 Item" (for 2010) and "Rev1 Item" (for 2005 counterpart) columns identify the AMR item numbers in the GALL Report, Chapters II and III, presenting detailed information summarized by this row.

Table 3.5-1 of this SRP-LR may identify a plant-specific aging management program. If the applicant chooses to use a plant-specific program that is not a GALL AMP, the NRC reviewer should confirm that the plant-specific program satisfies the criteria of Branch Technical Position RLSB-1 (Appendix A.1.2.3 of this SRP-LR).

3.5.3.5 FSAR Supplement

The reviewer confirms that the applicant has provided in its FSAR supplement information equivalent to that in Table 3.0-1 for aging management of the containments, structures, and component supports. Table 3.5-2 lists the AMPs that are applicable for this SRP-LR subsection. The reviewer also confirms that the applicant has provided information for Subsection 3.5.3.3, "AMR Results Not Consistent with or Not Addressed in the GALL Report," equivalent to that in Table 3.0-1.

The staff expects to impose a license condition on any renewed license to require the applicant to update its FSAR to include this FSAR supplement at the next update required pursuant to 10 CFR 50.71(e)(4). As part of the license condition until the FSAR update is complete, the applicant may make changes to the programs described in its FSAR supplement without prior NRC approval, provided that the applicant evaluates each such change and finds it acceptable pursuant to the criteria set forth in 10 CFR 50.59. If the applicant updates the FSAR to include the final FSAR supplement before the license is renewed, no condition will be necessary.

As noted in Table 3.0-1, an applicant need not incorporate the implementation schedule into its FSAR. However, the reviewer should confirm that the applicant has identified and committed in the license renewal application to any future aging management activities, including enhancements and commitments, to be completed before the period of extended operation. The staff expects to impose a license condition on any renewed license to ensure that the applicant will complete these activities no later than the committed date.

3.5.4 Evaluation Findings

If the reviewer determines that the applicant has provided information sufficient to satisfy the provisions of this section, then an evaluation finding similar to the following text should be included in the staff's safety evaluation report:

> On the basis of its review, as discussed above, the staff concludes that the applicant has demonstrated that the aging effects associated with the containments, structures, and component supports components will be adequately managed so that the intended functions will be maintained consistent with the CLB for the period of extended operation, as required by 10 CFR 54.21(a)(3).
>
> The staff also reviewed the applicable FSAR Supplement program summaries and concludes that they adequately describe the AMPs credited for managing

aging of the containments, structures, and component supports, as required by 10 CFR 54.21(d).

3.5.5 Implementation

Except in those cases in which the applicant proposes an acceptable alternative method for complying with specified portions of the NRC's regulations, the method described herein will be used by the staff in its evaluation of conformance with NRC regulations.

3.5.6 References

1. 10 CFR Part 50, Appendix B, "Quality Assurance Criteria for Nuclear Power Plants," Office of the Federal Register, National Archives and Records Administration, 2009.

2. 10 CFR Part 50.55a, "Codes and Standards," Office of the Federal Register, National Archives and Records Administration, 2009.

3. 10 CFR Part 50.59, "Changes, Tests, and Experiments," Office of the Federal Register, National Archives and Records Administration, 2009.

4. 10 CFR Part 50, Appendix J, "Primary Reactor Containment Leakage Testing for Water-Cooled Power Reactors," Office of the Federal Register, National Archives and Records Administration, 2009.

5. 10 CFR Part 50.71, "Maintenance of Record, Making of Reports," Office of the Federal Register, National Archives and Records Administration, 2009.

6. 10 CFR Part 50.65, "Requirements for Monitoring the Effectiveness of Maintenance at Nuclear Power Plants," Office of the Federal Register, National Archives and Records Administration, 2009.

7. 10 CFR 54.4, "Scope, Office of the Federal Register," National Archives and Records Administration, 2009.

8. NRC Regulatory Guide 1.127, "Inspection of Water-Control Structures Associated with Nuclear Power Plants," Revision 1, U.S. Nuclear Regulatory Commission, March 1978.

9. NUREG-0800, "Standard Review Plan for the Review of Safety Analysis Reports for Nuclear Power Plants," U.S. Nuclear Regulatory Commission, March 2007.

10. NUREG-1801, "Generic Aging Lessons Learned (GALL)," U.S. Nuclear Regulatory Commission, Revision 2, 2010.

11. NEI 95-10, "Industry Guideline for Implementing the Requirements of 10 CFR Part 54 – The License Renewal Rule," Nuclear Energy Institute, Revision 6.

12. ASME Section XI, "Rules for Inservice Inspection of Nuclear Power Plant Components," Subsection IWL, "Requirements for Class CC Concrete Components of Light-Water Cooled Power Plants," The ASME Boiler and Pressure Vessel Code, 2004 edition as approved in 10 CFR 50.55a, The American Society of Mechanical Engineers, New York, NY.

13. ASME Section XI, "Rules for Inservice Inspection of Nuclear Power Plant Components," Subsection IWE, "Requirements for Class MC and Metallic Liners of Class CC Components of Light-Water Cooled Power Plants," The ASME Boiler and Pressure Vessel Code, 2004 edition as approved in 10 CFR 50.55a, the American Society of Mechanical Engineers, New York, NY.

14. ASME Section XI, "Rules for Inservice Inspection of Nuclear Power Plant Components," Subsection IWF, "Requirements for Class 1, 2, 3, and MC Component Supports of Light-Water Cooled Power Plants," The ASME Boiler and Pressure Vessel Code, 2004 edition as approved in 10 CFR 50.55a, The American Society of Mechanical Engineers, New York, NY.

15. NUMARC 93-01, Rev. 2, "Industry Guideline for Monitoring the Effectiveness of Maintenance at Nuclear Power Plants" [Line-In/Line-Out Version], Nuclear Energy Institute, April 1996.

16. NRC Regulatory Guide 1.160, Revision 2, "Monitoring the Effectiveness of Maintenance at Nuclear Power Plants," March 1997.

17. NUREG-1557, October 1996, "Summary of Technical Information and Agreements from Nuclear Management and Resource Council Industry Report addressing License Renewal."

18. ACI Standard 318, "Building Code Requirements for Reinforced Concrete and Commentary," American Concrete Institute.

Table 3.5-1 Summary of Aging Management Programs for Containments, Structures and Component Supports Evaluated in Chapters II and III of the GALL Report

PWR Concrete (Reinforced and Prestressed) and Steel Containments

BWR Concrete and Steel (Mark I, II, and III) Containments

ID	Type	Component	Aging Effect/Mechanism	Aging Management Programs	Further Evaluation Recommended	Rev2 Item	Rev1 Item
1	BWR/PWR	Concrete: dome; wall; basemat; ring girders; buttresses, Concrete elements, all	Cracking and distortion due to increased stress levels from settlement	Chapter XI.S2, "ASME Section XI, Subsection IWL" or Chapter XI.S6, "Structure Monitoring" If a de-watering system is relied upon for control of settlement, then the licensee is to ensure proper functioning of the de-watering system through the period of extended operation.	Yes, if a de-watering system is relied upon to control settlement (See subsection 3.5.2.1.1)	II.A1.CP-101 II.A2.CP-69 II.B1.2.CP-105 II.B2.2.CP-105 II.B3.1.CP-69 II.B3.2.CP-105	II.A1-5(C-37) II.A2-5(C-36) II.B1.2-1(C-06) II.B2.2-1(C-06) II.B3.1-2(C-36) II.B3.2-1(C-06)
2	BWR/PWR	Concrete: foundation; subfoundation	Reduction of foundation strength and cracking due to differential settlement and erosion of porous concrete subfoundation	Chapter XI.S6, "Structures Monitoring" If a de-watering system is relied upon for control of erosion, then the licensee is to ensure proper functioning of the de-watering system through the period of extended operation.	Yes, if a de-watering system is relied upon to control settlement (See subsection 3.5.2.1.1)	II.A1.C-07 II.A2.C-07 II.B1.2.C-07 II.B2.2.C-07 II.B3.1.C-07 II.B3.2.C-07	II.A1-8(C-07) II.A2-8(C-07) II.B1.2-7(C-07) II.B2.2-7(C-07) II.B3.1-7(C-07) II.B3.2-8(C-07)

Table 3.5-1 Summary of Aging Management Programs for Containments, Structures and Component Supports Evaluated in Chapters II and III of the GALL Report

ID	Type	Component	Aging Effect/Mechanism	Aging Management Programs	Further Evaluation Recommended	Rev2 Item	Rev1 Item
3	BWR/PWR	Concrete: dome; wall; basemat; ring girders; buttresses, Concrete: containment; wall; basemat, Concrete: basemat, concrete fill-in annulus	Reduction of strength and modulus due to elevated temperature (>150°F general; >200°F local)	A plant-specific aging management program is to be evaluated.	Yes, if temperature limits are exceeded (See subsection 3.5.2.1.2)	II.A1.CP-34 II.B1.2.CP-57 II.B2.2.CP-57 II.B3.1.CP-65 II.B3.2.CP-108	II.A1-1(C-08) II.B1.2-3(C-35) II.B2.2-3(C-35) II.B3.1-4(C-50) II.B3.2-2(C-33)
4	BWR	Steel elements (inaccessible areas): drywell shell; drywell head; and drywell shell	Loss of material due to general, pitting, and crevice corrosion	Chapter XI.S1, "ASME Section XI, Subsection IWE," and Chapter XI.S4, "10 CFR Part 50, Appendix J"	Yes, if corrosion is indicated from the IWE examinations (See subsection 3.5.2.1.3.1)	II.B3.1.CP-113	II.B3.1-8(C-19)
5	BWR/PWR	Steel elements (inaccessible areas): liner; liner anchors; integral attachments, Steel elements (inaccessible areas): suppression chamber; drywell; drywell head; embedded shell; region shielded by diaphragm floor (as applicable)	Loss of material due to general, pitting, and crevice corrosion	Chapter XI.S1, "ASME Section XI, Subsection IWE" and Chapter XI.S4, "10 CFR Part 50, Appendix J"	Yes, if corrosion is indicated from the IWE examinations (See subsection 3.5.2.1.3.1)	II.A1.CP-98 II.A2.CP-98 II.B1.2.CP-63 II.B2.1.CP-63 II.B2.2.CP-63 II.B3.2.CP-98	II.A1-11(C-09) II.A2-9(C-09) II.B1.2-8(C-46) II.B2.1-1(C-46) II.B2.2-10(C-46) II.B3.2-9(C-09)
6	BWR	Steel elements: torus shell	Loss of material due to general, pitting, and crevice corrosion	Chapter XI.S1, "ASME Section XI, Subsection IWE" and Chapter XI.S4, "10 CFR Part 50, Appendix J"	Yes, if corrosion is significant Recoating of the torus is recommended. (See subsection 3.5.2.1.3.2)	II.B1.1.CP-48	II.B1.1-2(C-19)

Table 3.5-1 Summary of Aging Management Programs for Containments, Structures and Component Supports Evaluated in Chapters II and III of the GALL Report

ID	Type	Component	Aging Effect/Mechanism	Aging Management Programs	Further Evaluation Recommended	Rev2 Item	Rev1 Item
7	BWR	Steel elements: torus ring girders; downcomers;, Steel elements: suppression chamber shell (interior surface)	Loss of material due to general, pitting, and crevice corrosion	Chapter XI.S1, "ASME Section XI, Subsection IWE"	Yes, if corrosion is significant (See subsection 3.5.2.1.3.3)	II.B1.1.CP-109 II.B3.1.CP-158	II.B1.1-2(C-19) II.B3.1-8(C-19)
8	BWR/PWR	Prestressing system: tendons	Loss of prestress due to relaxation; shrinkage; creep; elevated temperature	Yes, TLAA	Yes, TLAA (See subsection 3.5.2.2.1.4)	II.A1.C-11 II.B2.2.C-11	II.A1-9(C-11) II.B2.2-8(C-11)
9	BWR/PWR	Penetration sleeves; penetration bellows, Steel elements: torus; vent line; vent header; vent line bellows; downcomers, Suppression pool shell; unbraced downcomers, Steel elements: vent header; downcomers	Cumulative fatigue damage due to fatigue (Only if CLB fatigue analysis exists)	Yes, TLAA	Yes, TLAA (See subsection 3.5.2.2.1.5)	II.A3.C-13 II.B1.1.C-21 II.B2.1.C-45 II.B2.2.C-48 II.B4.C-13	II.A3-4(C-13) II.B1.1-4(C-21) II.B2.1-4(C-45) II.B2.2-14(C-48) II.B4-4(C-13)
10	PWR	Penetration sleeves; penetration bellows	Cracking due to stress corrosion cracking	Chapter XI.S1, "ASME Section XI, Subsection IWE", and Chapter XI.S4, "10 CFR Part 50, Appendix J"	Yes, detection of aging effects is to be evaluated (See subsection 3.5.2.2.1.6)	II.A3.CP-38 II.B4.CP-38	II.A3-2(C-15) II.B4-2(C-15)
11	BWR/PWR	Concrete (inaccessible areas): dome; wall; basemat; ring girders; buttresses, Concrete	Loss of material (spalling, scaling) and cracking due to freeze-thaw	Further evaluation is needed for plants that are located in moderate to severe weathering	Yes, for plants located in moderate to severe weathering	II.A1.CP-147 II.A2.CP-70 II.B3.2.CP-135	II.A1-2(C-01) II.A2-2(C-28) II.B3.2-3(C-29)

Table 3.5-1 Summary of Aging Management Programs for Containments, Structures and Component Supports Evaluated in Chapters II and III of the GALL Report

ID	Type	Component	Aging Effect/Mechanism	Aging Management Programs	Further Evaluation Recommended	Rev2 Item	Rev1 Item
		(inaccessible areas): basemat, Concrete (inaccessible areas): dome; wall; basemat		conditions (weathering index >100 day-inch/yr) (NUREG-1557).	conditions (See subsection 3.5.2.1.7)		
12	BWR/PWR	Concrete (inaccessible areas): dome; wall; basemat; ring girders; buttresses, Concrete (inaccessible areas): basemat, Concrete (inaccessible areas): containment; wall; basemat, Concrete (inaccessible areas): basemat, concrete fill-in annulus	Cracking due to expansion from reaction with aggregates	Further evaluation is required to determine if a plant-specific aging management program is needed.	Yes, if concrete is not constructed as stated function (See subsection 3.5.2.1.8)	II.A1.CP-67 II.A2.CP-104 II.B1.2.CP-99 II.B2.2.CP-99 II.B3.1.CP-83 II.B3.2.CP-121	II.A1-3(C-04) II.A2-3(C-38) II.B1.2-4(C-39) II.B2.2-4(C-39) II.B3.1-5(C-51) II.B3.2-4(C-40)
13	BWR/PWR	Concrete (inaccessible areas): basemat, Concrete (inaccessible areas): dome; wall; basemat	Increase in porosity and permeability; loss of strength due to leaching of calcium hydroxide and carbonation	Further evaluation is required to determine if a plant-specific aging management program is needed.	Yes, if leaching is observed in accessible areas that impact intended function (See subsection 3.5.2.1.9)	II.A2.CP-53 II.B3.1.CP-53 II.B3.2.CP-122	II.A2-6(C-30) II.B3.1-3(C-30) II.B3.2-6(C-32)
14	BWR/PWR	Concrete (inaccessible areas): dome; wall; basemat; ring girders; buttresses, Concrete (inaccessible areas): containment; wall; basemat	Increase in porosity and permeability; loss of strength due to leaching of calcium hydroxide and carbonation	Further evaluation is required to determine if a plant-specific aging management program is needed.	Yes, if leaching is observed in accessible areas that impact intended function (See subsection 3.5.2.1.9)	II.A1.CP-102 II.B1.2.CP-110 II.B2.2.CP-110	II.A1-6(C-02) II.B1.2-6(C-31) II.B2.2-6(C-31)

Table 3.5-1 Summary of Aging Management Programs for Containments, Structures and Component Supports Evaluated in Chapters II and III of the GALL Report

ID	Type	Component	Aging Effect/Mechanism	Aging Management Programs	Further Evaluation Recommended	Rev2 Item	Rev1 Item
15	BWR/PWR	Concrete (accessible areas): basemat	Increase in porosity and permeability; loss of strength due to leaching of calcium hydroxide and carbonation	Chapter XI.S2, "ASME Section XI, Subsection IWL"	No	II.A2.CP-155 II.B3.1.CP-156	II.A2-6(C-30) II.B3.1-3(C-30)
16	BWR/PWR	Concrete (accessible areas): basemat, Concrete: containment; wall; basemat	Increase in porosity and permeability; cracking; loss of material (spalling, scaling) due to aggressive chemical attack	Chapter XI.S2, "ASME Section XI, Subsection IWL," or Chapter XI.S6, "Structures Monitoring"	No	II.A2.CP-72 II.B1.2.CP-106 II.B2.2.CP-106 II.B3.1.CP-72	II.A2-4(C-25) II.B1.2-5(C-26) II.B2.2-5(C-26) II.B3.1-1(C-25)
17	BWR	Concrete (accessible areas): dome; wall; basemat; ring girders; buttresses	Increase in porosity and permeability; cracking; loss of material (spalling, scaling) due to aggressive chemical attack	Chapter XI.S2, "ASME Section XI, Subsection IWL"	No	II.A1.CP-87	II.A1-4(C-03)
18	BWR/PWR	Concrete (accessible areas): dome; wall; basemat; ring girders; buttresses, Concrete (accessible areas): basemat	Loss of material (spalling, scaling) and cracking due to freeze-thaw	Chapter XI.S2, "ASME Section XI, Subsection IWL"	No	II.A1.CP-31 II.A2.CP-51 II.B3.2.CP-52	II.A1-2(C-01) II.A2-2(C-28) II.B3.2-3(C-29)
19	BWR/PWR	Concrete (accessible areas): dome; wall; basemat; ring girders; buttresses, Concrete (accessible areas):	Cracking due to expansion from reaction with aggregates	Chapter XI.S2, "ASME Section XI, Subsection IWL"	No	II.A1.CP-33 II.A2.CP-58 II.B1.2.CP-59 II.B2.2.CP-59 II.B3.1.CP-66 II.B3.2.CP-60	II.A1-3(C-04) II.A2-3(C-38) II.B1.2-4(C-39) II.B2.2-4(C-39) II.B3.1-5(C-51) II.B3.2-4(C-40)

Table 3.5-1 Summary of Aging Management Programs for Containments, Structures and Component Supports Evaluated in Chapters II and III of the GALL Report

ID	Type	Component	Aging Effect/Mechanism	Aging Management Programs	Further Evaluation Recommended	Rev2 Item	Rev1 Item
		basemat, Concrete (accessible areas): containment; wall; basemat, Concrete (accessible areas): basemat, concrete fill-in annulus					
20	BWR/PWR	Concrete (accessible areas): dome; wall; basemat; ring girders; buttresses, Concrete (accessible areas): containment; wall; basemat	Increase in porosity and permeability; loss of strength due to leaching of calcium hydroxide and carbonation	Chapter XI.S2, "ASME Section XI, Subsection IWL"	No	II.A1.CP-32 II.B1.2.CP-54 II.B2.2.CP-54 II.B3.2.CP-55	II.A1-6(C-02) II.B1.2-6(C-31) II.B2.2-6(C-31) II.B3.2-6(C-32)
21	BWR/PWR	Concrete (accessible areas): dome; wall; basemat; ring girders; buttresses; reinforcing steel, Concrete (accessible areas): basemat; reinforcing steel, Concrete (accessible areas): dome; wall; basemat; reinforcing steel	Cracking; loss of bond; and loss of material (spalling, scaling) due to corrosion of embedded steel	Chapter XI.S2, "ASME Section XI, Subsection IWL"	No	II.A1.CP-68 II.A2.CP-74 II.B1.2.CP-79 II.B2.2.CP-79 II.B3.1.CP-74 II.B3.2.CP-88	II.A1-7(C-05) II.A2-7(C-43) II.B1.2-2(C-41) II.B2.2-2(C-41) II.B3.1-6(C-43) II.B3.2-7(C-42)
22	BWR	Concrete (inaccessible areas): basemat; reinforcing steel	Cracking; loss of bond; and loss of material (spalling, scaling) due to corrosion of embedded steel	Chapter XI.S6, "Structures Monitoring"	No	II.B1.2.CP-80 II.B2.2.CP-80	II.B1.2(C-41) II.B2.2-2(C-41)

Table 3.5-1 Summary of Aging Management Programs for Containments, Structures and Component Supports Evaluated in Chapters II and III of the GALL Report

ID	Type	Component	Aging Effect/Mechanism	Aging Management Programs	Further Evaluation Recommended	Rev2 Item	Rev1 Item
23	BWR/PWR	Concrete (inaccessible areas): basemat; reinforcing steel, Concrete (inaccessible areas): dome; wall; basemat; reinforcing steel	Cracking; loss of bond; and loss of material (spalling, scaling) due to corrosion of embedded steel	Chapter XI.S2, "ASME Section XI, Subsection IWL," or Chapter XI.S6, "Structures Monitoring"	No	II.A2.CP-75 II.B3.1.CP-75 II.B3.2.CP-89	II.A2-7(C-43) II.B3.1-6(C-43) II.B3.2-7(C-42)
24	BWR/PWR	Concrete (inaccessible areas): dome; wall; basemat; ring girders; buttresses, Concrete (inaccessible areas): basemat, Concrete (accessible areas): dome; wall; basemat	Increase in porosity and permeability; cracking; loss of material (spalling, scaling) due to aggressive chemical attack	Chapter XI.S2, "ASME Section XI, Subsection IWL," or Chapter XI.S6, "Structures Monitoring"	No	II.A1.CP-100 II.A2.CP-71 II.B3.1.CP-71 II.B3.2.CP-73 II.B3.2.CP-84	II.A1-4(C-03) II.A2-4(C-25) II.B3.1-1(C-25) II.B3.2-5(C-27) II.B3.2-5(C-27)
25	PWR	Concrete (inaccessible areas): dome; wall; basemat; ring girders; buttresses; reinforcing steel	Cracking; loss of bond; and loss of material (spalling, scaling) due to corrosion of embedded steel	Chapter XI.S2, "ASME Section XI, Subsection IWL," or Chapter XI.S6, "Structures Monitoring"	No	II.A1.CP-97	II.A1-7(C-05)
26	BWR/PWR	Moisture barriers (caulking, flashing, and other sealants)	Loss of sealing due to wear, damage, erosion, tear, surface cracks, or other defects	Chapter XI.S1, "ASME Section XI, Subsection IWE"	No	II.A3.CP-40 II.B4.CP-40	II.A3-7(C-18) II.B4-7(C-18)
27	BWR/PWR	penetration sleeves; penetration bellows, Steel elements: torus; vent line; vent header; vent line bellows;	Cracking due to cyclic loading (CLB fatigue analysis does not exist)	Chapter XI.S1, "ASME Section XI, Subsection IWE," and Chapter XI.S4, "10 CFR Part 50, Appendix J"	No	II.A3.CP-37 II.B1.1.CP-49 II.B2.1.CP-107 II.B4.CP-37	II.A3-3(C-14) II.B1.1-3(C-20) II.B2.1-3(C-44) II.B4-3(C-14)

Table 3.5-1 Summary of Aging Management Programs for Containments, Structures and Component Supports Evaluated in Chapters II and III of the GALL Report

ID	Type	Component	Aging Effect/Mechanism	Aging Management Programs	Further Evaluation Recommended	Rev2 Item	Rev1 Item
		downcomers, Suppression pool shell					
28	BWR/PWR	Personnel airlock, equipment hatch, CRD hatch	Loss of material due to general, pitting, and crevice corrosion	Chapter XI.S1, "ASME Section XI, Subsection IWE," and Chapter XI.S4, "10 CFR Part 50, Appendix J"	No	II.A3.C-16 II.B4.C-16	II.A3-6(C-16) II.B4-6(C-16)
29	BWR/PWR	Personnel airlock, equipment hatch, CRD hatch: locks, hinges, and closure mechanisms	Loss of leak tightness due to mechanical wear of locks, hinges and closure mechanisms	Chapter XI.S1, "ASME Section XI, Subsection IWE," and Chapter XI.S4, "10 CFR Part 50, Appendix J"	No	II.A3.CP-39 II.B4.CP-39	II.A3-5(C-17) II.B4-5(C-17)
30	BWR/PWR	Pressure-retaining bolting	Loss of preload due to self-loosening	Chapter XI.S1, "ASME Section XI, Subsection IWE," and Chapter XI.S4, "10 CFR Part 50, Appendix J"	No	II.A3.CP-150 II.B4.CP-150	N/A N/A
31	BWR/PWR	Pressure-retaining bolting, Steel elements: downcomer pipes	Loss of material due to general, pitting, and crevice corrosion	Chapter XI.S1, "ASME Section XI, Subsection IWE"	No	II.A3.CP-148 II.B1.2.CP-117 II.B2.1.CP-117 II.B2.2.CP-117 II.B4.CP-148	N/A II.B1.2-8(C-46) II.B2.1-1(C-46) II.B2.2-10(C-46) N/A
32	BWR/PWR	Prestressing system: tendons; anchorage components	Loss of material due to corrosion	Chapter XI.S2, "ASME Section XI, Subsection IWL"	No	II.A1.C-10 II.B2.2.C-10	II.A1-10(C-10) II.B2.2-9(C-10)

Table 3.5-1 Summary of Aging Management Programs for Containments, Structures and Component Supports Evaluated in Chapters II and III of the GALL Report

ID	Type	Component	Aging Effect/Mechanism	Aging Management Programs	Further Evaluation Recommended	Rev2 Item	Rev1 Item
33	BWR/PWR	Seals and gaskets	Loss of sealing due to wear, damage, erosion, tear, surface cracks, or other defects	Chapter XI.S4, "10 CFR Part 50, Appendix J"	No	II.A3.CP-41 II.B4.CP-41	II.A3-7(C-18) II.B4-7(C-18)
34	BWR/PWR	Service Level I coatings	Loss of coating integrity due to blistering, cracking, flaking, peeling, or physical damage	Chapter XI.S8, "Protective Coating Monitoring and Maintenance"	No	II.A3.CP-152 II.B4.CP-152	N/A N/A
35	BWR/PWR	Steel elements (accessible areas): liner; liner anchors; integral attachments, Penetration sleeves, Steel elements (accessible areas): drywell shell; drywell head; drywell shell in sand pocket regions;, Steel elements (accessible areas): suppression chamber; drywell; drywell head; embedded shell; region shielded by diaphragm floor (as applicable), Steel elements (accessible areas): drywell shell; drywell head	Loss of material due to general, pitting, and crevice corrosion	Chapter XI.S1, "ASME Section XI, Subsection IWE," and Chapter XI.S4, "10 CFR Part 50, Appendix J"	No	II.A1.CP-35 II.A2.CP-35 II.A3.CP-36 II.B1.1.CP-43 II.B1.2.CP-46 II.B2.1.CP-46 II.B2.2.CP-46 II.B3.1.CP-43 II.B3.2.CP-35 II.B4.CP-36	II.A1-11(C-09) II.A2-9(C-09) II.A3-1(C-12) II.B1.1-2(C-19) II.B1.2-8(C-46) II.B2.1-1(C-46) II.B2.2-10(C-46) II.B3.1-8(C-19) II.B3.2-9(C-09) II.B4-1(C-12)

Table 3.5-1 Summary of Aging Management Programs for Containments, Structures and Component Supports Evaluated in Chapters II and III of the GALL Report

ID	Type	Component	Aging Effect/Mechanism	Aging Management Programs	Further Evaluation Recommended	Rev2 Item	Rev1 Item
36	BWR	Steel elements: drywell head; downcomers	Fretting or lockup due to mechanical wear	Chapter XI.S1, "ASME Section XI, Subsection IWE"	No	II.B1.1.C-23 II.B1.2.C-23 II.B2.1.C-23 II.B2.2.C-23	II.B1.1-1(C-23) II.B1.2-9(C-23) II.B2.1-2(C-23) II.B2.2-11(C-23)
37	BWR	Steel elements: suppression chamber (torus) liner (interior surface)	Loss of material due to general (steel only), pitting, and crevice corrosion	Chapter XI.S1, "ASME Section XI, Subsection IWE," and Chapter XI.S4, "10 CFR Part 50, Appendix J"	No	II.B1.2.C-49 II.B2.2.C-49	II.B1.2-10(C-49) II.B2.2-12(C-49)
38	BWR	Steel elements: suppression chamber shell (interior surface)	Cracking due to stress corrosion cracking	Chapter XI.S1, "ASME Section XI, Subsection IWE," and Chapter XI.S4, "10 CFR Part 50, Appendix J"	No	II.B3.1.C-24 II.B3.2.C-24	II.B3.1-9(C-24) II.B3.2-10(C-24)
39	BWR	Steel elements: vent line bellows	Cracking due to stress corrosion cracking	Chapter XI.S1, "ASME Section XI, Subsection IWE," and Chapter XI.S4, "10 CFR Part 50, Appendix J"	No	II.B1.1.CP-50	II.B1.1-5(C-22)
40	BWR	Unbraced downcomers, Steel elements: vent header; downcomers	Cracking due to cyclic loading (CLB fatigue analysis does not exist)	Chapter XI.S1, "ASME Section XI, Subsection IWE"	No	II.B2.1.CP-142 II.B2.2.CP-64	II.B2.1-3(C-44) II.B2.2-13(C-47)
41	BWR	Steel elements: drywell support skirt, Steel elements (inaccessible areas): support skirt	None	None	NA - No AEM or AMP	II.B1.1.CP-44 II.B1.2.CP-114 II.B2.1.CP-114 II.B2.2.CP-114	N/A N/A N/A N/A

Table 3.5-1 Summary of Aging Management Programs for Containments, Structures and Component Supports Evaluated in Chapters II and III of the GALL Report

Safety-Related and Other Structures; and Component Supports

ID	Type	Component	Aging Effect/Mechanism	Aging Management Programs	Further Evaluation Recommended	Rev2 Item	Rev1 Item
42	BWR/PWR	Groups 1-3, 5, 7-9: Concrete (inaccessible areas): foundation	Loss of material (spalling, scaling) and cracking due to freeze-thaw	Further evaluation is required for plants that are located in moderate to severe weathering conditions (weathering index >100 day-inch/yr) (NUREG-1557)	Yes, for plants located in moderate to severe weathering conditions (See subsection 3.5.2.2.1.1)	III.A1.TP-108 III.A2.TP-108 III.A3.TP-108 III.A5.TP-108 III.A7.TP-108 III.A8.TP-108 III.A9.TP-108	III.A1-6(T-01) III.A2-6(T-01) III.A3-6(T-01) III.A5-6(T-01) III.A7-5(T-01) III.A8-5(T-01) III.A9-5(T-01)
43	BWR/PWR	All Groups except Group 6: Concrete (inaccessible areas): all	Cracking due to expansion from reaction with aggregates	Further evaluation is required to determine if a plant-specific aging management program is needed.	Yes, if concrete is not constructed as stated (See subsection 3.5.2.2.1.2)	III.A1.TP-204 III.A2.TP-204 III.A3.TP-204 III.A4.TP-204 III.A5.TP-204 III.A7.TP-204 III.A8.TP-204 III.A9.TP-204	III.A1-2(T-03) III.A2-2(T-03) III.A3-2(T-03) III.A4-2(T-03) III.A5-2(T-03) III.A7-1(T-03) III.A8-1(T-03) III.A9-1(T-03)
44	BWR/PWR	All Groups: concrete: all	Cracking and distortion due to increased stress levels from settlement	Chapter XI.S6, "Structures Monitoring" If a de-watering system is relied upon for control of settlement, then the licensee is to ensure proper functioning of the de-watering system through the period of extended operation.	Yes, if a de-watering system is relied upon to control settlement (See subsection 3.5.2.2.1.3)	III.A1.TP-30 III.A2.TP-30 III.A3.TP-30 III.A4.TP-304 III.A5.TP-30 III.A6.TP-30 III.A7.TP-30 III.A8.TP-30 III.A9.TP-30	III.A1-3(T-08) III.A2-3(T-08) III.A3-3(T-08) N/A III.A5-3(T-08) III.A6-4(T-08) III.A7-2(T-08) III.A8-2(T-08) III.A9-2(T-08)

ID	Type	Component	Aging Effect/Mechanism	Aging Management Programs	Further Evaluation Recommended	Rev2 Item	Rev1 Item
45	BWR	Groups 1-3, 5-9: concrete: foundation; subfoundation	Reduction in foundation strength, cracking due to differential settlement, erosion of porous concrete subfoundation	Chapter XI.S6, "Structures Monitoring" If a de-watering system is relied upon for control of settlement, then the licensee is to ensure proper functioning of the de-watering system through the period of extended operation.	Yes, if a de-watering system is relied upon to control settlement (See subsection 3.5.2.2.1.3)	III.A9.TP-31	III.A9-7(T-09)
46	BWR/PWR	Groups 1-3, 5-9: concrete: foundation; subfoundation	Reduction of foundation strength and cracking due to differential settlement and erosion of porous concrete subfoundation	Chapter XI.S6, "Structures Monitoring" If a de-watering system is relied upon for control of settlement, then the licensee is to ensure proper functioning of the de-watering system through the period of extended operation.	Yes, if a de-watering system is relied upon to control settlement (See subsection 3.5.2.2.1.3)	III.A1.TP-31 III.A2.TP-31 III.A3.TP-31 III.A5.TP-31 III.A6.TP-31 III.A7.TP-31 III.A8.TP-31	III.A1-8(T-09) III.A2-8(T-09) III.A3-8(T-09) III.A5-8(T-09) III.A6-8(T-09) III.A7-7(T-09) III.A8-7(T-09)
47	BWR/PWR	Groups 1-5, 7-9: concrete (inaccessible areas): exterior above- and below-grade; foundation	Increase in porosity and permeability; loss of strength due to leaching of calcium hydroxide and carbonation	Further evaluation is required to determine if a plant-specific aging management program is needed.	Yes, if leaching is observed in accessible areas that impact intended function (See subsection 3.5.2.2.1.4)	III.A1.TP-67 III.A2.TP-67 III.A3.TP-67 III.A4.TP-305 III.A5.TP-67 III.A7.TP-67 III.A8.TP-67 III.A9.TP-67	III.A1-7(T-02) III.A2-7(T-02) III.A3-7(T-02) N/A III.A5-7(T-02) III.A7-6(T-02) III.A8-6(T-02) III.A9-6(T-02)

Table 3.5-1 Summary of Aging Management Programs for Containments, Structures and Component Supports Evaluated in Chapters II and III of the GALL Report

ID	Type	Component	Aging Effect/Mechanism	Aging Management Programs	Further Evaluation Recommended	Rev2 Item	Rev1 Item
48	BWR/PWR	Groups 1-5: concrete: all	Reduction of strength and modulus due to elevated temperature (>150°F general; >200°F local)	A plant-specific aging management program is to be evaluated.	Yes, if temperature limits are exceeded (See subsection 3.5.2.2.2)	III.A1.TP-114 III.A2.TP-114 III.A3.TP-114 III.A4.TP-114 III.A5.TP-114	III.A1-1(T-10) III.A2-1(T-10) III.A3-1(T-10) III.A4-1(T-10) III.A5-1(T-10)
49	BWR/PWR	Groups 6 - concrete (inaccessible areas): exterior above- and below-grade; foundation; interior slab	Loss of material (spalling, scaling) and cracking due to freeze-thaw	Further evaluation is required for plants that are located in moderate to severe weathering conditions (weathering index >100 day-inch/yr) (NUREG-1557)	Yes, for plants located in moderate to severe weathering conditions (See subsection 3.5.2.2.3.1)	III.A6.TP-110	III.A6-5(T-15)
50	BWR/PWR	Groups 6: concrete (inaccessible areas): all	Cracking due to expansion from reaction with aggregates	Further evaluation is required to determine if a plant-specific aging management program is needed.	Yes, if concrete is not constructed as stated See subsection 3.5.2.2.3.2	III.A6.TP-220	III.A6-2(T-17)
51	BWR/PWR	Groups 6: concrete (inaccessible areas): exterior above- and below-grade; foundation; interior slab	Increase in porosity and permeability; loss of strength due to leaching of calcium hydroxide and carbonation	Further evaluation is required to determine if a plant-specific aging management program is needed.	Yes, if leaching is observed in accessible areas that impact intended function (See subsection 3.5.2.2.3.3)	III.A6.TP-109	III.A6-6(T-16)
52	BWR/PWR	Groups 7, 8 - steel components: tank liner	Cracking due to stress corrosion cracking; Loss of material due to pitting and	A plant-specific aging management program is to be evaluated.	Yes, plant-specific (See subsection 3.5.2.2.4)	III.A7.T-23 III.A8.T-23	III.A7-11(T-23) III.A8-9(T-23)

Table 3.5-1 Summary of Aging Management Programs for Containments, Structures and Component Supports Evaluated in Chapters II and III of the GALL Report

ID	Type	Component	Aging Effect/Mechanism	Aging Management Programs	Further Evaluation Recommended	Rev2 Item	Rev1 Item
			crevice corrosion				
53	BWR/PWR	Support members; welds; bolted connections; support anchorage to building structure	Cumulative fatigue damage due to fatigue (Only if CLB fatigue analysis exists)	Yes, TLAA	Yes, TLAA (See subsection 3.5.2.2.5)	III.B1.1.T-26 III.B1.2.T-26 III.B1.3.T-26	III.B1.1-12(T-26) III.B1.2-9(T-26) III.B1.3-9(T-26)
54	BWR/PWR	All groups except 6: concrete (accessible areas): all	Cracking due to expansion from reaction with aggregates	Chapter XI.S6, "Structures Monitoring"	No	III.A1.TP-25 III.A2.TP-25 III.A3.TP-25 III.A4.TP-25 III.A5.TP-25 III.A7.TP-25 III.A8.TP-25 III.A9.TP-25	III.A1-2(T-03) III.A2-2(T-03) III.A3-2(T-03) III.A4-2(T-03) III.A5-2(T-03) III.A7-1(T-03) III.A8-1(T-03) III.A9-1(T-03)
55	BWR/PWR	Building concrete at locations of expansion and grouted anchors; grout pads for support base plates	Reduction in concrete anchor capacity due to local concrete degradation/ service-induced cracking or other concrete aging mechanisms	Chapter XI.S6, "Structures Monitoring"	No	III.B1.1.TP-42 III.B1.2.TP-42 III.B1.3.TP-42 III.B2.TP-42 III.B3.TP-42 III.B4.TP-42 III.B5.TP-42	III.B1.1-1(T-29) III.B1.2-1(T-29) III.B1.3-1(T-29) III.B2-1(T-29) III.B3-1(T-29) III.B4-1(T-29) III.B5-1(T-29)
56	BWR/PWR	Concrete: exterior above- and below-grade; foundation; interior slab	Loss of material due to abrasion; cavitation	Chapter XI.S7, "Regulatory Guide 1.127, Inspection of Water-Control Structures Associated with Nuclear Power Plants" or the	No	III.A6.T-20	III.A6-7(T-20)

Table 3.5-1 Summary of Aging Management Programs for Containments, Structures and Component Supports Evaluated in Chapters II and III of the GALL Report

ID	Type	Component	Aging Effect/Mechanism	Aging Management Programs	Further Evaluation Recommended	Rev2 Item	Rev1 Item
				FERC/US Army Corp of Engineers dam inspections and maintenance programs.			
57	BWR/PWR	Constant and variable load spring hangers; guides; stops	Loss of mechanical function due to corrosion, distortion, dirt, overload, fatigue due to vibratory and cyclic thermal loads	Chapter XI.S3, "ASME Section XI, Subsection IWF"	No	III.B1.1.T-28 III.B1.2.T-28 III.B1.3.T-28	III.B1.1-2(T-28) III.B1.2-2(T-28) III.B1.3-2(T-28)
58	BWR/PWR	Earthen water-control structures: dams; embankments; reservoirs; channels; canals and ponds	Loss of material; loss of form due to erosion, settlement, sedimentation, frost action, waves, currents, surface runoff, seepage	Chapter XI.S7, "Regulatory Guide 1.127, Inspection of Water-Control Structures Associated with Nuclear Power Plants" or the FERC/US Army Corp of Engineers dam inspections and maintenance programs.	No	III.A6.T-22	III.A6-9(T-22)
59	BWR/PWR	Group 6: concrete (accessible areas): all	Cracking; loss of bond; and loss of material (spalling, scaling) due to corrosion of embedded steel	Chapter XI.S7, "Regulatory Guide 1.127, Inspection of Water-Control Structures Associated with Nuclear Power Plants" or the FERC/US Army Corp of Engineers dam inspections and maintenance programs.	No	III.A6.TP-38	III.A6-1(T-18)

Table 3.5-1 Summary of Aging Management Programs for Containments, Structures and Component Supports Evaluated in Chapters II and III of the GALL Report

ID	Type	Component	Aging Effect/Mechanism	Aging Management Programs	Further Evaluation Recommended	Rev2 Item	Rev1 Item
60	BWR/PWR	Group 6: concrete (accessible areas): exterior above- and below-grade; foundation	Loss of material (spalling, scaling) and cracking due to freeze-thaw	Chapter XI.S7, "Regulatory Guide 1.127, Inspection of Water-Control Structures Associated with Nuclear Power Plants" or the FERC/US Army Corp of Engineers dam inspections and maintenance programs.	No	III.A6.TP-36	III.A6-5(T-15)
61	BWR/PWR	Group 6: concrete (accessible areas): exterior above- and below-grade; foundation; interior slab	Increase in porosity and permeability; loss of strength due to leaching of calcium hydroxide and carbonation	Chapter XI.S7, "Regulatory Guide 1.127, Inspection of Water-Control Structures Associated with Nuclear Power Plants" or the FERC/US Army Corp of Engineers dam inspections and maintenance programs.	No	III.A6.TP-37	III.A6-6(T-16)
62	BWR/PWR	Group 6: Wooden Piles; sheeting	Loss of material; change in material properties due to weathering, chemical degradation, and insect infestation repeated wetting and drying, fungal decay	Chapter XI.S7, "Regulatory Guide 1.127, Inspection of Water-Control Structures Associated with Nuclear Power Plants" or the FERC/US Army Corp of Engineers dam inspections and maintenance programs.	No	III.A6.TP-223	N/A

Table 3.5-1 Summary of Aging Management Programs for Containments, Structures and Component Supports Evaluated in Chapters II and III of the GALL Report

ID	Type	Component	Aging Effect/Mechanism	Aging Management Programs	Further Evaluation Recommended	Rev2 Item	Rev1 Item
63	BWR/PWR	Groups 1-3, 5, 7-9: concrete (accessible areas): exterior above- and below-grade; foundation	Increase in porosity and permeability; loss of strength due to leaching of calcium hydroxide and carbonation	Chapter XI.S6, "Structures Monitoring"	No	III.A1.TP-24 III.A2.TP-24 III.A3.TP-24 III.A5.TP-24 III.A7.TP-24 III.A8.TP-24 III.A9.TP-24	III.A1-7(T-02) III.A2-7(T-02) III.A3-7(T-02) III.A5-7(T-02) III.A7-6(T-02) III.A8-6(T-02) III.A9-6(T-02)
64	BWR/PWR	Groups 1-3, 5, 7-9: concrete (accessible areas): exterior above- and below-grade; foundation	Loss of material (spalling, scaling) and cracking due to freeze-thaw	Chapter XI.S6, "Structures Monitoring"	No	III.A1.TP-23 III.A2.TP-23 III.A3.TP-23 III.A5.TP-23 III.A7.TP-23 III.A8.TP-23 III.A9.TP-23	III.A1-6(T-01) III.A2-6(T-01) III.A3-6(T-01) III.A5-6(T-01) III.A7-5(T-01) III.A8-5(T-01) III.A9-5(T-01)
65	BWR/PWR	Groups 1-3, 5, 7-9: concrete (inaccessible areas): below-grade exterior, foundation; Groups 1-3, 5, 7-9: concrete (accessible areas): below-grade exterior, foundation; Groups 6: concrete (inaccessible areas): all	Cracking; loss of bond; and loss of material (spalling, scaling) due to corrosion of embedded steel	Chapter XI.S6, "Structures Monitoring"	No	III.A1.TP-212 III.A1.TP-27 III.A2.TP-212 III.A2.TP-27 III.A3.TP-212 III.A3.TP-27 III.A5.TP-212 III.A5.TP-27 III.A6.TP-104 III.A6.TP-212 III.A7.TP-212 III.A7.TP-27 III.A8.TP-212 III.A8.TP-27 III.A9.TP-212 III.A9.TP-27	III.A1-4(T-05) III.A1-4(T-05) III.A2-4(T-05) III.A2-4(T-05) III.A3-4(T-05) III.A3-4(T-05) III.A5-4(T-05) III.A5-4(T-05) III.A6-1(T-18) III.A7-3(T-05) III.A7-3(T-05) III.A8-3(T-05) III.A8-3(T-05) III.A9-3(T-05) III.A9-3(T-05)
66	BWR/PWR	Groups 1-5, 7, 9: concrete (accessible areas): interior and above-grade exterior	Cracking; loss of bond; and loss of material (spalling, scaling) due to corrosion of	Chapter XI.S6, "Structures Monitoring"	No	III.A1.TP-26 III.A2.TP-26 III.A3.TP-26 III.A4.TP-26 III.A5.TP-26 III.A7.TP-26	III.A1-9(T-04) III.A2-9(T-04) III.A3-9(T-04) III.A4-3(T-04) III.A5-9(T-04) III.A7-8(T-04)

Table 3.5-1 Summary of Aging Management Programs for Containments, Structures and Component Supports Evaluated in Chapters II and III of the GALL Report

ID	Type	Component	Aging Effect/Mechanism	Aging Management Programs	Further Evaluation Recommended	Rev2 Item	Rev1 Item
			embedded steel			III.A9.TP-26	III.A9-8(T-04)
67	BWR/PWR	Groups 1-5, 7, 9: Concrete: interior, above-grade exterior, Groups 1-3, 5, 7-9 - concrete (inaccessible areas): below-grade exterior, foundation, Group 6: concrete (inaccessible areas): all	Increase in porosity and permeability; cracking; loss of material (spalling, scaling) due to aggressive chemical attack	Chapter XI.S6, "Structures Monitoring"	No	III.A1.TP-28 III.A1.TP-29 III.A2.TP-28 III.A2.TP-29 III.A3.TP-28 III.A3.TP-29 III.A4.TP-28 III.A5.TP-28 III.A5.TP-29 III.A6.TP-107 III.A7.TP-28 III.A7.TP-29 III.A8.TP-29 III.A9.TP-28 III.A9.TP-29	III.A1-10(T-06) III.A1-5(T-07) III.A2-10(T-06) III.A2-5(T-07) III.A3-10(T-06) III.A3-5(T-07) III.A4-4(T-06) III.A5-10(T-06) III.A5-5(T-07) III.A6-3(T-19) III.A7-9(T-06) III.A7-4(T-07) III.A8-4(T-07) III.A9-9(T-06) III.A9-4(T-07)
68	BWR/PWR	High-strength structural bolting	Cracking due to stress corrosion cracking	Chapter XI.S3, "ASME Section XI, Subsection IWF"	No	III.B1.1.TP-41	III.B1.1-3(T-27)
69	BWR/PWR	High-strength structural bolting	Cracking due to stress corrosion cracking	Chapter XI.S6, "Structures Monitoring" Note: ASTM A 325, F 1852, and ASTM A 490 bolts used in civil structures have not shown to be prone to SCC. SCC potential need	No	III.A1.TP-300 III.A2.TP-300 III.A3.TP-300 III.A4.TP-300 III.A5.TP-300 III.A7.TP-300 III.A8.TP-300 III.A9.TP-300 III.B2.TP-300 III.B3.TP-300 III.B4.TP-300 III.B5.TP-300	N/A N/A N/A N/A N/A N/A N/A N/A N/A N/A N/A N/A

Table 3.5-1 Summary of Aging Management Programs for Containments, Structures and Component Supports Evaluated in Chapters II and III of the GALL Report

ID	Type	Component	Aging Effect/Mechanism	Aging Management Programs	Further Evaluation Recommended	Rev2 Item	Rev1 Item
				not be evaluated for these bolts.			
70	BWR/PWR	Masonry walls: all	Cracking due to restraint shrinkage, creep, and aggressive environment	Chapter XI.S5, "Masonry Walls"	No	III.A1.T-12 III.A2.T-12 III.A3.T-12 III.A5.T-12 III.A6.T-12	III.A1-11(T-12) III.A2-11(T-12) III.A3-11(T-12) III.A5-11(T-12) III.A6-10(T-12)
71	BWR/PWR	Masonry walls: all	Loss of material (spalling, scaling) and cracking due to freeze-thaw	Chapter XI.S5, "Masonry Walls"	No	III.A5.TP-34	N/A
72	BWR/PWR	Seals; gasket; moisture barriers (caulking, flashing, and other sealants)	Loss of sealing due to deterioration of seals, gaskets, and moisture barriers (caulking, flashing, and other sealants)	Chapter XI.S6, "Structures Monitoring"	No	III.A6.TP-7	III.A6-12(TP-7)
73	BWR/PWR	Service Level I coatings	Loss of coating integrity due to blistering, cracking, flaking, peeling, physical damage	Chapter XI.S8, "Protective Coating Monitoring and Maintenance"	No	III.A4.TP-301	N/A
74	BWR/PWR	Sliding support bearings; sliding support surfaces	Loss of mechanical function due to corrosion, distortion, dirt, debris, overload, wear	Chapter XI.S6, "Structures Monitoring"	No	III.B2.TP-46 III.B2.TP-47 III.B4.TP-46 III.B4.TP-47	III.B2-2(TP-1) III.B2-3(TP-2) III.B4-2(TP-1) III.B4-3(TP-2)

ID	Type	Component	Aging Effect/Mechanism	Aging Management Programs	Further Evaluation Recommended	Rev2 Item	Rev1 Item
75	BWR/PWR	Sliding surfaces	Loss of mechanical function due to corrosion, distortion, dirt, debris, overload, wear	Chapter XI.S3, "ASME Section XI, Subsection IWF"	No	III.B1.1.TP-45 III.B1.2.TP-45 III.B1.3.TP-45	III.B1.1-5(T-32) III.B1.2-3(T-32) III.B1.3-3(T-32)
76	BWR/PWR	Sliding surfaces: radial beam seats in BWR drywell	Loss of mechanical function due to corrosion, distortion, dirt, overload, wear	Chapter XI.S6, "Structures Monitoring"	No	III.A4.TP-35	III.A4-6(T-13)
77	BWR/PWR	Steel components: all structural steel	Loss of material due to corrosion	Chapter XI.S6, "Structures Monitoring" If protective coatings are relied upon to manage the effects of aging, the structures monitoring program is to include provisions to address protective coating monitoring and maintenance.	No	III.A1.TP-302 III.A2.TP-302 III.A3.TP-302 III.A4.TP-302 III.A5.TP-302 III.A7.TP-302 III.A8.TP-302	III.A1-12(T-11) III.A2-12(T-11) III.A3-12(T-11) III.A4-5(T-11) III.A5-12(T-11) III.A7-10(T-11) III.A8-8(T-11)
78	BWR/PWR	Steel components: fuel pool liner	Cracking due to stress corrosion cracking; Loss of material due to pitting and crevice corrosion	Chapter XI.M2, "Water Chemistry," and Monitoring of the spent fuel pool water level in accordance with technical specifications and leakage from the leak chase channels.	No, unless leakages have been detected through the SFP liner that cannot be accounted for from the leak chase channels	III.A5.T-14	III.A5-13(T-14)

ID	Type	Component	Aging Effect/Mechanism	Aging Management Programs	Further Evaluation Recommended	Rev2 Item	Rev1 Item
79	BWR/PWR	Steel components: piles	Loss of material due to corrosion	Chapter XI.S6, "Structures Monitoring"	No	III.A3.TP-219	N/A
80	BWR/PWR	Structural bolting	Loss of material due to general, pitting and crevice corrosion	Chapter XI.S6, "Structures Monitoring"	No	III.A1.TP-248 III.A2.TP-248 III.A3.TP-248 III.A4.TP-248 III.A5.TP-248 III.A6.TP-248 III.A7.TP-248 III.A8.TP-248 III.A9.TP-248 III.B2.TP-248 III.B3.TP-248 III.B4.TP-248 III.B5.TP-248	N/A N/A N/A N/A N/A N/A N/A N/A N/A N/A N/A N/A N/A
81	BWR/PWR	Structural bolting	Loss of material due to general, pitting, and crevice corrosion	Chapter XI.S3, "ASME Section XI, Subsection IWF"	No	III.B1.1.TP-226 III.B1.2.TP-226 III.B1.3.TP-226	N/A N/A N/A
82	BWR/PWR	Structural bolting	Loss of material due to general, pitting, and crevice corrosion	Chapter XI.S6, "Structures Monitoring"	No	III.A1.TP-274 III.A2.TP-274 III.A3.TP-274 III.A4.TP-274 III.A5.TP-274 III.A7.TP-274 III.A8.TP-274 III.A9.TP-274 III.B2.TP-274 III.B3.TP-274 III.B4.TP-274 III.B5.TP-274	N/A N/A N/A N/A N/A N/A N/A N/A N/A N/A N/A N/A

Table 3.5-1 Summary of Aging Management Programs for Containments, Structures and Component Supports Evaluated in Chapters II and III of the GALL Report

ID	Type	Component	Aging Effect/Mechanism	Aging Management Programs	Further Evaluation Recommended	Rev2 Item	Rev1 Item
83	BWR/PWR	Structural bolting	Loss of material due to general, pitting, and crevice corrosion	Chapter XI.S7, "Regulatory Guide 1.127, Inspection of Water-Control Structures Associated with Nuclear Power Plants" or the FERC/US Army Corp of Engineers dam inspections and maintenance programs.	No	III.A6.TP-221	N/A
84	BWR/PWR	Structural bolting	Loss of material due to pitting and crevice corrosion	Chapter XI.M2, "Water Chemistry," and Chapter XI.S3, "ASME Section XI, Subsection IWF"	No	III.B1.3.TP-232	N/A
85	BWR/PWR	Structural bolting	Loss of material due to pitting and crevice corrosion	Chapter XI.M2, "Water Chemistry," for BWR water, and Chapter XI.S3, "ASME Section XI, Subsection IWF"	No	III.B1.1.TP-232 III.B1.2.TP-232	N/A N/A
86	BWR/PWR	Structural bolting	Loss of material due to pitting and crevice corrosion	Chapter XI.S3, "ASME Section XI, Subsection IWF"	No	III.B1.1.TP-235 III.B1.2.TP-235 III.B1.3.TP-235	N/A N/A N/A
87	BWR/PWR	Structural bolting	Loss of preload due to self-loosening	Chapter XI.S3, "ASME Section XI, Subsection IWF"	No	III.B1.1.TP-229 III.B1.2.TP-229 III.B1.3.TP-229	N/A N/A N/A

Table 3.5-1 Summary of Aging Management Programs for Containments, Structures and Component Supports Evaluated in Chapters II and III of the GALL Report

ID	Type	Component	Aging Effect/Mechanism	Aging Management Programs	Further Evaluation Recommended	Rev2 Item	Rev1 Item
88	BWR/PWR	Structural bolting	Loss of preload due to self-loosening	Chapter XI.S6, "Structures Monitoring"	No	III.A1.TP-261 III.A2.TP-261 III.A3.TP-261 III.A4.TP-261 III.A5.TP-261 III.A6.TP-261 III.A7.TP-261 III.A8.TP-261 III.A9.TP-261 III.B2.TP-261 III.B3.TP-261 III.B4.TP-261 III.B5.TP-261	N/A N/A N/A N/A N/A N/A N/A N/A N/A N/A N/A N/A N/A
89	PWR	Support members; welds; bolted connections; support anchorage to building structure	Loss of material due to boric acid corrosion	Chapter XI.M10, "Boric Acid Corrosion"	No	III.B1.1.T-25 III.B1.1.TP-3 III.B1.2.T-25 III.B1.3.TP-3 III.B1.2.TP-3 III.B2.T-25 III.B2.TP-3 III.B3.T-25 III.B3.TP-3 III.B4.T-25 III.B4.TP-3 III.B5.T-25 III.B5.TP-3	III.B1.1-14(T-25) III.B1.1-8(TP-3) III.B1.2-11(T-25) III.B1.3-6(TP-3) III.B1.2-6(TP-3) III.B2-11(T-25) III.B2-6(TP-3) III.B3-8(T-25) III.B3-4(TP-3) III.B4-11(T-25) III.B4-6(TP-3) III.B5-8(T-25) III.B5-4(TP-3)
90	BWR/PWR	Support members; welds; bolted connections; support anchorage to building structure	Loss of material due to general (steel only), pitting, and crevice corrosion	Chapter XI.M2, "Water Chemistry," for BWR water, and Chapter XI.S3, "ASME Section XI, Subsection IWF"	No	III.B1.1.TP-10	III.B1.1-11(TP-10)

Table 3.5-1 Summary of Aging Management Programs for Containments, Structures and Component Supports Evaluated in Chapters II and III of the GALL Report

ID	Type	Component	Aging Effect/Mechanism	Aging Management Programs	Further Evaluation Recommended	Rev2 Item	Rev1 Item
91	BWR/PWR	Support members; welds; bolted connections; support anchorage to building structure	Loss of material due to general and pitting corrosion	Chapter XI.S3, "ASME Section XI, Subsection IWF"	No	III.B1.1.T-24 III.B1.2.T-24 III.B1.3.T-24	III.B1.1-13(T-24) III.B1.2-10(T-24) III.B1.3-10(T-24)
92	BWR/PWR	Support members; welds; bolted connections; support anchorage to building structure	Loss of material due to general and pitting corrosion	Chapter XI.S6, "Structures Monitoring"	No	III.B2.TP-43 III.B3.TP-43 III.B4.TP-43 III.B5.TP-43	III.B2-10(T-30) III.B3-7(T-30) III.B4-10(T-30) III.B5-7(T-30)
93	BWR/PWR	Support members; welds; bolted connections; support anchorage to building structure	Loss of material due to pitting and crevice corrosion	Chapter XI.S6, "Structures Monitoring"	No	III.B2.TP-6 III.B4.TP-6	III.B2-7(TP-6) III.B4-7(TP-6)
94	BWR/PWR	Vibration isolation elements	Reduction or loss of isolation function due to radiation hardening, temperature, humidity, sustained vibratory loading	Chapter XI.S3, "ASME Section XI, Subsection IWF"	No	III.B1.1.T-33 III.B1.2.T-33 III.B1.3.T-33 III.B4.TP-44	III.B1.1-15(T-33) III.B1.2-12(T-33) III.B1.3-11(T-33) III.B4-12(T-31)
95	BWR/PWR	Aluminum, galvanized steel and stainless steel Support members; welds; bolted connections; support anchorage to building structure exposed to Air –	None	None	NA - No AEM or AMP	III.B1.1.TP-4 III.B1.1.TP-8 III.B1.2.TP-4 III.B1.2.TP-8 III.B1.3.TP-4 III.B1-3.TP-8 III.B2.TP-4 III.B2.TP-8 III.B3.TP-4 III.B3.TP-8 III.B4.TP-4	III.B1.1-10(TP-4) III.B1.1-6(TP-8) III.B1.1-7(TP-11) III.B.1.1-9(TP-5) III.B1.2-8(TP-4) III.B1.2-4(TP-8) III.B1.2-5(TP-11) III.B1.2-7(TP-5)

Table 3.5-1 Summary of Aging Management Programs for Containments, Structures and Component Supports Evaluated in Chapters II and III of the GALL Report

ID	Type	Component	Aging Effect/Mechanism	Aging Management Programs	Further Evaluation Recommended	Rev2 Item	Rev1 Item
		indoor, uncontrolled				III.B4.TP-8 III.B5.TP-4 III.B5.TP-8	III.B1.3-8(TP-4) III.B1.3-4(TP-8) III.B1.3-5(TP-11) III.B1.3-7(TP-5) III.B2-9(TP-4) III.B2-4(TP-8) III.B2-8(TP-5) III.B2-5(TP-11) III.B3-6(TP-4) III.B3-2(TP-8) III.B3-5(TP-5) III.B3-3(TP-11) III.B4-9(TP-4) III.B4-4(TP-8) III.B4-8(TP-5) III.B4-5(TP-11) III.B5-6(TP-4) III.B5-2(TP-8) III.B5-5(TP-5) III.B5-3(TP-11)

Table 3.5-2   Aging Management Programs Recommended for Containments, Structures, and Component Supports

GALL Report Chapter/AMP	Program Name
Chapter XI.M2	Water Chemistry
Chapter XI.M10	Boric Acid Corrosion
Chapter XI.M18	Bolting Integrity
Chapter XI.S1	ASME Section XI, Subsection IWE Inservice Inspection (IWE)
Chapter XI.S2	ASME Section XI, Subsection IWL Inservice Inspection (IWL)
Chapter XI.S3	ASME Section XI, Subsection IWF Inservice inspection (IWF)
Chapter XI.S4	10 CFR Part 50, Appendix J
Chapter XI.S5	Masonry Walls
Chapter XI.S6	Structures Monitoring
Chapter XI.S7	R.G. 1.127, Inspection of Water-Control Structures Associated with Nuclear Power Plants
Chapter XI.S8	Protective Coating Monitoring and Maintenance
Appendix for GALL	Quality Assurance for Aging Management Programs
SRP-LR Appendix A	Plant-specific AMP

3.6 AGING MANAGEMENT OF ELECTRICAL AND INSTRUMENTATION AND CONTROLS

Review Responsibilities

Primary - Branches assigned responsibility by PM as described in SRP-LR Section 3.0 of this SRP-LR

3.6.1 Areas of Review

This section addresses the AMR and the associated AMP of the electrical and instrumentation and controls (I&C). For a recent vintage plant, the information related to the electrical and I&C is contained in Chapter 7, "Instrumentation and Controls," and Chapter 8, "Electric Power," of the plant's FSAR, consistent with the "Standard Review Plan for the Review of Safety Analysis Reports for Nuclear Power Plants" (NUREG-0800) (Ref. 1). For older plants, the location of applicable information is plant-specific because an older plant's FSAR may have predated NUREG-0800. Typical electrical and I&C components that are subject to an AMR for license renewal are electrical cables and connections, metal enclosed buses, fuse holders, high-voltage insulators, transmission conductors and connections, and switchyard bus and connections.

The responsible review organization is to review the following LRA AMR and AMP items assigned to it, per SRP-LR Section 3.0:

AMRs
- AMR results consistent with the GALL Report
- AMR results for which further evaluation is recommended by the GALL Report
- AMR results not consistent with or not addressed in the GALL Report

AMPs
- Consistent with GALL Report AMPs
- Plant-specific AMPs

FSAR Supplement
- The responsible review organization is to review the FSAR Supplement associated with each assigned AMP.

3.6.2 Acceptance Criteria

The acceptance criteria for the areas of review describe methods for determining whether the applicant has met the requirements of the NRC's regulations in 10 CFR 54.21.

3.6.2.1 AMR Results Consistent with the GALL Report

The AMRs and the AMPs applicable to the electrical and I&C components are described and evaluated in Chapter VI of NUREG-1801, "Generic Aging Lessons Learned (GALL) Report" (Ref. 2).

The applicant's LRA should provide sufficient information for the NRC reviewer to confirm that the specific LRA AMR item and the associated LRA AMP are consistent with the cited GALL

Report AMR item. The staff reviewer should then confirm that the LRA AMR item is consistent with the GALL Report AMR item to which it is compared.

When the applicant is crediting a different AMP than recommended in the GALL Report, the reviewer should confirm that the alternate AMP is valid to use for aging management and will be capable of managing the effects of aging as adequately as the AMP recommended by the GALL Report.

3.6.2.2 AMR Results for Which Further Evaluation is Recommended by the GALL Report

The basic acceptance criteria defined in Section 3.6.2.1 need to be applied first for all of the AMRs and AMPs reviewed as part of this section. In addition, if the GALL Report AMR item to which the LRA AMR item is compared identifies that "further evaluation is recommended," then additional criteria apply as identified by the GALL Report for each of the following aging effect/aging mechanism combinations. Refer to Table 3.6-1, comparing the "Further Evaluation Recommended" and the "Rev2 Item" columns, for the AMR items that reference the following subsections. The 2005 AMR item counterpart is provided in the "Rev1 Item" column.

3.6.2.2.1 Electrical Equipment Subject to Environmental Qualification

Environmental qualification is a TLAA as defined in 10 CFR 54.3. TLAAs are required to be evaluated in accordance with 10 CFR 54.21(c)(1). The evaluation of this TLAA is addressed separately in Section 4.4, "Environmental Qualification (EQ) of Electrical Equipment," of this SRP-LR.

3.6.2.2.2 Reduced Insulation Resistance due to Presence of Any Salt Deposits and Surface Contamination, and Loss of Material due to Mechanical Wear Caused by Wind Blowing on Transmission Conductors

Reduced insulation resistance due to presence of any salt deposits and surface contamination could occur in high-voltage insulators. The GALL Report recommends further evaluation of a plant-specific AMP for plants located such that the potential exists for salt deposits or surface contamination (e.g., in the vicinity of salt water bodies or industrial pollution). Loss of material due to mechanical wear caused by wind blowing on transmission conductors could occur in high-voltage insulators. The GALL Report recommends further evaluation of a plant-specific AMP to ensure that this aging effect is adequately managed. Acceptance criteria are described in Branch Technical Position RLSB-1 (Appendix A.1 of this SRP-LR).

3.6.2.2.3 Loss of Material due to Wind-Induced Abrasion, Loss of Conductor Strength due to Corrosion, and Increased Resistance of Connection due to Oxidation or Loss of Pre-load

Loss of material due to wind-induced abrasion, loss of conductor strength due to corrosion, and increased resistance of connection due to oxidation or loss of pre-load could occur in transmission conductors and connections, and in switchyard bus and connections. The GALL Report recommends further evaluation of a plant-specific AMP to ensure that this aging effect is adequately managed. Acceptance criteria are described in Branch Technical Position RLSB-1 (Appendix A.1 of this SRP-LR).

3.6.2.2.4   Quality Assurance for Aging Management of Nonsafety-Related Components

Acceptance criteria are described in Branch Technical Position IQMB-1 (Appendix A.2 of this SRP-LR).

3.6.2.3   AMR Results Not Consistent With or Not Addressed in the GALL Report

Acceptance criteria are described in Branch Technical Position RLSB-1 (Appendix A.1 of this SRP-LR).

3.6.2.4   Aging Management Programs

For those AMPs that will be used for aging management and that are based on the program elements of an AMP in the GALL Report, the NRC reviewer performs an audit of AMPs credited in the LRA to confirm consistency with the GALL AMPs identified in the GALL Report, Chapters X and XI.

If the applicant identifies an exception to any of the program elements of the cited GALL Report AMP, the LRA AMP should include a basis demonstrating how the criteria of 10 CFR 54.21(a)(3) would still be met. The NRC reviewer should then confirm that the LRA AMP, with all exceptions, would satisfy the criteria of 10 CFR 54.21(a)(3). If, while reviewing the LRA AMP, the reviewer identifies a difference between the LRA AMP and the GALL Report AMP that should have been identified as an exception to the GALL Report AMP, the difference should be reviewed and properly dispositioned. The reviewer should document the disposition of all LRA-defined exceptions and staff-identified differences.

The LRA should identify any enhancements that are needed to permit an existing AMP to be declared consistent with the GALL Report AMP to which the LRA AMP is compared. The reviewer is to confirm both that the enhancement, when implemented, would allow the existing plant AMP to be consistent with the GALL Report AMP and also that the applicant has a commitment in the FSAR supplement to implement the enhancement prior to the period of extended operation. The reviewer should review and document the disposition of all enhancements.

If the applicant chooses to use a plant-specific program that is not a GALL AMP, the NRC reviewer should confirm that the plant-specific program satisfies the criteria of Branch Technical Position RLSB-1 (Appendix A.1 of this SRP-LR).

3.6.2.5   FSAR Supplement

The summary description of the programs and activities for managing the effects of aging for the period of extended operation in the FSAR supplement should be sufficiently comprehensive, such that later changes can be controlled by 10 CFR 50.59. The description should contain information associated with the bases for determining that aging effects are managed during the period of extended operation. The description should also contain any future aging management activities, including enhancements and commitments, to be completed before entering the period of extended operation. Table 3.0-1 of this SRP-LR provides examples of the type of information to be included in the FSAR Supplement. Table 3.6-2 lists the programs that are applicable for this SRP-LR subsection.

3.6.3 Review Procedures

For each area of review, the following review procedures are to be followed:

3.6.3.1 AMR Results Consistent with the GALL Report

The applicant may reference the GALL Report in its LRA, as appropriate, and demonstrate that the AMRs and AMPs at its facility are consistent with those reviewed and approved in the GALL Report. The reviewer should not conduct a re-review of the substance of the matters described in the GALL Report. If the applicant has provided the information necessary to adopt the finding of program acceptability as described and evaluated in the GALL Report, the reviewer should find acceptable the applicant's reference to the GALL Report in its LRA. In making this determination, the reviewer confirms that the applicant has provided a brief description of the system, components, materials, and environment. The reviewer also confirms that the applicant has stated that the applicable aging effects, and that industry and plant-specific operating experience have been reviewed by the applicant and are evaluated in the GALL Report.

Furthermore, the reviewer should confirm that the applicant has addressed operating experience identified after the issuance of the GALL Report. Performance of this review includes confirming that the applicant has identified those aging effects for the electrical and I&C components that are contained in the GALL Report as applicable to its plant.

3.6.3.2 AMR Results for Which Further Evaluation is Recommended by the GALL Report

The basic review procedures defined in Section 3.6.3.1 need to be applied first for all of the AMRs and AMPs provided in this section. In addition, if the GALL AMR item to which the LRA AMR item is compared identifies that "further evaluation is recommended," then additional criteria apply as identified by the GALL Report for each of the following aging effect/aging mechanism combinations.

3.6.3.2.1 Electrical Equipment Subject to Environmental Qualification

Environmental qualification is a TLAA as defined in 10 CFR 54.3. TLAAs are required to be evaluated in accordance with 10 CFR 54.21(c)(1). The staff reviews the evaluation of this TLAA separately following the guidance in Section 4.4 of this SRP-LR.

3.6.3.2.2 Reduced Insulation Resistance due to Presence of Any Salt Deposits and Surface Contamination, and Loss of Material due to Mechanical Wear Caused by Wind Blowing on Transmission Conductors

The GALL Report recommends a plant-specific AMP for the management of reduced insulation resistance due to presence of any salt deposits and surface contamination for plants located such that the potential exists for salt deposits or surface contamination (e.g., in the vicinity of salt water bodies or industrial pollution), and loss of material due to mechanical wear caused by wind blowing on transmission conductors in high-voltage insulators. The reviewer reviews the applicant's proposed program on a case-by-case basis to ensure that an adequate program will be in place for the management of these aging effects.

3.6.3.2.3 Loss of Material due to Wind-Induced Abrasion, Loss of Conductor Strength due to Corrosion, and Increased Resistance of Connection due to Oxidation or Loss of Pre-load

The GALL Report recommends a plant-specific AMP for the management of loss of material due to wind-induced abrasion, loss of conductor strength due to corrosion, and increased resistance of connection due to oxidation or loss of pre-load in transmission conductors and connections, and in switchyard bus and connections. The reviewer reviews the applicant's proposed program on a case-by-case basis to ensure that an adequate program will be in place for the management of these aging effects.

3.6.3.2.4 Quality Assurance for Aging Management of Nonsafety-Related Components

The applicant's AMPs for license renewal should contain the elements of corrective actions, the confirmation process, and administrative controls. Safety-related components are covered by 10 CFR Part 50, Appendix B, which is adequate to address these program elements. However, Appendix B does not apply to nonsafety-related components that are subject to an AMR for license renewal. Nevertheless, the applicant has the option to expand the scope of its 10 CFR Part 50, Appendix B program to include these components and address these program elements. If the applicant chooses this option, the reviewer confirms that the applicant has documented such a commitment in the FSAR supplement. If the applicant chooses alternative means, the branch responsible for quality assurance should be requested to review the applicant's proposal on a case-by-case basis.

3.6.3.3 AMR Results Not Consistent With or Not Addressed in the GALL Report

The reviewer should confirm that the applicant, in the license renewal application, has identified applicable aging effects, listed the appropriate combination of materials and environments, and has credited AMPs that will adequately manage the aging effects. The AMP credited by the applicant could be an AMP that is described and evaluated in the GALL Report or in a plant-specific program. Review procedures are described in Branch Technical Position RLSB-1 (Appendix A.1 of this SRP-LR).

3.6.3.4 Aging Management Programs

The reviewer confirms that the applicant has identified the appropriate AMPs as described and evaluated in the GALL Report. If the applicant commits to an enhancement to make its LRA AMP consistent with a GALL Report AMP, then the reviewer is to confirm that this enhancement, when implemented, will make the LRA AMP consistent with the GALL Report AMP. If the applicant identifies, in the LRA AMP, an exception to any of the program elements of the GALL Report AMP, the reviewer is to confirm that the LRA AMP with the exception will satisfy the criteria of 10 CFR 54.21(a)(3). If the reviewer identifies a difference, not identified by the LRA, between the LRA AMP and the GALL Report AMP with which the LRA claims to be consistent, the reviewer should confirm that the LRA AMP with this difference satisfies 10 CFR 54.21(a)(3). The reviewer should document the basis for accepting enhancements, exceptions, or differences. The AMPs evaluated in the GALL Report pertinent to the electrical and I&C components are summarized in Table 3.6-1 of this SRP-LR. The "Rev 2 Item" (for 2010) and "Rev1 Item" (for 2005 counterpart) columns identify the AMR item numbers in the GALL Report, Chapters VI, presenting detailed information summarized by this row.

Table 3.6-1 of this SRP-LR may identify a plant-specific AMP. If the applicant chooses to use a plant-specific program that is not a GALL AMP, the NRC reviewer should confirm that the plant-specific program satisfies the criteria of Branch Technical Position RLSB-1 (Appendix A.1 of this SRP-LR).

3.6.3.5   FSAR Supplement

The reviewer confirms that the applicant has provided in its FSAR supplement information equivalent to that in Table 3.0-1 for aging management of the Electrical and I&C System. Table 3.6-2 lists the AMPs that are applicable for this SRP-LR subsection. The reviewer also confirms that the applicant has provided information for Subsection 3.6.3.3, "AMR Results Not Consistent With or Not Addressed in the GALL Report," equivalent to that in Table 3.0-1.

The staff expects to impose a license condition on any renewed license to require the applicant to update its FSAR to include this FSAR supplement at the next update required pursuant to 10 CFR 50.71(e)(4). As part of the license condition, until the FSAR update is complete, the applicant may make changes to the programs described in its FSAR supplement without prior NRC approval, provided that the applicant evaluates each such change pursuant to the criteria set forth in 10 CFR 50.59. If the applicant updates the FSAR to include the final FSAR supplement before the license is renewed, no condition will be necessary.

As noted in Table 3.0-1, an applicant need not incorporate the implementation schedule into its FSAR. However, the reviewer should confirm that the applicant has identified and committed in the license renewal application to any future aging management activities, including enhancements and commitments to be completed before the period of extended operation. The staff expects to impose a license condition on any renewed license to ensure that the applicant will complete these activities no later than the committed date.

3.6.4   Evaluation Findings

If the reviewer determines that the applicant has provided information sufficient to satisfy the provisions of this section, then an evaluation finding similar to the following text should be included in the staff's safety evaluation report:

> On the basis of its review, as discussed above, the staff concludes that the applicant has demonstrated that the aging effects associated with the electrical and instrumentation and controls components will be adequately managed so that the intended functions will be maintained consistent with the CLB for the period of extended operation, as required by 10 CFR 54.21(a)(3).
>
> The staff also reviewed the applicable FSAR Supplement program summaries and concludes that they adequately describe the AMPs credited for managing aging of electrical and instrumentation and controls, as required by 10 CFR 54.21(d).

3.6.5   Implementation

Except in those cases in which the applicant proposes an acceptable alternative method for complying with specified portions of the NRC's regulations, the method described herein will be used by the staff in its evaluation of conformance with NRC regulations.

3.6.6   References

1. NUREG-0800, "Standard Review Plan for the Review of Safety Analysis Reports for Nuclear Power Plants, LWR Edition" U.S. Nuclear Regulatory Commission, March 2007.

2. NUREG-1801, "Generic Aging Lessons Learned (GALL) Report," U.S. Nuclear Regulatory Commission, Revision 2, 2010.

3. NEI 95-10, "Industry Guideline for Implementing the Requirements of 10 CFR Part 54 – The License Renewal Rule," Nuclear Energy Institute, Revision 6.

Table 3.6-1 Summary of Aging Management Programs for the Electrical Components Evaluated in Chapter VI of the GALL Report

ID	Type	Component	Aging Effect/Mechanism	Aging Management Programs	Further Evaluation Recommended	Rev2 Item	Rev1 Item
1	BWR/PWR	Electrical equipment subject to 10 CFR 50.49 EQ requirements composed of Various polymeric and metallic materials exposed to Adverse localized environment caused by heat, radiation, oxygen, moisture, or voltage	Various aging effects due to various mechanisms in accordance with 10 CFR 50.49	EQ is a time-limited aging analysis (TLAA) to be evaluated for the period of extended operation. See the Standard Review Plan, Section 4.4, "Environmental Qualification (EQ) of Electrical Equipment," for acceptable methods for meeting the requirements of 10 CFR 54.21(c)(1)(i) and (ii). See Chapter X.E1, "Environmental Qualification (EQ) of Electric Components," of this report for meeting the requirements of 10 CFR 54.21(c)(1)(iii).	Yes, TLAA (See subsection 3.6.2.2.1)	VI.B.L-05	VI.B-1(L-05)
2	BWR/PWR	High-voltage insulators composed of Porcelain; malleable iron; aluminum; galvanized steel; cement exposed to Air – outdoor	Loss of material due to mechanical wear caused by wind blowing on transmission conductors	A plant-specific aging management program is to be evaluated	Yes, plant-specific (See subsection 3.6.2.2.2)	VI.A.LP-32	VI.A-10(LP-11)

Table 3.6-1 Summary of Aging Management Programs for the Electrical Components Evaluated in Chapter VI of the GALL Report

ID	Type	Component	Aging Effect/Mechanism	Aging Management Programs	Further Evaluation Recommended	Rev2 Item	Rev1 Item
3	BWR/PWR	High-voltage insulators composed of Porcelain; malleable iron; aluminum; galvanized steel; cement exposed to Air – outdoor	Reduced insulation resistance due to presence of salt deposits or surface contamination	A plant-specific aging management program is to be evaluated for plants located such that the potential exists for salt deposits or surface contamination (e.g., in the vicinity of salt water bodies or industrial pollution)	Yes, plant-specific (See subsection 3.6.2.2.2)	VI.A.LP-28	VI.A-9(LP-07)
4	BWR/PWR	Transmission conductors composed of Aluminum; steel exposed to Air – outdoor	Loss of conductor strength due to corrosion	A plant-specific aging management program is to be evaluated for ACSR	Yes, plant-specific (See subsection 3.6.2.2.3)	VI.A.LP-38	VI.A-16(LP-08)
5	BWR/PWR	Transmission connectors composed of Aluminum; steel exposed to Air – outdoor	Increased resistance of connection due to oxidation or loss of pre-load	A plant-specific aging management program is to be evaluated	Yes, plant-specific (See subsection 3.6.2.2.3)	VI.A.LP-48	VI.A-16(LP-08)
6	BWR/PWR	Switchyard bus and connections composed of Aluminum; copper; bronze; stainless steel; galvanized steel exposed to Air – outdoor	Loss of material due to wind-induced abrasion; Increased resistance of connection due to oxidation or loss of pre-load	A plant-specific aging management program is to be evaluated	Yes, plant-specific (See subsection 3.6.2.2.3)	VI.A.LP-39	VI.A-15(LP-09)
7	BWR/PWR	Transmission conductors composed of Aluminum; Steel exposed to Air – outdoor	Loss of material due to wind-induced abrasion	A plant-specific aging management program is to be evaluated for ACAR and ACSR	Yes, plant-specific (See subsection 3.6.2.2.3)	VI.A.LP-47	VI.A-16(LP-08)

Table 3.6-1 Summary of Aging Management Programs for the Electrical Components Evaluated in Chapter VI of the GALL Report

ID	Type	Component	Aging Effect/Mechanism	Aging Management Programs	Further Evaluation Recommended	Rev2 Item	Rev1 Item
8	BWR/PWR	Insulation material for electrical cables and connections (including terminal blocks, fuse holders, etc.) composed of Various organic polymers (e.g., EPR, SR, EPDM, XLPE) exposed to Adverse localized environment caused by heat, radiation, or moisture	Reduced insulation resistance due to thermal/thermoxidative degradation of organics, radiolysis, and photolysis (UV sensitive materials only) of organics; radiation-induced oxidation; moisture intrusion	Chapter XI.E1, "Insulation Material for Electrical Cables and Connections Not Subject to 10 CFR 50.49 Environmental Qualification Requirements"	No	VI.A.LP-33	VI.A-2(L-01)
9	BWR/PWR	Insulation material for electrical cables and connections used in instrumentation circuits that are sensitive to reduction in conductor insulation resistance (IR) composed of Various organic polymers (e.g., EPR, SR, EPDM, XLPE) exposed to Adverse localized environment caused by heat, radiation, or moisture	Reduced insulation resistance due to thermal/thermoxidative degradation of organics, radiolysis, and photolysis (UV sensitive materials only) of organics; radiation-induced oxidation; moisture intrusion	Chapter XI.E2, "Insulation Material for Electrical Cables and Connections Not Subject to 10 CFR 50.49 Environmental Qualification Requirements Used in Instrumentation Circuits"	No	VI.A.LP-34	VI.A-3(L-02)

Table 3.6-1 Summary of Aging Management Programs for the Electrical Components Evaluated in Chapter VI of the GALL Report

ID	Type	Component	Aging Effect/Mechanism	Aging Management Programs	Further Evaluation Recommended	Rev2 Item	Rev1 Item
10	BWR/PWR	Conductor insulation for inaccessible power cables greater than or equal to 400 volts (e.g., installed in conduit or direct buried) composed of Various organic polymers (e.g., EPR, SR, EPDM, XLPE) exposed to Adverse localized environment caused by significant moisture	Reduced insulation resistance due to moisture	Chapter XI.E3, "Inaccessible Power Cables Not Subject to 10 CFR 50.49 Environmental Qualification Requirements"	No	VI.A.LP-35	VI.A-4(L-03)
11	BWR/PWR	Metal enclosed bus: enclosure assemblies composed of Elastomers exposed to Air – indoor, controlled or uncontrolled or Air – outdoor	Surface cracking, crazing, scuffing, dimensional change (e.g. "ballooning" and "necking"), shrinkage, discoloration, hardening and loss of strength due to elastomer degradation	Chapter XI.E4, "Metal Enclosed Bus," or Chapter XI.M38, "Inspection of Internal Surfaces in Miscellaneous Piping and Ducting Components"	No	VI.A.LP-29	VI.A-12(LP-10)
12	BWR/PWR	Metal enclosed bus: bus/connections composed of Various metals used for electrical bus and connections exposed to Air – indoor, controlled or uncontrolled or Air – outdoor	Increased resistance of connection due to the loosening of bolts caused by thermal cycling and ohmic heating	Chapter XI.E4, "Metal Enclosed Bus"	No	VI.A.LP-25	VI.A-11(LP-04)

Table 3.6-1 Summary of Aging Management Programs for the Electrical Components Evaluated in Chapter VI of the GALL Report

ID	Type	Component	Aging Effect/Mechanism	Aging Management Programs	Further Evaluation Recommended	Rev2 Item	Rev1 Item
13	BWR/PWR	Metal enclosed bus: insulation; insulators composed of Porcelain; xenoy; thermo-plastic organic polymers exposed to Air – indoor, controlled or uncontrolled or Air – outdoor	Reduced insulation resistance due to thermal/thermoxidative degradation of organics/thermoplastics, radiation-induced oxidation, moisture/debris intrusion, and ohmic heating	Chapter XI.E4, "Metal Enclosed Bus"	No	VI.A.LP-26	VI.A-14(LP-05)
14	BWR/PWR	Metal enclosed bus: external surface of enclosure assemblies composed of Steel exposed to Air – indoor, uncontrolled or Air – outdoor	Loss of material due to general, pitting, and crevice corrosion	Chapter XI.E4, "Metal Enclosed Bus," or Chapter XI.S6, "Structures Monitoring"	No	VI.A.LP-43	VI.A-13(LP-06)
15	BWR/PWR	Metal enclosed bus: external surface of enclosure assemblies composed of Galvanized steel; aluminum exposed to Air – outdoor	Loss of material due to pitting and crevice corrosion	Chapter XI.E4, "Metal Enclosed Bus," or Chapter XI.S6, "Structures Monitoring"	No	VI.A.LP-42	VI.A-13(LP-06)

Table 3.6-1 Summary of Aging Management Programs for the Electrical Components Evaluated in Chapter VI of the GALL Report

ID	Type	Component	Aging Effect/Mechanism	Aging Management Programs	Further Evaluation Recommended	Rev2 Item	Rev1 Item
16	BWR/PWR	Fuse holders (not part of active equipment): metallic clamps composed of Various metals used for electrical connections exposed to Air – indoor, uncontrolled	Increased resistance of connection due to chemical contamination, corrosion, and oxidation (in an air, indoor controlled environment, increased resistance of connection due to chemical contamination, corrosion and oxidation do not apply); fatigue due to ohmic heating, thermal cycling, electrical transients	Chapter XI.E5, "Fuse Holders"	No	VI.A.LP-23	VI.A-8(LP-01)
17	BWR/PWR	Fuse holders (not part of active equipment): metallic clamps composed of Various metals used for electrical connections exposed to Air – indoor, controlled or uncontrolled	Increased resistance of connection due to fatigue caused by frequent manipulation or vibration	Chapter XI.E5, "Fuse Holders" No aging management program is required for those applicants who can demonstrate these fuse holders are located in an environment that does not subject them to environmental aging mechanisms or fatigue caused by frequent manipulation or vibration	No	VI.A.LP-31	VI.A-8(LP-01)

Table 3.6-1 Summary of Aging Management Programs for the Electrical Components Evaluated in Chapter VI of the GALL Report

ID	Type	Component	Aging Effect/Mechanism	Aging Management Programs	Further Evaluation Recommended	Rev2 Item	Rev1 Item
18	BWR/PWR	Cable connections (metallic parts) composed of Various metals used for electrical contacts exposed to Air – indoor, controlled or uncontrolled or Air – outdoor	Increased resistance of connection due to thermal cycling, ohmic heating, electrical transients, vibration, chemical contamination, corrosion, and oxidation	Chapter XI.E6, "Electrical Cable Connections Not Subject to 10 CFR 50.49 Environmental Qualification Requirements"	No	VI.A.LP-30	VI.A-1(LP-12)
19	PWR	Connector contacts for electrical connectors exposed to borated water leakage composed of Various metals used for electrical contacts exposed to Air with borated water leakage	Increased resistance of connection due to corrosion of connector contact surfaces caused by intrusion of borated water	Chapter XI.M10, "Boric Acid Corrosion"	No	VI.A.LP-36	VI.A-5(L-04)
20	BWR/PWR	Transmission conductors composed of Aluminum exposed to Air – outdoor	Loss of conductor strength due to corrosion	None – for Aluminum Conductor Aluminum Alloy Reinforced (ACAR)	None	VI.A.LP-46	VI.A-16(LP-08)

Table 3.6-1 Summary of Aging Management Programs for the Electrical Components Evaluated in Chapter VI of the GALL Report

ID	Type	Component	Aging Effect/Mechanism	Aging Management Programs	Further Evaluation Recommended	Rev2 Item	Rev1 Item
21	BWR/PWR	Fuse holders (not part of active equipment): insulation material, Metal enclosed bus: external surface of enclosure assemblies composed of Insulation material: bakelite; phenolic melamine or ceramic; molded polycarbonate; other, Galvanized steel; aluminum, Steel exposed to Air – indoor, controlled or uncontrolled	None	None	NA - No AEM or AMP	VI.A.LP-24 VI.A.LP-41 VI.A.LP-44	VI.A-7(LP-02) VI.A-13(LP-06) VI.A-13(LP-06)

Table 3.6-2   Aging Management Programs Recommended for Electrical and Instrumentation and Control Systems

GALL Report Chapter/AMP	Program Name
Chapter X.E1	Environmental Qualification (EQ) of Electric Components (TLAA)
Chapter XI.E1	Insulation Material for Electrical Cables and Connections Not Subject to 10 CFR 50.49 Environmental Qualification Requirements
Chapter XI.E2	Insulation Material for Electrical Cables and Connections Not Subject to 10 CFR 50.49 Environmental Qualification Requirements Used in Instrumentation Circuits
Chapter XI.E3	Inaccessible Power Cables Not Subject to 10 CFR 50.49 Environmental Qualification Requirements
Chapter XI.E4	Metal Enclosed Bus
Chapter XI.E5	Fuse Holders
Chapter XI.E6	Electrical Cable Connections Not Subject to 10 CFR 50.49 Environmental Qualification Requirements
Chapter XI.M10	Boric Acid Corrosion
Chapter XI.M38	Inspection of Internal Surfaces in Miscellaneous Piping and Ducting Components
Chapter XI.S6	Structures Monitoring
Appendix for GALL	Quality Assurance for Aging Management Programs
SRP-LR Appendix A	Plant-specific AMP

CHAPTER 4

TIME-LIMITED AGING ANALYSES

4.1 IDENTIFICATION OF TIME-LIMITED AGING ANALYSES

Review Responsibilities

Primary - Branch responsible for the TLAA issues

Secondary - Other branches responsible for engineering, as appropriate

4.1.1 Areas of Review

This review plan section addresses the identification of time-limited aging analyses (TLAAs). The technical review of TLAAs is addressed in Sections 4.2 through 4.7. As explained in more detail below, the list of TLAAs are certain plant-specific safety analyses that are based on an explicitly assumed 40-year plant life (for example, aspects of the reactor vessel design). Pursuant to 10 CFR 54.21(c)(1), a license renewal applicant is required to provide a list of TLAAs, as defined in 10 CFR 54.3. The area relating to the identification of TLAAs is reviewed.

TLAAs may have developed since issuance of a plant's operating license. As indicated in 10 CFR 54.30, the adequacy of the plant's CLB, which includes TLAAs, is not an area within the scope of the license renewal review. Any questions regarding the adequacy of the CLB are addressed under the backfit rule (10 CFR 50.109) and are separate from the license renewal process.

In addition, pursuant to 10 CFR 54.21(c)(2), an applicant must provide a list of plant-specific exemptions granted under 10 CFR 50.12 that are based on TLAAs. However, the initial license renewal applicants have found no such exemptions for their plants. It is an applicant's option to include more analyses than those required by 10 CFR 54.21(c)(1). The staff should focus its review to confirm that the applicant did not omit any TLAAs, as defined in 10 CFR 54.3.

Pursuant to 10 CFR 54.21(d), each application includes an FSAR supplement summary description for each TLAA that is identified in accordance with 10 CFR 54.3.

4.1.2 Acceptance Criteria

The acceptance criteria for the areas of review described in Subsection 4.1.1 of this review plan section delineate acceptable methods for meeting the requirements of the NRC's regulations in 10 CFR 54.21(c)(1). For the applicant's list of exemptions to be acceptable, the staff should have reasonable assurance that there has been no omission of TLAAs from that list.

Pursuant to 10 CFR 54.3, TLAAs are those licensee calculations and analyses that:

1. *Involve systems, structures, and components within the scope of license renewal, as delineated in 10 CFR 54.4(a);*

2. *Consider the effects of aging;*

3. *Involve time-limited assumptions defined by the current operating term, for example, 40 years;*

4. *Were determined to be relevant by the licensee in making a safety determination;*

5. *Involve conclusions or provide the basis for conclusions related to the capability of the*

system, structure, or component to perform its intended function(s), as delineated in 10 CFR 54.4(b); and

6. *Are contained or incorporated by reference in the CLB.*

The reviewer reviews the FSAR supplement for each TLAA identified as being within the scope of the LRA, as defined in 10 CFR 54.3.

4.1.3 Review Procedures

For each area of review described in Subsection 4.1.1, the reviewer adheres to the following review procedures:

The reviewer uses the plant UFSAR and other CLB documents, such as staff SERs, to perform the review. The reviewer selects analyses that the applicant did not identify as TLAAs that are likely to meet the six criteria identified in Subsection 4.1.2. The reviewer verifies that the selected analyses, not identified by the applicant as TLAAs, do not meet at least one of the following criteria (Ref. 1).

Sections 4.2 through 4.6 identify typical types of TLAAs for most plants. Information on the applicant's methodology for identifying TLAAs also may be useful in identifying calculations that did not meet the six criteria below.

1. *Involve systems, structures, and components within the scope of license renewal,* as delineated in 10 CFR 54.4(a). Chapter 2 of this SRP-LR provides the reviewer guidance on the scoping and screening methodology, and on plant-level and various system-level scoping results.

2. *Consider the effects of aging.* The effects of aging include but are not limited to loss of material, change in dimension, change in material properties, loss of toughness, loss of prestress, settlement, cracking, and loss of dielectric properties.

3. *Involve time-limited assumptions defined by the current operating term* (e.g., 40 years). The defined operating term should be explicit in the analysis. Simply asserting that a component is designed for a service life or plant life is not sufficient. The assertion is supported by calculations or other analyses that explicitly include a time limit.

4. *Were determined to be relevant by the licensee in making a safety determination.* Relevancy is a determination that the applicant makes based on a review of the information available. A calculation or analysis is relevant if it can be shown to have a direct bearing on the action taken as a result of the analysis performed. Analyses are also relevant if they provide the basis for a licensee's safety determination, and, in the absence of the analyses, the applicant might have reached a different safety conclusion.

5. *Show capability of the system, structure, or component to perform its intended function(s),* as delineated. Involve conclusions or provide the basis for conclusions related to 10 CFR 54.4(b). Analyses that do not affect the intended functions of systems, structures, or components are not TLAAs.

6. *Are contained or incorporated by reference in the CLB.* The CLB includes the technical specifications as well as design basis information (as defined in 10 CFR 50.2), or

licensee commitments documented in the plant-specific documents contained or incorporated by reference in the CLB, including but not limited to the FSAR, NRC SERs, the fire protection plan/hazards analyses, correspondence to and from the NRC, the quality assurance plan, and topical reports included as references to the FSAR. Calculations and analyses that are not in the CLB or not incorporated by reference in the CLB are not TLAAs. If a code of record is in the FSAR for particular groups of structures or components, reference material includes all calculations called for by that code of record for those structures and components.

TLAAs that need to be addressed are not necessarily those analyses that have been previously reviewed or approved by the NRC. The following examples illustrate TLAAs that need to be addressed that were not previously reviewed and approved by the NRC:

- The FSAR states that the design complies with a certain national code and standard. A review of the code and standard reveals that it calls for an analysis or calculation. Some of these calculations or analysis will be TLAAs. The actual calculation was performed by the applicant to meet the code and standard. The specific calculation was not referenced in the FSAR. The NRC had not reviewed the calculation. In response to a generic letter, a licensee submitted a letter to the NRC committing to perform a TLAA that would address the concern in the generic letter. The NRC had not documented a review of the applicant's response and had not reviewed the actual analysis.

The following examples illustrate analyses that are *not* TLAAs and need not be addressed under 10 CFR 54.21(c):

- Population projections (Section 2.1.3 of NUREG-0800) (Ref. 2).
- Cost-benefit analyses for plant modifications.
- Analysis with time-limited assumptions defined short of the current operating term of the plant, for example, an analysis for a component based on a service life that would not reach the end of the current operating term.

The number and type of TLAAs vary depending on the plant-specific CLB. All six criteria set forth in 10 CFR 54.3 (and repeated in Subsection 4.1.2) must be satisfied to conclude that a calculation or analysis is a TLAA. Table 4.1-1 provides examples of how these six criteria may be applied (Ref. 1). Table 4.1-2 provides a list of generic TLAAs that are included in the SRP-LR. Table 4.1-3 provides a list of other potential plant-specific TLAAs that have been identified by license renewal applicants. It is not expected that all applicants would identify all the analyses in these tables as TLAAs for their plants. Also, an applicant may perform specific TLAAs for its plant that are not shown in these tables.

As appropriate, staff members from other branches of NRR review the application in their assigned areas without examining the identification of TLAAs. However, they may come across situations in which they may question why the applicant did not identify certain analyses as TLAAs. The reviewer coordinates the resolution of any such questions with these other staff members to determine whether these analyses should be evaluated as TLAAs.

In order to determine whether there is reasonable assurance that the applicant has identified the TLAAs for its plant, the reviewer should find that the analyses omitted from the applicant's list are not TLAAs. Should an applicant identify a TLAA that is also a basis for a plant-specific exemption granted pursuant to 10 CFR 50.12 and the exemption is in effect, the reviewer verifies that the applicant also has identified that exemption pursuant to 10 CFR 54.21(c)(2). However, the initial license renewal applicants have found no such exemptions for their plants.

4.1.4   Evaluation Findings

The reviewer determines whether the applicant has provided sufficient information to satisfy the provisions of this section, and whether the staff's evaluation supports conclusions of the following type, to be included in the staff's safety evaluation report:

> On the basis of its review, as discussed above, the staff concludes that the applicant has provided an acceptable list of TLAAs as defined in 10 CFR 54.3, and that no 10 CFR 50.12 exemptions have been granted on the basis of a TLAA, as defined in 10 CFR 54.3.

4.1.5   Implementation

Except in those cases in which the applicant proposes an acceptable alternative method, the method described herein are used by the staff to evaluate conformance with NRC regulations.

4.1.6   REFERENCES

1. NEI 95-10, "Industry Guideline for Implementing the Requirements of 10 CFR Part 54 – The License Renewal Rule," Nuclear Energy Institute, Revision 6.

2. NUREG-0800, "Standard Review Plan for the Review of Safety Analysis Reports Nuclear Power Plants," U.S. Nuclear Regulatory Commission, March 2007.

Table 4.1-1   Sample Process for Identifying Potential Time-Limited Aging Analyses and Basis for Disposition

Example	Disposition
NRC correspondence requests a utility to justify that unacceptable cumulative wear did not occur during the design life of control rods.	Does not qualify as a TLAA because the design life of control rods is less than 40 years. Therefore, does not meet criterion (3) of the TLAA definition in 10 CFR 54.3.
Maximum wind speed of 100 mph is expected to occur once per 50 years.	Not a TLAA because it does not involve an aging effect.
Correspondence from the utility to the NRC states that the membrane on the containment basemat is certified by the vendor to last for 40 years.	The membrane was not credited in any safety evaluation, and therefore the analysis is not considered a TLAA. This example does not meet criterion (4) of the TLAA definition in 10 CFR 54.3.
Fatigue usage factor for the pressurizer surge line was determined not to be an issue for the current license period in response to NRC Bulletin 88-11.	This example is a TLAA because it meets all 6 criteria in the definition of TLAA in 10 CFR 54.3. The utility's fatigue design basis relies on assumptions defined by the 40-year operating life for this component, which is the current operating term.
Containment tendon lift-off forces are calculated for the 40-year life of the plant. These data are used during Technical Specification surveillance for comparing measured to predicted lift-off forces.	This example is a TLAA because it meets all 6 criteria of the TLAA definition in 10 CFR 54.3. The lift-off force curves are currently limited to 40-year values, and are needed to perform a required Technical Specification surveillance.

Table 4.1-2   Generic Time-Limited Aging Analyses

Reactor vessel neutron embrittlement (Subsection 4.2)
Metal fatigue (Subsection 4.3)
Environmental qualification of electrical equipment (Subsection 4.4)
Concrete containment tendon prestress (Subsection 4.5)
Inservice local metal containment corrosion analyses (Subsection 4.6)

Table 4.1-3   Examples of Potential Plant-Specific TLAAs

Intergranular separation in the heat-affected zone (HAZ) of reactor vessel low-alloy steel under austenitic SS cladding
Low-temperature overpressure protection (LTOP) analyses
Fatigue analysis for the main steam supply lines to the turbine-driven auxiliary feedwater pumps
Fatigue analysis of the reactor coolant pump flywheel
Fatigue analysis of polar crane
Flow-induced vibration endurance limit for the reactor vessel internals
Transient cycle count assumptions for the reactor vessel internals
Ductility reduction of fracture toughness for the reactor vessel internals
Leak before break
Fatigue analysis for the containment liner plate
Containment penetration pressurization cycles
Metal corrosion allowance
High-energy line-break postulation based on fatigue cumulative usage factor
Inservice flaw growth analyses that demonstrate structure stability for 40 years

4.2 REACTOR VESSEL NEUTRON EMBRITTLEMENT ANALYSIS

Review Responsibilities

Primary - Branch responsible for the TLAA issues

Secondary - Branch responsible for reactor systems

4.2.1 Areas of Review

During plant service, neutron irradiation reduces the fracture toughness of ferritic steel in the reactor vessel beltline region of light-water nuclear power reactors. Areas of review to ensure that the reactor vessel has adequate fracture toughness to prevent brittle failure during normal and off-normal operating conditions are (a) upper-shelf energy, (b) pressurized thermal shock (PTS) for pressurized water reactors (PWRs), (c) heat-up and cool-down (pressure-temperature limits) curves, (d) Boiling Water Reactor Vessel and Internals Project (BWRVIP)-05 analysis for elimination of circumferential weld inspection and analysis of the axial welds, and (e) other plant-specific TLAAs on reactor vessel neutron embrittlement.

The adequacy of the analyses for these five areas is reviewed for the period of extended operation.

The branch responsible for reactor systems reviews neutron fluence and dosimetry information in the application.

4.2.2 Acceptance Criteria

The acceptance criteria for the areas of review described in Subsection 4.2.1 of this review plan section delineate acceptable methods for meeting the requirements of the U.S. Nuclear Regulatory Commission's (NRC's) regulation in 10 CFR 54.21(c)(1).

4.2.2.1 Time-Limited Aging Analysis

Pursuant to 10 CFR 54.21(c)(1)(i) - (iii), an applicant must demonstrate one of the following:

(i) The analyses remain valid for the period of extended operation;

(ii) The analyses have been projected to the end of the extended period of operation; or

(iii) The effects of aging on the intended function(s) will be adequately managed for the period of extended operation.

For the first three areas of review for the analysis of reactor vessel neutron embrittlement, the specific acceptance criteria depend on the applicant's choice of 10 CFR 54.21(c)(1)(i), (ii), or (iii).

4.2.2.1.1 Upper-Shelf Energy (USE)

10 CFR Part 50 Appendix G (Ref. 1) paragraph IV.A.1 requires that the reactor vessel beltline materials have a Charpy upper-shelf energy of no less than 68 J (50 ft-lb) throughout the life of the reactor vessel, unless otherwise approved by the NRC. An applicant may take any one of the following three approaches.

4.2.2.1.1.1   10 CFR 54.21(c)(1)(i)

The reactor vessel components evaluated in the existing upper-shelf energy analysis or NRC-approved equivalent margins analysis (EMA) are re-evaluated to demonstrate that the existing analysis remains valid during the period of extended operation because the neutron fluence projected to the end of the period of extended operation is bound by the fluence assumed in the existing analysis.

4.2.2.1.1.2   10 CFR 54.21(c)(1)(ii)

The reactor vessel components evaluated in the existing upper-shelf energy analysis or NRC-approved EMA are re-evaluated to consider the period of extended operation in accordance with 10 CFR Part 50, Appendix G.

10 CFR Part 50, Appendix G, Section IV.A.1 (the rule) requires applicants to take further corrective actions for those cases where the 75 ft-lbs (102 joules) unirradiated USE (UUSE) criterion or 50 ft-lbs (68 joules) end-of-life (EOL) USE criterion cannot be met (i.e., when the respective UUSE value falls below 75 ft-lbs or the EOL USE falls below 50 ft-lbs). When this occurs, the rule requires a licensee to submit a supplemental analysis for NRC approval for any case where the UUSE value is less than 75 ft-lbs (102 joules) or where the projected EOL USE value for a given material is projected to be less than the 50 ft-lbs (68 joules) acceptance criteria at the expiration of the operating license. Thus, if the USE value for a PWR reactor vessel (RV) material, as projected to the expiration of the period of extended operation, falls below either the 50 ft-lbs (68 joules) acceptance criterion or the USE value criterion specified in a previously NRC-approved EMA, or where the percent-drop in USE value for a BWR RV material, as projected to the expiration of the period of extended operation, falls below that percent-drop in USE value approved by the NRC in its safety evaluation of the BWRVIP's generic EMA for BWRs, an applicant will need to submit a plant-specific engineering analysis (usually an EMA) for NRC approval as supplemental information for license renewal. Otherwise, failure to meet the USE requirements of 10 CFR Part 50, Appendix G for the RV materials as evaluated using the neutron fluence that are projected for the period of extended operation mandates imposition of additional commitments or license condition on USE for the license renewal application.

4.2.2.1.1.3   10 CFR 54.21(c)(1)(iii)

Acceptance criteria under 10 CFR 54.21(c)(1)(iii) have yet to be developed. They will be evaluated on a case-by-case basis to ensure that the aging effects will be managed such that the intended function(s) will be maintained during the period of extended operation.

4.2.2.1.2   Pressurized Thermal Shock (for PWRs)

For PWRs, 10 CFR 50.61 (Ref. 2) requires that the "reference temperature" for reactor vessel beltline materials evaluated at EOL fluence, RT_{PTS}, be less than the "PTS screening criteria" at the expiration date of the operating license, unless otherwise approved by the NRC. The "PTS screening criteria" are 132°C (270°F) for plates, forgings, and axial weld materials, and 149°C (300°F) for circumferential weld materials. The regulations require updating of the PTS assessment upon a request for a change in the expiration date of a facility's operating license, or change of the projected material neutron fluence or change in the material properties in any of the reactor vessel beltline materials. Therefore, the RT_{PTS} value must be calculated for the entire life of the facility, including the period of extended operation. The PTS TLAA may be handled as follows.

4.2.2.1.2.1   10 CFR 54.21(c)(1)(i)

The existing PTS analysis remains valid during the period of extended operation because the neutron fluence projected to the end of the period of extended operation is bound by the fluence assumed in the existing analysis.

4.2.2.1.2.2   10 CFR 54.21(c)(1)(ii)

The PTS analysis is re-evaluated to consider the period of extended operation in accordance with 10 CFR 50.61. An analysis may be performed in accordance with NRC Regulatory Guide (RG) 1.154 (Ref. 3) or 10 CFR 50.61a (Ref. 16) if the PTS screening criteria in 10 CFR 50.61 are projected to be exceeded during the period of extended operation.

4.2.2.1.2.3   10 CFR 54.21(c)(1)(iii)

The staff position for license renewal on this option is described in a May 27, 2004 letter from L.A. Reyes (EDO) to the Commission (Ref. 4), which states that if the applicant does not extend the TLAA, the applicant provides an assessment of the current licensing basis TLAA for PTS, a discussion of the flux reduction program implemented in accordance with 10 CFR 50.61(b)(3), if necessary, and an identification of the viable options that exist for managing the aging effect in the future.

4.2.2.1.3   Pressure-Temperature (P-T) Limits

10 CFR Part 50, Appendix G (Ref. 1) requires that the reactor pressure vessel (RPV) be maintained within established pressure-temperature (P-T) limits, including during any condition of normal operation. This includes heatup and cooldown. These limits specify the maximum allowable pressure as a function of reactor coolant temperature. As the reactor pressure vessel becomes embrittled and its fracture toughness is reduced, the allowable pressure (given the required minimum temperature) is reduced.

P-T limits are TLAAs for the application if the plant currently has P-T limit curves approved for the expiration of the current period of operation (i.e., 32 EFPY or other licensed EFPY values at expiration of the current license). However, the P-T limits for the period of extended operation need not be submitted as part of the LRA since the P-T limits need to be updated through the 10 CFR 50.90 licensing process when necessary for P-T limits that are located in the limiting conditions of operation (LCOs) of the Technical Specifications (TS). For those plants that have approved pressure-temperature limit reports (PTLRs), the P-T limits for the period of extended operation will be updated at the appropriate time through the plant's Administrative Section of the TS and the plant's PTLR process. In either case, the 10 CFR 50.90 or the PTLR processes, which constitute the current licensing basis, will ensure that the P-T limits for the period of extended operation will be updated prior to expiration of the P-T limit curves for the current period of operation.

P-T limits may be handled as follows.

4.2.2.1.3.1   10 CFR 54.21(c)(1)(i)

The existing P-T limits are valid during the period of extended operation because the neutron fluence projected to the end of the period of extended operation is bound by the fluence assumed in the existing analysis.

4.2.2.1.3.2   10 CFR 54.21(c)(1)(ii)

The P-T limits are reevaluated to consider the period of extended operation in accordance with 10 CFR Part 50, Appendix G (Ref. 1).

4.2.2.1.3.3   10 CFR 54.21(c)(1)(iii)

Updated P-T limits for the period of extended operation must be available prior to entering the period of extended operation. The 10 CFR 50.90 process for P-T limits located in the LCOs or the Administrative Controls Process for P-T limits that are administratively amended through a PTLR process can be considered adequate aging management programs within the scope of 10 CFR 54.21(c)(1)(iii), such that P-T limits will be maintained through the period of extended operation.

4.2.2.1.4   Elimination of Circumferential Weld Inspection (for BWRs)

Some BWRs have an approved technical alternative, which eliminates the reactor vessel circumferential shell weld inspections for the current license term because they satisfy the limiting conditional failure probability for the circumferential welds at the expiration of the current license, based on BWRVIP-05 and the extent of neutron embrittlement (Refs. 5-7). These assessments are performed through the 10 CFR 50.55a process. If the applicant indicates that relief from circumferential weld examination will be made under 10 CFR 50.55a(a)(3), the applicant will manage this TLAA in accordance with 10 CFR 54.21(c)(1)(iii).

4.2.2.1.4.1   10 CFR 54.21(c)(1)(iii)

An applicant for renewal of a license should address this issue by noting that it will be handled through a re-application under 10 CFR 50.55a(a)(3). An applicant for a license renewal to operate, such a BWR may provide justification to extend this relief into the period of extended operation in accordance with BWRVIP-74-A (Ref. 8), which is the revised and NRC-approved version of BWRVIP-74 (Ref. 9). The staff's review of BWRVIP-74 (Ref. 9) is contained in an October 18, 2001 letter to C. Terry, BWRVIP Chairman (Ref. 10). Appendix E of the staff's final safety evaluation report (FSER) (Ref. 10) conservatively evaluated BWR RPV's to have 64 effective full-power years (EFPY), which is 10 EFPY greater than the maximum of what is realistically expected for the end of the license renewal period. Since this is a generic analysis, a licensee relying on BWRVIP-74-A should provide plant-specific information to demonstrate that at the end of the renewal period, the circumferential beltline weld materials meet the limiting conditional failure probability for circumferential welds specified in Appendix E of the FSER (Ref. 10) and that operator training and procedures are utilized during the license renewal term to limit the frequency for cold over-pressure events to the amount specified in the NRC FSER (Ref. 10).

4.2.2.1.5   Axial Welds (for BWRs)

The staff's SER contained in a letter to Carl Terry dated March 7, 2000, "Supplement to Final Safety Evaluation of the BWR Vessel and Internals Project BWRVIP-05 Report" (Ref. 11), discussed the staff's concern related to RPV failure frequency for axial welds and the BWRVIP's analysis of the RPV failure frequency of axial welds. These discussions are also presented in the staff's FSER of BWRVIP-74 (Ref. 10). The SER indicates that the RPV failure frequency due to failure of the limiting axial welds in the BWR fleet at the end of 40 years of operation is less than 5×10^{-6} per reactor year, given the assumptions on flaw density, distribution, and location described in the SER. Since the BWRVIP analysis was generic, a licensee relying on BWRVIP-74-A should monitor axial beltline weld embrittlement. The applicant may provide plant-specific information to demonstrate that the axial beltline weld materials at the extended period of operation meet the criteria specified in the report or have a program to monitor axial weld embrittlement relative to the values specified by the staff in its March 7, 2000 (Ref. 11) letter.

4.2.2.2   FSAR Supplement

The specific criterion for meeting 10 CFR 54.21(d) is:

> The summary description of the evaluation of TLAAs for the period of extended operation in the FSAR supplement is appropriate, such that later changes can be controlled by 10 CFR 50.59. The description contains information associated with the TLAAs regarding the basis for determining that the applicant has made the demonstration required by 10 CFR 54.21(c)(1).

4.2.3   Review Procedures

For each area of review described in Subsection 4.2.1, the following review procedures should be followed.

4.2.3.1   Time-Limited Aging Analysis

For the first three areas of review for the analysis of reactor vessel neutron embrittlement, the review procedures depend on the applicant's choice of 10 CFR 54.21(c)(1)(i), (ii), or (iii). For each area, the applicant's three options under section 54.21(c)(1) are discussed in turn, as follows.

4.2.3.1.1   Upper-Shelf Energy

4.2.3.1.1.1   10 CFR 54.21(c)(1)(i)

The projected ¼T neutron fluence at the end of the period of extended operation is reviewed to verify that it is bound by the fluence assumed in the existing upper-shelf energy analysis.

Neutron Fluence: The applicant identifies (a) the neutron fluence at the ¼T location for each beltline material at the expiration of the license renewal period, (b) the staff-approved methodology used to determine the neutron fluence or submits the methodology for staff review, and (c) whether the methodology follows the guidance in NRC RG 1.190 (Ref. 15).

4.2.3.1.1.2   10 CFR 54.21(c)(1)(ii)

The documented results of the revised upper-shelf energy analysis based on the projected neutron fluence at the end of the period of extended operation are reviewed for compliance with 10 CFR Part 50, Appendix G. The applicant may use NRC RG 1.99 Rev. 2 (Ref. 12) to project upper-shelf energy to the end of the period of extended operation. The applicant also may use ASME Code Section XI Appendix K (Ref. 13) for the purpose of performing an equivalent margins analysis to demonstrate that adequate protection for ductile failure is maintained to the end of the period of extended operation. The staff reviews the applicant's methodology for this evaluation. Branch Position MTEB 5-3, "Fracture Toughness Requirements," in Standard

Review Plan (Ref. 14), Section 5.3.2, "Pressure Temperature Limits, Upper-Shelf Energy, and Pressurized Thermal Shock," provides additional NRC positions on estimations of USE values for reactor vessel beltline materials.

The staff confirms that the applicant has provided sufficient information for all Upper Shelf Energy (USE) and/or equivalent margins analysis calculations for the period of extended operation as follows:

Neutron Fluence: The applicant identifies (a) the neutron fluence at the ¼T location for each beltline material at the expiration of the license renewal period, (b) the staff-approved methodology used to determine the neutron fluence or submits the methodology for staff review, and (c) whether the methodology follows the guidance in NRC RG 1.190 (Ref. 15).

To confirm that the USE analysis meets the requirements of Appendix G of 10 CFR Part 50 at the end of the license renewal period, the staff determines whether:

1. For each beltline material, the applicant provides the unirradiated Charpy USE, and the projected Charpy USE at the end of the license renewal period, and whether the drop in Charpy USE was determined using the limit lines in Figure 2 of NRC RG 1.99, Revision 2, or from surveillance data and the percentage copper.

2. If an equivalent margins analysis is used to demonstrate compliance with the USE requirements in Appendix G of 10 CFR Part 50, the applicant provides the analysis or identifies an approved topical report that contains the analysis. Information the staff considers to assess the equivalent margins analysis includes the unirradiated USE (if available) for the limiting material, its copper content, the fluence (¼T and at 1-inch depth), the EOLE USE (if available), the operating temperature in the downcomer at full power, the vessel radius, the vessel wall thickness, the J-applied analysis for Service Level C and D, the vessel accumulation pressure, and the vessel bounding heatup/cooldown rate during normal operation.

For Boiling Water Reactors, the staff confirms that the beltline materials are evaluated in accordance with Renewal Applicant Action Item 10 in the staff's SER for BWRVIP-74 (Letter to C. Terry dated October 18, 2001) (Ref.10). Action Item 10: To demonstrate that the beltline materials meet the Charpy USE criteria specified in Appendix B of BWRVIP-74-A or the NRC FSER (Ref. 10), the applicant demonstrates that the percent reduction in Charpy USE for their beltline materials is less than that specified for the limiting BWR/3-6 plates and the non-Linde 80 submerged arc welds and that the percent reduction in Charpy USE for their surveillance weld and plate are less than or equal to the values projected using the methodology in NRC RG 1.99, Revision 2.

The applicant identifies whether there are two or more surveillance material samples available that are relevant to the RPV beltline materials. If there are two or more data points for a surveillance material, the applicant provides analyses of the data to determine whether the data are consistent with the NRC RG 1.99, Revision 2 methodology that was utilized in the BWRVIP-74-A analyses.

4.2.3.1.1.3   10 CFR 54.21(c)(1)(iii)

The applicant's proposal to demonstrate that the effects of aging on the intended function(s) will be adequately managed for the period of extended operation is reviewed on a case-by-case basis.

4.2.3.1.2   Pressurized Thermal Shock (for PWRs)

4.2.3.1.2.1   10 CFR 54.21(c)(1)(i)

The projected clad-to-base metal interface neutron fluence at the end of the period of extended operation is reviewed to verify that it is bound by the fluence assumed in the existing PTS analysis.

Neutron Fluence: The applicant identifies the neutron fluence at the clad-to-base metal interface for each beltline material at the expiration of the license renewal period. The applicant identifies the staff-approved methodology used in determining the neutron fluence or submits the methodology for staff review and identifies whether the methodology followed the guidance in NRC RG 1.190 (Ref. 15).

4.2.3.1.2.2   10 CFR 54.21(c)(1)(ii)

The documented results of the revised PTS analysis based on the projected neutron fluence at the end of the period of extended operation are reviewed for compliance with 10 CFR 50.61.

The staff confirms that the applicant has provided sufficient information for Pressurized Thermal Shock for the period of extended operation as follows:

Neutron Fluence: Identified the neutron fluence at the clad-to-base metal interface for each beltline material at the expiration of the license renewal period. Identified the staff-approved methodology used in determining the neutron fluence or submit the methodology for staff review and identified whether the methodology followed the guidance in NRC RG 1.190 (Ref. 15).

There are two methodologies from 10 CFR 50.61 that can be used in the PTS analysis, based on the projected neutron fluence at the end of the period of extended operation. RT_{NDT} is the reference temperature (NDT means nil-ductility temperature) used as an indexing parameter to determine the fracture toughness and the amount of embrittlement of a material. RT_{PTS} is the reference temperature used in the PTS analysis and is related to RT_{NDT} at the end of the facility's operating license.

The first methodology does not rely on plant-specific surveillance data to calculate delta RT_{NDT} (i.e., the mean value of the adjustment or shift in reference temperature caused by irradiation). The delta RT_{NDT} is determined by multiplying a chemistry factor from the tables in 10 CFR 50.61 by a fluence factor calculated from the neutron flux using an equation.

The second methodology relies on plant-specific surveillance data to determine the delta RT_{NDT}. In this methodology, two or more sets of surveillance data are needed. A surveillance datum consists of a measured delta RT_{NDT} for corresponding neutron fluence. 10 CFR 50.61 specifies a procedure and a criterion for determining whether the surveillance data are credible. For the surveillance data to be defined as credible, the difference in the predicted value and the measured value for delta RT_{NDT} must be less than 28°F for weld metal. When a credible surveillance data set exists, the chemistry factor can be determined from these data in lieu of a value from the table in 10 CFR 50.61. Then the standard deviation of the increase in the RT_{NDT} can be reduced from 28°F to 14°F for welds.

To confirm that the Pressurized Thermal Shock analysis results in RT_{PTS} values below the screening criteria in 10 CFR 50.61 at the end of the license renewal period, the applicant provides the following:

1. For each beltline material, provide the unirradiated RT_{NDT}, the method of calculating the unirradiated RT_{NDT} (either generic or plant-specific), the margin, the chemistry factor, the method of calculating the chemistry factor, the mean value for the shift in transition temperature, and the RT_{PTS} value.

2. If there are two or more data for a surveillance material that is from the same heat of material as the beltline material, provide analyses to determine whether the data are credible in accordance with NRC RG 1.99, Revision 2 and whether the margin value used in the analysis is appropriate.

3. If a surveillance program does not include the vessel beltline controlling material but two or more data seta are available from other beltline materials, then provide an analysis of the data in accordance with Regulatory Guide 1.99, Revision 2, Regulatory Position C.2.1, to show that the results either bound or are comparable to the values that would be calculated for the same materials using Regulatory Position C.1.1.

If the PTS screening criteria in 10 CFR 50.61 are projected to be exceeded during the period of extended operation, an analysis based on NRC RG 1.154 (Ref. 3) or 10 CFR 50.61a may be submitted for review.

4.2.3.1.2.3   10 CFR 54.21(c)(1)(iii)

The applicant's proposal to demonstrate that the effects of aging on the intended function(s) will be adequately managed for the period of extended operation will be reviewed on a case-by-case basis.

The license renewal application provides an assessment of the current licensing basis TLAA for PTS, a discussion of the flux reduction program implemented in accordance with §50.61(b)(3), if necessary, and an identification of the viable options that exist for managing the aging effect in the future.

A. The applicant explains its core management plans (e.g., operation with a low leakage core design and/or integral burnable neutron absorbers) from now through the end of the period of extended operation. Based on this core management strategy, the applicant:

   (1) Identifies the material in the RPV which has limiting RT_{PTS} value,
   (2) Provides the projected fluence value for the limiting material at end of license extended (EOLE),
   (3) Provides the projected RT_{PTS} value for the limiting material at EOLE, and
   (4) Provides the projected date and fluence values at which the limiting material will exceed the screening criteria in §50.61.

B. The applicant discusses aging management programs that it intends to implement which actively "manage" the condition of the facility's RPV and hence, the risk associated with PTS. This discussion is expected to address, at least, the facility's reactor pressure vessel material surveillance program.

C. The applicant briefly discusses the options that it is considering with respect to "resolving" the PTS issue through EOLE. It is anticipated that this discussion includes some or all of the following:

   (1) Plant modifications (e.g., heating of ECCS injection water) which could limit the risk associated with postulated PTS events [see §50.61(b)(4) and/or (b)(6)],
   (2) More detailed safety analyses (e.g., using Regulatory Guide 1.154) which may be performed to show that the PTS risk for the facility is acceptably low through EOLE [see §50.61(b)(4)],
   (3) More advanced material property evaluation (e.g., use of Master Curve technology) to demonstrate greater fracture resistance for the limiting material [applies to §50.61(b)(4)],
   (4) The potential for RPV thermal annealing in accordance with §50.66 [see §50.61(b)(7)], and/or
   (5) Use of the alternative PTS Rule (Ref. 16).

4.2.3.1.3   Pressure-Temperature (P-T) Limits

4.2.3.1.3.1   10 CFR 54.21(c)(1)(i)

The projected neutron fluence for the ¼T and ¾T locations at the end of the period of extended operation are bounded by the neutron fluences used to develop the existing P-T limit analysis.

Neutron Fluence: The applicant identifies (a) the neutron fluence at the ¼T and ¾T locations for each beltline material at the expiration of the license renewal period, (b) the staff-approved methodology used to determine the neutron fluence or submits the methodology for staff review, and (c) whether the methodology follows the guidance in NRC RG 1.190 (Ref. 15).

4.2.3.1.3.2   10 CFR 54.21(c)(1)(ii)

The documented results of the revised P-T limit analysis based on the projected reduction in fracture toughness at the end of the period of extended operation is reviewed for compliance with 10 CFR Part 50, Appendix G. The P-T limit evaluations are dependent upon the neutron fluence.

Neutron Fluence: The applicant identifies (a) the neutron fluence at the ¼T and ¾T locations for each beltline material at the expiration of the license renewal period, (b) the staff-approved methodology used to determine the neutron fluence or submits the methodology for staff review, and (c) whether the methodology follows the guidance in NRC RG 1.190 (Ref. 15).

4.2.3.1.3.3   10 CFR 54.21(c)(1)(iii)

Updated P-T limits for the period of extended operation must be available prior to entering the period of extended operation. The 10 CFR 50.90 process for P-T limits located in the LCOs or the Administrative Controls Process for P-T limits that are administratively amended through a PTLR process can be considered adequate aging management programs within the scope of

10 CFR 54.21(c)(1)(iii), such that P-T limits will be maintained through the period of extended operation.

For Boiling Water Reactors, the staff confirms that the applicant addresses the following Renewal Applicant Action Item in the staff's SER for BWRVIP-74 (Letter to C. Terry dated October 18, 2001) (Ref.10).

> Action Item 9: Appendix A of the BWRVIP-74-A Report indicates that a set of P-T curves should be developed for the heat-up and cool-down operating conditions in the plant at a given EFPY in the license renewal period.

This means that, for this action item, the applicant has not provided updated curves, but shall have a procedure for updating P-T limits in accordance with 10 CFR Part 50, Appendix G, that will cover 60 years.

4.2.3.1.4   Elimination of Circumferential Weld Inspection (for BWRs)

The staff verifies that the applicant has identified that, should the inspection relief be desired for the period of extended operation, an application will be made under 10 CFR 50.55a(a)(3) prior to entering the period of extended operation. If the applicant indicates that relief from circumferential weld examination will be made under 10 CFR 50.55a(a)(3), the applicant will manage this TLAA in accordance with 10 CFR 54.21(c)(1)(iii).

4.2.3.1.5   Axial Welds (for BWRs)

To demonstrate that the vessel has not been embrittled beyond the basis for the staff and BWRVIP analyses, the applicant should provide (a) a comparison of the neutron fluence, initial RT_{NDT}, chemistry factor amounts of copper and nickel, delta RT_{NDT}, and mean RT_{NDT} of the limiting axial weld at the end of the license renewal period to the reference case in the BWRVIP and staff analyses and (b) an estimate of conditional failure probability of the RPV at the end of the license renewal term based on the comparison of the mean RT_{NDT} for the limiting axial welds and the reference case. If this comparison does not indicate that the RPV failure frequency for axial welds is less than 5×10^{-6} per reactor year, the applicant should provide a probabilistic analysis to determine the RPV failure frequency for axial welds. Consistent with the staff's supplemental safety evaluation report (SER) of BWR Vessel and Internals Project BWRVIP-05 Report, dated May 7, 2000 (Ref. 11), the staff should ensure that the applicant's plant is bounded by the BWRVIP-05 analysis or that the applicant has committed to a program to monitor axial weld embrittlement relative to the values specified by the staff in its May 7, 2000 SER. The staff also confirms that the applicant has addressed the following Renewal Applicant Action Item in the staff's SER for BWRVIP-74 (Letter to C. Terry dated October 18, 2001) (Ref.10).

Action Item 12: As indicated in the staff's March 7, 2000 letter to Carl Terry, a license renewal (LR) applicant shall monitor axial beltline weld embrittlement. One acceptable method is to determine the mean RT_{NDT} of the limiting axial beltline weld at the end of the extended period of operation is less than the values specified in Table 1 of the staff's Oct. 18, 2001 FSER (Ref. 10).

4.2.3.2   FSAR Supplement

The reviewer verifies that the applicant has provided information to be included in the FSAR supplement that includes a summary description of the evaluation of the reactor vessel neutron embrittlement TLAA. Table 4.2-1 of this review plan section contains examples of acceptable FSAR supplement information for this TLAA. The reviewer verifies that the applicant has provided an FSAR supplement with information equivalent to that in Table 4.2-1.

The staff expects to impose a license condition on any renewed license to require the applicant to update its FSAR to include this FSAR supplement at the next update required pursuant to 10 CFR 50.71(e)(4). As part of the license condition, until the FSAR update is complete, the applicant may make changes to the programs described in its FSAR supplement without prior NRC approval, provided that the applicant evaluates each such change pursuant to the criteria set forth in 10 CFR 50.59. If the applicant updates the FSAR to include the final FSAR supplement before the license is renewed, no condition will be necessary.

As noted in Table 4.2-1, an applicant need not incorporate the implementation schedule into its FSAR. However, the reviewer should verify that the applicant has identified and committed in the license renewal application to any future aging management activities, including enhancements and commitments to be completed before the period of extended operation. The staff expects to impose a license condition on any renewed license to ensure that the applicant will complete these activities no later than the committed date.

4.2.4   Evaluation Findings

The reviewer determines whether the applicant has provided sufficient information to satisfy the provisions of this section and whether the staff's evaluation supports conclusions of the following type, depending on the applicant's choice of 10 CFR 54.21(c)(1)(i), (ii), or (iii), to be included in the staff's safety evaluation report:

> On the basis of its review, as discussed above, the staff concludes that the applicant has provided an acceptable demonstration, pursuant to 10 CFR 54.21(c)(1), that, for the reactor vessel neutron embrittlement TLAA, [choose which is appropriate] (i) the analyses remain valid for the period of extended operation, (ii) the analyses have been projected to the end of the period of extended operation, or (iii) the effects of aging on the intended function(s) will be adequately managed for the period of extended operation. The staff also concludes that the FSAR supplement contains an appropriate summary description of the reactor vessel neutron embrittlement TLAA evaluation for the period of extended operation as reflected in the license condition.

4.2.5   IMPLEMENTATION

Except in those cases in which the applicant proposes an acceptable alternative method, the method described herein will be used by the staff in its evaluation of conformance with NRC regulations.

4.2.6   References

1. 10 CFR Part 50 Appendix G, "Fracture Toughness Requirements," Office of the Federal Register, National Archives and Records Administration, 2010.

2. 10 CFR 50.61, "Fracture Toughness Requirements for Protection Against Pressurized Thermal Shock Events," Office of the Federal Register, National Archives and Records Administration, 2010.

3. Regulatory Guide 1.154, "Format and Content of Plant-Specific Pressurized Thermal Shock Safety Analysis Reports for Pressurized Water Reactors," U.S. Nuclear Regulatory Commission, January 1987.

4. Letter to the Commission from L.A. Reyes (EDO), dated May 27, 2004 (ADAMS accession number ML041190564)

5. BWRVIP-05, "BWR Vessel and Internals Project, BWR Reactor Pressure Vessel Shell Weld Inspection Recommendations," EPRI TR-105697, September 1995.

6. Letter to Carl Terry of Niagara Mohawk Power Company, BWRVIP Chairman, from Gus C. Lainas of NRC, "Final Safety Evaluation of the BWR Vessel and Internals Project BWRVIP-05 Report (TAC No. M93925)," July 28, 1998.

7. Generic Letter 98-05, "Boiling Water Reactor Licensees Use of the BWRVIP-05 Report to Request Relief from Augmented Examination Requirements on Reactor Pressure Vessel Circumferential Shell Welds," U.S. Nuclear Regulatory Commission, November 10, 1998.

8. BWRVIP-74-A, "BWR Vessels and Internals Project, BWR Reactor Pressure Vessel Inspection and Flaw Evaluation Guidelines for License Renewal," EPRI TR-1008872, June 2003.

9. BWRVIP-74, "BWR Vessels and Internals Project, BWR Reactor Pressure Vessel Inspection and Flaw Evaluation Guidelines," EPRI TR-113596, September 1999.

10. Letter to Carl Terry of Niagara Mohawk Power Company, BWRVIP Chairman, from Christopher Grimes, of NRC, "Acceptance for Referencing of EPRI Proprietary Report TR-113596, BWR Vessel and Internals Project, BWR Reactor Pressure Vessel Inspection and Flaw Evaluation Guidelines BWRVIP-74), and Appendix A, Demonstration of Compliance with the Technical Information Requirements of the License Renewal Rule (10CFR54.21)," dated October 18, 2001.

11. Letter to Carl Terry of Niagara Mohawk Power Company, BWRVIP Chairman, from Jack R. Strosnider, Jr., of NRC, "Supplement to Final Safety Evaluation of the BWR Vessel and Internals Project BWRVIP-05 Report (TAC No. MA3395)," dated March 7, 2000.12.

12. Regulatory Guide 1.99 Rev. 2, "Radiation Embrittlement of Reactor Vessel Materials," May, 1988.

13. Appendix K of ASME Code, Section XI, "Rules for Inservice Inspection of Nuclear Power Plant Components," 2004 Edition.

14. NUREG-0800, "U.S. Nuclear Regulatory Commission, Standard Review Plan," U.S. Nuclear Regulatory Commission, March 2007.

15. Regulatory Guide 1.190 Rev. 0, "Calculational and Dosimetry Methods for Determining Pressure Vessel Neutron Fluence," U.S. Nuclear Regulatory Commission, March 2001.

16. 75 FR 23, "Alternative Fracture Toughness Requirements for Protection Against Pressurized Thermal Shock Events," January 4, 2010.

Table 4.2-1   Examples of FSAR Supplement for Reactor Vessel Neutron Embrittlement TLAA Evaluation

TLAA	Description of Evaluation	Implementation Schedule*
Upper-shelf energy	10 CFR Part 50 Appendix G paragraph IV.A.1 requires that the reactor vessel beltline materials must have Charpy upper-shelf energy of no less than 50 ft-lb (68 J) throughout the life of the reactor vessel unless otherwise approved by the NRC. The upper-shelf energy has been determined to exceed 50 ft-lb (68 J) to the end of the period of extended operation.	Completed
Pressurized thermal shock (for PWRs)	For PWRs, 10 CFR 50.61 requires the "reference temperature RT_{PTS}" for reactor vessel beltline materials to be less than the "PTS screening criteria" at the expiration date of the operating license unless otherwise approved by the NRC. The "PTS screening criteria" are 270°F (132°C) for plates, forgings, and axial weld materials, or 300°F (149°C) for circumferential weld materials. The "reference temperature" has been determined to be less than the "PTS screening criteria" at the end of the period of extended operation.	Completed
Pressure-temperature (P-T) limits	10 CFR Part 50 Appendix G requires that heatup and cooldown of the RPV be accomplished within established P-T limits. These limits specify the maximum allowable pressure as a function of reactor coolant temperature. As the RPV becomes embrittled and its fracture toughness is reduced, the allowable pressure is reduced. 10 CFR Part 50 Appendix G requires periodic update of P-T limits based on projected embrittlement and data from a material surveillance program. The P-T limits will be updated to consider the period of extended operation.	Update should be completed before the period of extended operation
Elimination of circumferential weld inspection and analysis of axial welds (for BWRs)	NRC has granted relief from the reactor vessel circumferential shell weld inspections because the applicant has demonstrated through plant-specific analysis that the plant meets the staff-approved BWRVIP-74-A Report and has provided sufficient information that the probability of vessel failure due to embrittlement of axial welds is low. If the applicant indicates that relief from circumferential weld examination will be made under 10 CFR 50.55a(a)(3), the applicant will manage this TLAA in accordance with 10 CFR 54.21(c)(1)(iii).	Re-submittal under 10 CFR 50.55a(a)(3) should be completed before the period of extended operation
Other miscellaneous TLAAs on RV neutron embrittlement	Provide sufficient information on how the calculations for plant-specific TLAAs were performed, what the limiting TLAA parameter was calculated to be in accordance with the neutron fluence projected for the period of extended operation, and why the TLAA is acceptable under either	

| 10 CFR 54.21 (c)(1)(i), (ii), or (iii). | |

* An applicant need not incorporate the implementation schedule into its FSAR. However, the reviewer should verify that the applicant has identified and committed in the license renewal application to any future aging management activities to be completed before the period of extended operation. The staff expects to impose a license condition on any renewed license to ensure that the applicant will complete these activities by no later than the committed date.

4.3   METAL FATIGUE

Review Responsibilities

Primary - Branch responsible for the TLAA issues

Secondary - None

4.3.1   Area of Review

A metal component may progressively degrade and lose its structural integrity when it is subjected to fluctuating stresses, even at magnitudes less than the design static loads, due to a well-known degradation mechanism - fatigue. This mechanism of degradation can occur in flaw-free components by developing cracks during services. ASME Section III (Ref. 1) requires a fatigue analysis for Class 1 components unless allowed by the Code to be exempted under applicable ASME Section III provisions. The analysis considers all transient loads based on the anticipated number of thermal and pressure transients, and includes calculation of a parameter "cumulative usage factor" (CUF) that is used for estimating the extent of fatigue damage in the component. The ASME Code limits the CUF to a value of less than or equal to one for acceptable fatigue design. A CUF below a value of one provides assurance that no crack has been formed. A CUF above a value of one allows for the possibility that a crack may form, and that if left untreated, the crack could propagate exponentially under fatigue loading and eventually lead to coolant leakage in reactor pressure boundary components, or even general structural failure. Metal fatigue of components may have been evaluated based on an assumed number of transients or cycles for the current operating term. The validity of such metal fatigue analysis is reviewed for the period of extended operation.

Areas of review to ensure that the metal fatigue or flaw growth/tolerance evaluations are valid for the period of extended operation include:

1. CUF calculations for ASME Code Class 1 components designed to ASME Section III requirements, and other Codes that are based on a CUF calculation [the 1969 edition of ANSI B31.7 (Ref. 3) for Class 1 piping, ASME NC-3200 vessels, ASME NE 3200 Class MC components, and metal bellows designed to ASME NC-3649.4(e)(3), ND 3649.4(e)(3), or NE-3366.2(e)(3)]. ASME Class 1 components, which include core support structures, are analyzed for metal fatigue.

2. Implicit fatigue-based maximum allowable stress calculations for piping components designed to USAS ANSI B31.1 (Ref. 2) requirements, and ASME Code Class 2 and 3 components designed to ASME Section III design requirements that are similar to the guidance in ANSI B31.1.

    ANSI B31.1 applies only to piping and does not call for an explicit fatigue analysis. It specifies allowable stress levels based on the number of anticipated full thermal range transient cycles. The specific stress range reduction factors due to full thermal cycles are listed in Table 4.3-1.

3. Environmental fatigue calculations for ASME Code Class 1 reactor coolant pressure boundary components.

Generic Safety Issue: The fatigue design criteria for nuclear power plant components have changed as the industry consensus codes and standards have developed. The fatigue design criteria for a specific component depend on the version of the design code that applied to that component, i.e., the code of record. There is a concern that the effects of the reactor coolant environment on the fatigue life of components were not adequately addressed by the code of record.

The NRC has decided that the adequacy of the code of record relating to metal fatigue is a potential safety issue to be addressed by the current regulatory process for operating reactors (Refs. 4 and 5). The effects of fatigue for the initial 40-year reactor license period were studied and resolved under Generic Safety Issue (GSI)-78, "Monitoring of Fatigue Transient Limits for reactor coolant system," and GSI-166, "Adequacy of Fatigue Life of Metal Components" (Ref. 6). GSI-78 addressed whether fatigue monitoring was necessary at operating plants. As part of the resolution of GSI-166, an assessment was made of the significance of the more recent fatigue test data on the fatigue life of a sample of components in plants where Code fatigue design analysis had been performed. The efforts on fatigue life estimation and ongoing issues under GSI-78 and GSI-166 for 40-year plant life were addressed separately under a staff generic task action plan (Refs. 7 and 8). The staff documented its completion of the fatigue action plan in SECY-95-245 (Ref. 9).

SECY-95-245 was based on a study described in NUREG/CR-6260, "Application of NUREG/CR-5999 Interim Fatigue Curves to Selected Nuclear Power Plant Components" (Ref. 10). In NUREG/CR-6260, sample locations with high fatigue usage were evaluated. Conservatisms in the original fatigue calculations, such as actual cycles versus assumed cycles, were removed, and the fatigue usage was recalculated using a fatigue curve considering the effects of the environment. The staff found that most of the locations would have a CUF of less than the ASME Code limit of 1.0 for 40 years. On the basis of the component assessments, supplemented by a 40-year risk study, the staff concluded that a backfit of the environmental fatigue data to operating plants could not be justified. However, because the staff was less certain that sufficient excessive conservatisms in the original fatigue calculations could be removed to account for an additional 20 years of operation for renewal, the staff recommended in SECY-95-245 that the samples in NUREG/CR-6260 should be evaluated considering environmental effects for license renewal. GSI-190, "Fatigue Evaluation of Metal Components for 60-year Plant Life," was established to address the residual concerns of GSI-78 and GSI-166 regarding the environmental effects on fatigue of pressure boundary components for 60 years of plant operation.

The scope of GSI-190 included design basis fatigue transients. It studied the probability of fatigue failure and its effect on core damage frequency (CDF) of selected metal components for 60-year plant life. The results showed that some components have cumulative probabilities of crack initiation and through-wall growth that approach one within the 40- to 60-year period. The maximum failure rate (through-wall cracks per year) was in the range of 10E-2 per year, and those failures were generally associated with high cumulative usage factor locations and components with thinner walls, i.e., pipes more vulnerable to through-wall cracks. In most cases, the leakage from these through-wall cracks is small and not likely to lead to core damage. It was concluded that no generic regulatory action is necessary and that GSI-190 is resolved based on results of probabilistic analyses and sensitivity studies, interactions with the industry (NEI and

EPRI), and different approaches available to licensees to manage the effects of aging (Refs. 11 and 12).

However, the calculations supporting resolution of this issue, which included consideration of environmental effects, indicate the potential for an increase in the frequency of pipe leaks as plants continue to operate. Thus, the staff concluded that licensees are to address the effects of coolant environment on component fatigue life as aging management programs are formulated in support of license renewal.

The applicant's consideration of the effects of coolant environment on component fatigue life for license renewal is an area of review.

4. Potential fatigue assessments for BWR vessel internals components (potential TLAAs based on applicable applicant action items identified in applicable BWRVIP reports)

    For Boiling Water Reactors, license renewal applications that reference the following BWR Vessels and Internals Project (BWRVIP) reports should identify and evaluate the projected fatigue CUFs as a potential TLAA issue, which may impact the structural integrity of the subject reactor pressure vessel internal components.

    - BWRVIP-18-A (Ref. 16, action item #4) for core spray internals
    - BWRVIP-27-A (Ref. 17, action item #4) for standby liquid control system/core plate ΔP
    - BWRVIP-47-A (Ref. 18, action item #4) for lower plenum.

    In addition, license renewal applications that reference the BWRVIP-74-A report (Ref. 19) for reactor pressure vessel, should address the following renewal applicant action items in the staff's SER for BWRVIP-74-A report.

    Item #8: For the license renewal period, verify that the original fatigue analysis is valid and also addresses environmental fatigue for components mentioned in NUREG/CR-6260. As a minimum, these components normally include reactor coolant pressure boundary components, such as reactor vessel head and shell components, reactor vessel flange bolting or stud components, reactor vessel nozzle components, and piping components, including safe-end locations.

    Item #14: Components that have indications that were previously analytically evaluated in accordance with ASME Section XI Subsection IWB-3600 until the end of the 40-year service period shall be re-evaluated for the period corresponding to the license renewal term.

5. Potential fatigue-based flaw growth analyses or fatigue-based fracture mechanics analyses, including those for high-energy line breaks, reactor coolant pump (RCP) flywheels, reactor vessel metal bellows, and reactor vessel underclad cracking analyses (applicable to reactor vessels fabricated from SA-508 Class 2 or 3 forgings), as appropriate

The validity of these analyses is reviewed for the period of extended operation. The design criteria used to determine the postulated high-energy line break design locations include the calculated fatigue CUF based on the number of design transients assumed for the 40-year life of the plant. The aging effect of concern for the RCP flywheel is fatigue crack initiation and growth in the flywheel bore keyway from stresses due to starting the motor during start/stop cycles of the RCP during the 40-year design. Similarly, the primary containment process metal bellows are designed for a specific number of cycles of expansion and contraction for 40 years of operation. The fracture toughness (including the effects of neutron irradiation) and flaw growth analyses for underclad cracks that are postulated in the internal cladding of SA-508 Class 2 and 3 alloy steel components are also based upon 40-year design transients.

4.3.2   Acceptance Criteria

The acceptance criteria for the areas of review described in Subsection 4.3.1 of this review plan section delineate acceptable methods for meeting the requirements of the NRC's regulations in 10 CFR 54.21(c)(1).

4.3.2.1   Time-Limited Aging Analysis

Pursuant to 10 CFR 54.21(c)(1)(i) - (iii), an applicant must demonstrate one of the following:

(i) the analyses remain valid for the period of extended operation,

(ii) the analyses have been projected to the end of the extended period of operation, or

(iii) the effects of aging on the intended function(s) will be adequately managed for the period of extended operation.

Specific acceptance criteria for metal fatigue are discussed in the following sub-sections.

4.3.2.1.1   ASME Code Class 1 Components Designed to ASME Section III and other Codes based on CUF

For components designed or analyzed to ASME Code Section III requirements for Class I components or other Codes that require a CUF calculation, the acceptance criteria, depending on the applicant's choice of 10 CFR 54.21(c)(1)(i), (ii), or (iii), are discussed in the following sub-sections.

4.3.2.1.1.1   10 CFR 54.21(c)(1)(i)

The existing CUF calculations remain valid for the period of extended operation because the number of accumulated cycles for the design basis transients would not exceed the limits established for these transients.

4.3.2.1.1.2   10 CFR 54.21(c)(1)(ii)

The CUF calculation analyses are projected to the end of the extended period of operation based on projecting the cumulative number of transient occurrences for design basis transients through to the expiration of the period of extended operation. The resulting CUF values are verified to remain less than or equal to a CUF value of one for the period of extended operation.

4.3.2.1.1.3   10 CFR 54.21(c)(1)(iii)

In Chapter X.M1 of the GALL Report (Ref. 13), the staff has evaluated a program for monitoring and tracking the number of critical thermal and pressure transients for the selected components. The staff has determined that this program is an acceptable aging management program to address metal fatigue of the reactor coolant system components according to 10 CFR 54.21(c)(1)(iii). The GALL Report may be referenced in a license renewal application and should be treated in the same manner as an approved topical report. In referencing the GALL Report, the applicant should indicate that the material referenced is applicable to the specific plant involved and should provide the information necessary to adopt the finding of program acceptability as described and evaluated in the report. The applicant also should verify that the approvals set forth in the GALL Report for the generic program apply to the applicant's program.

4.3.2.1.2 Piping Components Designed to USAS ANSI B31.1 Requirements and ASME Code Class 2 and 3 Components Designed to ASME Section III Requirements

For piping designed or analyzed to B31.1 guidance or ASME Code Class 2 and 3 components designed to ASME Section III cyclic design requirements, the acceptance criteria, depending on the applicant's choice of 10 CFR 54.21(c)(1)(i), (ii), or (iii), are discussed in the following sub-sections.

4.3.2.1.2.1 10 CFR 54.21(c)(1)(i)

The maximum allowable stress range values for the existing fatigue analysis remain valid because the allowable limit for the number of full thermal range transient cycles would not be exceeded during the period of extended operation.

4.3.2.1.2.2 10 CFR 54.21(c)(1)(ii)

The maximum allowable stress range values are re-evaluated based on the projected number of assumed full thermal range transient cycles above a value of 7000 (refer to Table 4.3-1). The current stress range value in the design basis for the component is then compared to the reduced maximum allowable stress range value to make sure it is within the value of the newly-reduced limit, and to ensure that the design basis value remains acceptable during the period of extended operation.

4.3.2.1.2.3 10 CFR 54.21(c)(1)(iii)

The effects of aging on the intended function(s) will be adequately managed for the period of extended operation. Chapter X.M1 in the GALL Report provides an acceptable method for dispositioning fatigue analyses under 10 CFR 54.21(c)(1)(iii). Alternatively, the component could be replaced and the allowable stresses for the replacement will be sufficient as specified by the code during the period of extended operation. The analyses will be evaluated on a case-by-case basis to ensure that the aging effects will be managed such that the intended functions(s) will be maintained during the period of extended operation.

4.3.2.1.3 Environmental Fatigue Calculations for Code Class 1 Components

The staff recommendation for the closure of GSI-190 is contained in a December 26, 1999 memorandum from Ashok Thadani to William Travers (Ref. 11). The staff recommended that licensees address the effects of the coolant environment on component fatigue life as aging management programs are formulated in support of license renewal. For reactor coolant pressure boundary components, one method acceptable to the staff for satisfying this recommendation is to assess the impact of the reactor coolant environment on a sample of critical components. These critical components should include, as a minimum, those selected in NUREG/CR-6260 (Ref. 10). Applicants should consider adding additional component locations if they are considered to be more limiting than those considered in NUREG/CR-6260. The sample of critical components can be evaluated by applying environmental correction factors to the existing ASME Code fatigue analyses.

Applicants should consider adding additional component locations if they are considered to be more limiting than those considered in NUREG/CR-6260. The sample of critical components can be evaluated by applying environmental correction factors to the existing ASME Code fatigue analyses. Applicants should add additional plant specific component locations if they may be more limiting than those considered in NUREG/CR-6260. Environmental effects on fatigue for these critical components can be evaluated using one of the following sets of formulae:

- Carbon and Low Alloy Steels
  - Those provided in NUREG/CR-6583 (Ref. 14), using the applicable ASME Section III fatigue design curve.
  - Those provided in Appendix A of NUREG/CR-6909 (Ref. 21), using either the applicable ASME Section III fatigue design curve or the fatigue design curve for carbon and low alloy steel provided in NUREG/CR-6909 (Figures A.1 and A.2, respectively, and Table A.1).
  - A staff approved alternative.

- Austenitic Stainless Steels
  - Those provided in NUREG/CR-5704 (Ref. 15), using the applicable ASME Section III fatigue design curve.
  - Those provided in NUREG/CR-6909 (Ref. 21), using the fatigue design curve for austenitic stainless steel provided in NUREG/CR-6909 (Figure A.3 and Table A.2).
  - A staff approved alternative.

- Nickel Alloys
  - Those provided in NUREG/CR-6909 (Ref. 21), using the fatigue design curve for austenitic stainless steel provided in NUREG/CR-6909 (Figure A.3 and Table A.2).
  - A staff approved alternative.

Any one option may be used for calculating the CUF_{en} for each material.

4.3.2.1.4   Potential Fatigue Assessments for BWR Vessel Internals Components

The acceptance criteria in Subsection 4.3.2.1.1 of this review plan section apply.

4.3.2.1.5   Potential Flaw Growth and Fracture Mechanics Analysis

Depending on the choice of 10 CFR 54.21(c)(1)(i), (ii), or (iii), the acceptance criteria are discussed in the following sub-sections.

4.3.2.1.5.1   10 CFR 54.21(c)(1)(i)

The existing analyses remain valid because the number of cycles assumed for the 40-year life would not be exceeded during the period of extended operation.

4.3.2.1.5.2   10 CFR 54.21(c)(1)(ii)

The degree of flaw growth or increase in the stress intensity factor are projected based on the cumulative number of transient occurrences to the end of the period of extended operation. The newly-projected values are compared to the acceptance criteria for these parameters that were set in the original analysis.

4.3.2.1.5.3   10 CFR 54.21(c)(1)(iii)

The acceptance criteria under 10 CFR 54.21(c)(1)(iii) will be evaluated on a case-by-case basis to ensure that the aging effects will be managed such that the intended functions(s) of the subject components will be maintained during the period of extended operation.

Chapter X.M1 in the GALL Report may not be used as a basis for accepting fatigue-based flaw growth analyses or fracture mechanics analyses in accordance with 10 CFR 54.21(c)(1)(iii).

4.3.2.2   FSAR Supplement

The specific criterion for meeting 10 CFR 54.21(d) is:

> The summary description of the evaluation of TLAAs for the period of extended operation in the FSAR supplement is appropriate such that later changes can be controlled by 10 CFR 50.59. The description should contain information associated with the TLAAs regarding the basis for determining that the applicant has made the demonstration required by 10 CFR 54.21(c)(1).

4.3.3   Review Procedures

For each area of review described in Subsection 4.3.1, the review procedures in the subsequent sub-sections should be followed.

4.3.3.1   Time-Limited Aging Analysis

4.3.3.1.1   ASME Code Class 1 Components Designed to ASME Section III and Other Codes based on CUF

For components designed or analyzed to ASME Class 1 requirements or other Codes that are based on a CUF calculation, the review procedures, depending on the applicant's choice of 10 CFR 54.21(c)(1)(i), (ii), or (iii), are discussed in the following sub-sections.

4.3.3.1.1.1   10 CFR 54.21(c)(1)(i)

The operating transient experience and a list of the assumed transients used in the existing CUF calculations for the current operating term are reviewed to ensure that the number of assumed transients would not be exceeded during the period of extended operation. For consistency purposes, the review also includes an assessment of the TLAA information against relevant design basis information and CLB information (including applicable cycle-counting requirements in technical specifications).

4.3.3.1.1.2   10 CFR 54.21(c)(1)(ii)

The operating transient experience is reviewed to ensure that the increased number of transients used for any re-analysis meets or exceeds the number of transients projected to the end of the period of extended operation. The revised CUF calculations are reviewed to ensure that the CUF remains less than or equal to one at the end of the period of extended operation. For consistency purposes, the review also includes an assessment of the TLAA information against relevant design basis information and CLB information (including applicable cycle-counting requirements in technical specifications).

The code of record should be used for the reevaluation, or the applicant may update to a later code edition pursuant to 10 CFR 50.55a. In the latter case, the reviewer verifies that the requirements in 10 CFR 50.55a are met.

4.3.3.1.1.3   10 CFR 54.21(c)(1)(iii)

The applicant may reference Chapter X.M1 of the GALL Report in its license renewal application and use this GALL chapter to accept the TLAA in accordance with 10 CFR 54.21(c)(1)(iii), as appropriate. The review should verify that the applicant has stated that the report is applicable to its plant with respect to its program that monitors and tracks the number of critical thermal and pressure transients for the selected reactor coolant system components. The reviewer verifies that the applicant has identified the appropriate program as described and evaluated in the GALL Report. The reviewer also ensures that the applicant has stated that its program contains the same program elements that the staff evaluated and relied upon in approving the corresponding generic program in the GALL Report. For consistency purposes, the review also includes an assessment of the TLAA information against relevant design basis and CLB information (including applicable cycle-counting requirements in technical specifications). No staff re-evaluation of the acceptability of the basis in the GALL Report Chapter X.M1 is necessary.

4.3.3.1.2   Piping Components Designed to USAS ANSI B31.1 Requirements and ASME Code Class 2 and 3 Components Designed to ASME Section III Requirements

For piping designed or analyzed to ANSI B31.1 guidance or ASME Code Class 2 and 3 components designed to ASME Section III cyclic design requirements, the review procedures, depending on the applicant's choice of 10 CFR 54.21(c)(1)(i), (ii), or (iii), are discussed in the following sub-sections.

4.3.3.1.2.1   10 CFR 54.21(c)(1)(i)

The staff reviews the relevant information in the TLAA, operating plant transient history, design basis, and CLB (including technical specification cycle counting requirements) to verify that the maximum allowable stress range values for the existing fatigue analysis remain valid for the period of extended operation and that the allowable limit for full thermal range transients will not be exceeded during the period of extended operation.

4.3.3.1.2.2   10 CFR 54.21(c)(1)(ii)

The maximum allowable stress range values are re-evaluated based on the projected number of full thermal range transient cycles above a value of 7000 (refer to Table 4.3-1). The reviewer confirms that (1) if the value of the number of equivalent full temperature cycles is above 7000, the applicant has reduced the maximum allowable stress range limit according to the reduction factors in the table, and (2) the design basis stress value for the component is within the newly-reduced allowable limit. The review includes all relevant design basis, operating history, CLB, and TLAA information, including technical specification cycle counting requirements.

4.3.3.1.2.3   10 CFR 54.21(c)(1)(iii)

The effects of aging on the intended function(s) will be adequately managed for the period of extended operation. Chapter X.M1 in the GALL Report provides an acceptable method to disposition these fatigue analyses under 10 CFR 54.21(c)(1)(iii). The staff reviews an applicant's program for dispositioning the TLAA in accordance with the requirements in 10 CFR 54.21(c)(1)(iii) and the guidance in GALL Chapter X.M1.

In addition to continued monitoring and tracking, the components could be re-analyzed, inspected for cracking, repaired, or replaced. Such approaches will be evaluated on a case-by-case basis to ensure that the aging effects will be managed such that the intended functions(s) will be maintained during the period of extended operation.

4.3.3.1.3   Other Evaluations Based on CUF

The reviewer verifies that the applicant has addressed the staff recommendation for the closure of GSI-190 contained in a December 26, 1999 memorandum from Ashok Thadani to William Travers (Ref. 11). The reviewer verifies that the applicant has addressed the effects of the coolant environment on component fatigue life as aging management programs are formulated in support of license renewal. If an applicant has chosen to assess the impact of the reactor coolant environment on a sample of critical components, the reviewer verifies the following:

1. The critical components include a sample of high-fatigue usage locations. This sample is to include the locations identified in NUREG/CR-6260 (Ref. 10), as a minimum, and proposed additional locations based on plant specific considerations.

2. The sample of critical components has been evaluated (1) by applying environmental correction factors to the existing ASME Code required CUF analyses, or (2) using the methodology provided in NUREG/CR-6909 (Ref. 21). If the Class 1 component was designed to a code not requiring CUF, a new environmental CUF calculation has been performed or is addressed in an appropriate license renewal commitment.

3. Environmental effects on fatigue for these critical components have been evaluated using one of the following sets of formulae:

    *   Carbon and Low Alloy Steels
        o   Those provided in NUREG/CR-6583 (Ref. 14), using the applicable ASME Section III fatigue design curve.
        o   Those provided in Appendix A of NUREG/CR-6909 (Ref. 21), using either the applicable ASME Section III fatigue design curve or the fatigue design curve for carbon and low alloy steel provided in NUREG/CR-6909 (Figures A.1 and A.2, respectively, and Table A.1).
        o   A staff approved alternative.

    *   Austenitic Stainless Steels

- Those provided in NUREG/CR-5704 (Ref. 15), using the applicable ASME Section III fatigue design curve.
- Those provided in NUREG/CR-6909 (Ref. 21), using the fatigue design curve for austenitic stainless steel provided in NUREG/CR-6909 (Figure A.3 and Table A.2).
- A staff approved alternative.

- **Nickel Alloys**
  - Those provided in NUREG/CR-6909 (Ref. 21), using the fatigue design curve for austenitic stainless steel provided in NUREG/CR-6909 (Figure A.3 and Table A.2).
  - A staff approved alternative.

Any one option may be used for calculating the CUF_{en} for each material.

4.3.3.1.4 Potential Fatigue Assessments for BWR Vessel Internals Components

The review procedures in Subsection 4.3.3.1.1 of this review plan section apply.

4.3.3.1.5 Potential Flaw Growth and Fracture Mechanics Analysis

Depending on the choice of 10 CFR 54.21(c)(1)(i), (ii), or (iii), the review procedures are discussed in the following sub-sections.

4.3.3.1.5.1 10 CFR 54.21(c)(1)(i)

The operating cyclic experience and a list of the assumed cycles used in the existing analyses are reviewed to ensure that the number of assumed cycles would not be exceeded during the period of extended operation.

4.3.3.1.5.2 10 CFR 54.21(c)(1)(ii)

The staff reviews the applicant's basis for projecting the amount of flaw growth or increase in the stress intensity factor value to the end of the period of extended operation. The staff ensures that the applicant's basis is valid and that the newly-projected values are compared to the acceptance criteria for these parameters that were set in the original analysis.

4.3.3.1.5.3 10 CFR 54.21(c)(1)(iii)

The staff reviews the applicant's proposed program to ensure that the effects of aging on the intended function(s) of the subject component will be adequately managed for the period of extended operation. The acceptance criteria under 10 CFR 54.21(c)(1)(iii) will be evaluated on a case-by-case basis. Chapter X.M1 in the GALL Report may not be used as a basis for accepting fatigue-based flaw growth analyses or fracture mechanics analyses in accordance with 10 CFR 54.21(c)(1)(iii).

4.3.3.2 FSAR Supplement

The reviewer verifies that the applicant has provided information to be included in the FSAR supplement that includes a summary description of the evaluation of the metal fatigue TLAA.

Table 4.3-2 contains examples of acceptable FSAR supplement information for this TLAA. The reviewer verifies that the applicant has provided a FSAR supplement with information equivalent to that in Table 4.3-2.

The staff expects to impose a license condition on any renewed license to require the applicant to update its FSAR to include this FSAR supplement at the next update required pursuant to 10 CFR 50.71(e)(4). As part of the license condition, until the FSAR update is complete, the applicant may make changes to the programs described in its FSAR supplement without prior NRC approval, provided that the applicant evaluates each such change pursuant to the criteria set forth in 10 CFR 50.59. If the applicant updates the FSAR to include the final FSAR supplement before the license is renewed, no condition will be necessary.

As noted in Table 4.3-2, an applicant need not incorporate the implementation schedule into its FSAR. However, the reviewer should verify that the applicant has identified and committed in the license renewal application to any future aging management activities, including enhancements and commitments to be completed before the period of extended operation. The staff expects to impose a license condition on any renewed license to ensure that the applicant will complete these activities no later than the committed date.

4.3.4   Evaluation Findings

The reviewer determines whether the applicant has provided sufficient information to satisfy the provisions of this section and whether the staff's evaluation supports conclusions of the following type, depending on the applicant's choice of 10 CFR 54.21(c)(1)(i), (ii), or (iii), to be included in the staff's safety evaluation report:

> On the basis of its review, as discussed above, the staff concludes that the applicant has provided an acceptable demonstration, pursuant to 10 CFR 54.21(c)(1), that, for the metal fatigue TLAA, [choose which is appropriate] (i) the analyses remain valid for the period of extended operation, (ii) the analyses have been projected to the end of the period of extended operation, or (iii) the effects of aging on the intended function(s) will be adequately managed for the period of extended operation. The staff also concludes that the FSAR Supplement contains an appropriate summary description of the metal fatigue TLAA evaluation for the period of extended operation as reflected in the license condition.

4.3.5   Implementation

Except in those cases in which the applicant proposes an acceptable alternative method, the method described herein will be used by the staff in its evaluation of conformance with NRC regulations.

4.3.6   References

1. ASME Section III, "Rules for Construction of Nuclear Power Plant Components," The ASME Boiler and Pressure Vessel Code, 2004 edition, The American Society of Mechanical Engineers, New York, NY.

2. ANSI/ASME B31.1, "Power Piping," American National Standards Institute.

3. ANSI/ASME B31.7-1969, "Nuclear Power Piping," American National Standards Institute.

4. SECY-93-049, "Implementation of 10 CFR Part 54, 'Requirements for Renewal of Operating Licenses for Nuclear Power Plants,'" March 1, 1993.

5. Staff Requirements Memorandum from Samuel J. Chilk, dated June 28, 1993.

6. NUREG-0933, "Resolution of Generic Safety Issues," Supplement 32, U.S. Nuclear Regulatory Commission, August 2008.

7. Letter from William T. Russell of NRC to William Rasin of the Nuclear Management and Resources Council, dated July 30, 1993.

8. SECY-94-191, "Fatigue Design of Metal Components," July 26, 1994.

9. SECY-95-245, "Completion of the Fatigue Action Plan," September 25, 1995.

10. NUREG/CR-6260, "Application of NUREG/CR-5999 Interim Fatigue Curves to Selected Nuclear Power Plant Components," March 1995.

11. Letter from Ashok C. Thadani of the Office of Nuclear Regulatory Research to William D. Travers, Executive Director of Operations, dated December 26, 1999.

12. NUREG/CR-6674, "Fatigue Analysis of Components for 60-Year Plant Life," June 2000.

13. NUREG-1801, "Generic Aging Lessons Learned (GALL) Report," U.S. Nuclear Regulatory Commission, Revision 2, 2010.

14. NUREG/CR-6583, "Effects of LWR Coolant Environments on Fatigue Design Curves of Carbon and Low–Alloy Steels," March 1998.

15. NUREG/CR-5704, "Effects of LWR Coolant Environments on Fatigue Design Curves of Austenitic Stainless Steels," April 1999.

16. NRC Acceptance for Referencing of BWR Vessel and Internals Project, BWR Core Spray Internals Inspection and Flaw Evaluation Guidelines for Compliance with the License Renewal Rule (10CFR Part 54), Dec. 7, 2000, in BWRVIP-18-A, "BWR Vessel and Internals Project BWR Core Spray Internals Inspection and Flaw Evaluation Guidelines," EPRI TR-1011469, February 2005.

17. NRC Acceptance for Referencing of Report, "BWR Vessel and Internals Project, BWR Standby Liquid Control System/Core Plate ΔP Inspection and Flaw Evaluation Guidelines," for Compliance with the License Renewal Rule (10CFR Part 54), Dec. 20, 1999, in BWRVIP-27-A, "BWR Vessel and Internals Project BWR Standby Liquid Control System/Core Plate $\Box P$ Inspection and Flaw Evaluation Guidelines," EPRI TR-1007279, August 2003.

18. NRC Acceptance for Referencing of Report, "BWR Vessel and Internals Project, BWR Lower Plenum Inspection and Flaw Evaluation Guidelines," for Compliance with the License Renewal Rule (10CFR Part 54), Dec. 7, 2000, in BWRVIP-47-A, "BWR Vessel and Internals

Project BWR Lower Plenum Inspection and Flaw Evaluation Guidelines," EPRI TR-1009947, June 2004.

19. NRC Acceptance for Referencing of EPRI Proprietary Report TR-113596, "BWR Vessel and Internals Project, BWR Reactor Pressure Vessel Inspection and Flaw Evaluation Guidelines," and Appendix A, "Demonstration of Compliance with the Technical Information Requirements of the License Renewal Rule (10CFR 54.21)," Oct. 18, 2001, in BWRVIP-74-A, "BWR Vessel and Internals Project BWR Reactor Pressure Vessel Inspection and Flaw Evaluation Guidelines," EPRI TR-1008872, June 2003.

20. Regulatory Guide RG-1.207, "Guidelines for Evaluating Fatigue Analyses Incorporating the Life Reduction of Metal Components due to the Effects of the Light-Water Reactor Environment for New Reactors," U.S. Nuclear Regulatory Commission, Washington, DC, March 2007.

21. NUREG/CR-6909, "Effect of LWR Coolant Environments on Fatigue Life of Reactor Materials" (Final Report), February 2007.

Table 4.3-1 Stress Range Reduction Factors

Number of Equivalent Full Temperature Cycles	Stress Range Reduction Factor
7,000 and less	1.0
7,000 to 14,000	0.9
14,000 to 22,000	0.8
22,000 to 45,000	0.7
45,000 to 100,000	0.6
100,000 and over	0.5

Table 4.3-2 Example of FSAR Supplement for Metal Fatigue TLAA Evaluation

10 CFR 54.21(c)(1)(iii) Example

TLAA	Description of Evaluation	Implementation Schedule*

Metal fatigue	The aging management program monitors and tracks the number of critical thermal and pressure test transients and monitors the cycles for the selected reactor coolant system components.
	
	The aging management program addresses the effects of the coolant environment on component fatigue life by assessing the impact of the reactor coolant environment on a sample of critical components that include, as a minimum, those components selected in NUREG/CR-6260. Environmental effects on fatigue for these critical components can be evaluated using one of the following sets of formulae:
	
	• Carbon and Low Alloy Steels
	  ○ Those provided in NUREG/CR-6583 (Ref. 14), using the applicable ASME Section III fatigue design curve.
	  ○ Those provided in Appendix A of NUREG/CR-6909 (Ref. 21), using either the applicable ASME Section III fatigue design curve or the fatigue design curve for carbon and low alloy steel provided in NUREG/CR-6909 (Figures A.1 and A.2, respectively, and Table A.1).
	  ○ A staff approved alternative.
	
	• Austenitic Stainless Steels
	  ○ Those provided in NUREG/CR-5704 (Ref. 15), using the applicable ASME Section III fatigue design curve.
	  ○ Those provided in NUREG/CR-6909 (Ref. 21), using the fatigue design curve for austenitic stainless steel provided in NUREG/CR-6909 (Figure A.3 and Table A.2).
	  ○ A staff approved alternative.
		Evaluation should be completed before the period of extended operation

Table 4.3-2   Example of FSAR Supplement for Metal Fatigue TLAA Evaluation

10 CFR 54.21(c)(1)(iii) Example

TLAA	Description of Evaluation	Implementation Schedule*
	• <u>Nickel Alloys</u> ○ Those provided in NUREG/CR-6909 (Ref. 21), using the fatigue design curve for austenitic stainless steel provided in NUREG/CR-6909 (Figure A.3 and Table A.2). ○ A staff approved alternative. Any one option may be used for calculating the CUF_{en} for each material.	

* An applicant need not incorporate the implementation schedule into its FSAR. However, the reviewer should verify that the applicant has identified and committed in the license renewal application to any future aging management activities to be completed before the period of extended operation. The staff expects to impose a license condition on any renewed license to ensure that the applicant will complete these activities no later than the commitment date.

4.4 ENVIRONMENTAL QUALIFICATION (EQ) OF ELECTRIC EQUIPMENT

Review Responsibilities

Primary - Branch responsible for electrical engineering

Secondary - None

4.4.1 Areas of Review

The NRC has established environmental qualification requirements in 10 CFR Part 50, Appendix A, Criterion 4, and 10 CFR 50.49. Section 50.49 specifically requires each nuclear power plant licensee to establish a program to qualify certain electric equipment (not including equipment located in mild environments) so that such equipment, in its end-of-life condition, will meet its performance specifications during and following design basis accidents under the most severe environmental conditions postulated at the equipment's location after such an accident. Such conditions include, among others, conditions resulting from loss of coolant accidents (LOCAs), high energy line breaks (HELBs), and post-LOCA radiation. Equipment qualified by test must be preconditioned by aging to its end-of-life condition (i.e., the condition at the end of the current operating term). Those components with a qualified life equal to or greater than the duration of the current operating term are covered by TLAAs.

In a related subject, some nuclear power plants have mechanical equipment that was qualified in accordance with the provisions of Criterion 4 of Appendix A to 10 CFR Part 50. If a plant has qualified mechanical equipment, it is typically documented in the plant's master EQ list. If this qualified mechanical equipment requires a performance of a TLAA, it should be performed in accordance with the provisions of SRP-LR Section 4.7, "Other Plant-Specific Time-Limited Aging Analyses." If a TLAA of qualified mechanical equipment is necessary, usually it will involve assessments of the environmental effects on components such as seals, gaskets, lubricants, fluids for hydraulic systems, or diaphragms.

4.4.1.1 Time-Limited Aging Analysis

All operating plants must meet the requirements of 10 CFR 50.49 for certain important-to-safety electrical components. 10 CFR 50.49 defines the scope of components to be included, requires the preparation and maintenance of a list of in-scope components, and requires the preparation and maintenance of a qualification file that includes component performance specifications, electrical characteristics, and environmental conditions. 10 CFR 50.49(e)(5) contains provisions for aging that require, in part, consideration of all significant types of aging degradation that can affect component functional capability. 10 CFR 50.49(e) also requires component replacement or refurbishment prior to the end of designated life, unless additional life is established through ongoing qualification. 10 CFR 50.49(f) establishes four methods of demonstrating qualification for aging and accident conditions. 10 CFR 50.49(k) and (l) permit different qualification criteria to apply based on plant and component vintage. Supplemental environmental qualification regulatory guidance for compliance with these different qualification criteria is provided in NRC RG 1.89, Rev. 1, "Environmental Qualification of Certain Electric Equipment Important to Safety for Nuclear Power Plants" (Ref. 1), the Division of Operating Reactors (DOR) Guidelines (Ref. 2), and NUREG-0588 (Ref. 3). The principal nuclear industry qualification standards for electric equipment are IEEE STD 323-1971 (Ref. 4) and IEEE STD 323-1974 (Ref. 5). These standards contain explicit environmental qualification considerations based on TLAAs. Compliance with 10 CFR 50.49 provides reasonable assurance that the component can perform its intended functions during accident conditions after experiencing the effects of inservice aging.

4.4.1.1.1 DOR Guidelines

The qualification of electric equipment that is subject to significant known degradation due to aging where a qualified life was previously required to be established in accordance with Section 5.2.4 of the DOR Guidelines is reviewed for the period of extended operation according to those requirements. If a qualified life was not previously established, the qualification is reviewed in accordance with Section 7 of the DOR Guidelines.

4.4.1.1.2 NUREG-0588, CATEGORY II (IEEE STD 323-1971)

The qualification of certain electric equipment important to safety for which qualification was required in accordance with NUREG-0588, Category II, is reviewed for conformance to those requirements for the period of extended operation to assess the validity of the extended qualification. These requirements include IEEE STD 382-1972 (Ref. 6) for valve operators and IEEE STD 334-1971 (Ref. 7).

4.4.1.1.3 NUREG-0588, CATEGORY I (IEEE STD 323-1974)

The qualification of certain electric equipment important to safety for which qualification was required in accordance with NUREG-0588, Category I, is reviewed for conformance to those requirements for the period of extended operation to assess the validity of the extended qualification.

4.4.1.2 Generic Safety Issue

Regulatory Issue Summary (RIS) 2003-09 was issued on May 2, 2003, (Ref. 8) to inform addressees of the results of the technical assessment of GSI-168, "Environmental Qualification of Electrical Equipment" (Ref. 9). This RIS requires no action on the part of the addressees.

4.4.1.3 FSAR Supplement

The detailed information on the evaluation of TLAAs is contained in the renewal application. A summary description of the evaluation of TLAAs for the period of extended operation is contained in the applicant's FSAR supplement. The FSAR supplement is an area of review.

4.4.2 Acceptance Criteria

The acceptance criteria for the areas of review described in Subsection 4.4.1 of this review plan section delineate acceptable methods for meeting the requirements of the NRC's regulations in 10 CFR 54.21(c)(1).

4.4.2.1 Time-Limited Aging Analysis

Pursuant to 10 CFR 54.21(c)(1)(i) - (iii), an applicant must demonstrate one of the following:

(i) the analyses remain valid for the period of extended operation,

(ii) the analyses have been projected to the end of the extended period of operation, or

(iii) the effects of aging on the intended function(s) will be adequately managed for the period of extended operation.

Specific acceptance criteria for environmental qualification of certain electric equipment important to safety analyzed to Section 5.2.4 of the DOR Guidelines; NUREG-0588, Category II (Section 4); or NUREG-0588, Category I, depend on the applicant's choice, that is, 10 CFR 54.21(c)(1)(i), (ii), or (iii), and are:

4.4.2.1.1   10 CFR 54.21(c)(1)(i)

The existing qualification is based on previous testing, analysis, or operating experience, or combinations thereof, that demonstrate that the equipment is qualified for the period of extended operation. For option (i), the aging evaluation existing at the time of the renewal application for the component remains valid for the period of extended operation, and no further evaluation is necessary.

4.4.2.1.2   10 CFR 54.21(c)(1)(ii)

Qualification of the equipment is extended for the period of extended operation by testing, analysis, or operating experience, or combinations thereof, in accordance with the current licensing basis. For option (ii), a reanalysis of the aging evaluation is performed in order to project the qualification of the component through the period of extended operation. Important reanalysis attributes of an aging evaluation include analytical methods, data collection and reduction methods, underlying assumptions, acceptance criteria, and corrective actions if acceptance criteria are not met. These reanalysis attributes are discussed in Table 4.4-1.

4.4.2.1.3   10 CFR 54.21(c)(1)(iii)

In Chapter X of the GALL Report (Ref. 10), the staff has evaluated the environmental qualification program (10 CFR 50.49) and determined that it is an acceptable aging management program to address environmental qualification according to 10 CFR 54.21(c)(1)(iii). The GALL Report may be referenced in a license renewal application and should be treated in the same manner as an approved topical report. However, the GALL Report contains one acceptable way and is not the only way to manage aging for license renewal.

In referencing the GALL Report, the applicant should indicate that the material referenced is applicable to the specific plant involved and should provide the information necessary to adopt the finding of program acceptability as described and evaluated in the report. The applicant should also verify that the approvals set forth in the GALL Report for the generic program apply to the applicant's program.

4.4.2.2   FSAR Supplement

The specific criterion for meeting 10 CFR 54.21(d) is:

> The summary description of the evaluation of TLAAs for the period of extended operation in the FSAR supplement is appropriate such that later changes can be controlled by 10 CFR 50.59. The description should contain information associated with the TLAA regarding the basis for determining that the applicant has made the demonstration required by 10 CFR 54.21(c)(1).

4.4.3 Review Procedures

For each area of review described in Subsection 4.4.1, the following review procedures should be followed:

4.4.3.1 Time-Limited Aging Analysis

For electric equipment qualified to the requirements of 10 CFR 50.49, the review procedures, depending on the applicant's choice of 10 CFR 54.21(c)(1)(i), (ii), or (iii), are:

4.4.3.1.1 10 CFR 54.21(c)(1)(i)

The documented results, test data, analyses, etc. of the previous qualification, which consisted of an appropriate combination of testing, analysis, and operating experience, are reviewed to confirm that the original qualified life remains valid for the period of extended operation.

4.4.3.1.2 10 CFR 54.21(c)(1)(ii)

The results of projecting the qualification to the end of the period of extended operation are reviewed. The qualification methods include testing, analysis, operating experience, or combinations thereof.

The reanalysis of an aging evaluation is normally performed to extend the qualification by reducing excess conservatisms incorporated in the prior evaluation. Such a reanalysis is performed on a routine basis as part of an environmental qualification program. A component life-limiting condition may be due to thermal, radiation, or cyclical aging; the vast majority of component aging limits are based on thermal conditions. Conservatisms may exist in aging evaluation parameters, such as the assumed ambient temperature of the component, unrealistically low activation energy, or in the application of a component (de-energized versus energized). The reanalysis of an aging evaluation is documented in accordance with the plant's quality assurance program, which provides for the verification of assumptions and conclusions. For reanalysis, the reviewer verifies that an applicant has completed its reanalysis, addressing attributes of analytical methods, data collection and reduction methods, underlying assumptions, acceptance criteria, and corrective actions if acceptance criteria are not met (see Table 4.4-1). The reviewer also verifies that the reanalysis has been completed in a timely manner prior to the end of qualified life.

4.4.3.1.3 10 CFR 54.21(c)(1)(iii)

The applicant may reference the GALL Report in its license renewal application, as appropriate. The review should verify that the applicant has stated that the report is applicable to its plant with respect to its environmental qualification program. The reviewer verifies that the applicant has identified the appropriate program as described and evaluated in the GALL Report. The reviewer also ensures that the applicant has stated that its environmental qualification program contains the same program elements that the staff evaluated and relied upon in approving the corresponding generic program in the GALL Report. No further staff evaluation is necessary.

If the applicant does not reference the GALL Report in its renewal application, additional staff evaluation is necessary to determine whether the applicant's program is acceptable for this area of review.

4.4.3.2   FSAR Supplement

The reviewer verifies that the applicant has provided information to be included in the FSAR supplement that includes a summary description of the TLAA evaluation of the environmental qualification of electric equipment. Table 4.4-2 contains examples of acceptable FSAR supplement information for this TLAA. The reviewer verifies that the applicant has provided a FSAR supplement with information equivalent to that in Table 4.4-2.

The staff expects to impose a license condition on any renewed license to require the applicant to update its FSAR to include this FSAR supplement at the next update required pursuant to 10 CFR 50.71(e)(4). As part of the license condition, until the FSAR update is complete, the applicant may make changes to the programs described in its FSAR supplement without prior NRC approval, provided that the applicant evaluates each such change pursuant to the criteria set forth in 10 CFR 50.59. The staff will review any such changes when the next update is submitted. If the applicant updates the FSAR to include the final FSAR supplement before the license is renewed, no condition will be necessary.

As noted in Table 4.4-2, an applicant need not incorporate the implementation schedule into its FSAR. However, the reviewer should verify that the applicant has identified and committed in the license renewal application to any future aging management activities, including enhancements and commitments to be completed before the period of extended operation. The staff expects to impose a license condition on any renewed license to ensure that the applicant will complete these activities no later than the committed date.

4.4.4   Evaluation of Findings

The reviewer determines whether the applicant has provided information sufficient to satisfy the provisions of this section and whether the staff's evaluation supports conclusions of the following type, depending on the applicant's choice of 10 CFR 54.21(c)(1)(i), (ii), or (iii), to be included in the staff's safety evaluation report:

> On the basis of its review, as discussed above, the staff concludes that the applicant has provided an acceptable demonstration, pursuant to 10 CFR 54.2 (c)(1), that, for the environmental qualification of Electric Equipment TLAA, [choose which is appropriate] (i) the analyses remain valid for the period of extended operation, (ii) the analyses have been projected to the end of the period of extended operation, or (iii) the effects of aging on the intended function(s) will be adequately managed for the period of extended operation. The staff also concludes that the FSAR supplement contains an appropriate summary description of the environmental qualification of electric equipment TLAA evaluation for the period of extended operation as reflected in the license condition.

4.4.5   Implementation

Except in those cases in which the applicant proposes an acceptable alternative method for complying with specific portions of the NRC's regulations, the method described herein will be used by the staff in its evaluation of conformance with NRC regulations.

4.4.6   References

1. Regulatory Guide 1.89, Rev. 1, "Environmental Qualification of Certain Electric Equipment Important to Safety for Nuclear Power Plants," June 1984.

2. "Guidelines for Evaluating Environmental Qualification of Class 1E Electrical Equipment in Operating Reactors," (DOR Guidelines), November 1979.

3. NUREG-0588, "Interim Staff Position on Environmental Qualification of Safety-Related Equipment," July 1981.

4. IEEE STD 323-1971, "IEEE Trial Use Standard; General Guide for Qualifying Class 1E Equipment for Nuclear Power Generating Stations."

5. IEEE STD 323-1974, "IEEE Standard for Qualifying Class 1E Equipment for Nuclear Power Generating Stations."

6. IEEE STD 382-1972, "Standard for Qualification of Actuators for Power Operated Valve Assemblies with Safety Related Functions for Nuclear Power Plants."

7. IEEE STD 334-1971, "IEEE Standard for Type Tests of Continuous Duty Class 1E Motors for Nuclear Power Generating Stations."

8. NRC Regulatory Issue Summary 2003-09, "Environmental Qualification of Low-Voltage Instrumentation and Control Cables," May 2, 2003.

9. Generic Safety Issue -168, "Environmental Qualification of Low-Voltage Instrumentation and Control Cables," February 2001.

10. NUREG-1801, "Generic Aging Lessons Learned (GALL) Report," U.S. Nuclear Regulatory Commission, Revision 2, 2010.

Table 4.4-1 Environmental Qualification Reanalysis Attributes

Reanalysis Attributes	Description
Analytical methods	The analytical models used in the reanalysis of an aging evaluation should be the same as those previously applied during the prior evaluation. The Arrhenius methodology is an acceptable thermal model for performing a thermal aging evaluation. The analytical method used for a radiation aging evaluation is to demonstrate qualification for the total integrated dose (i.e., normal radiation dose for the projected installed life plus accident radiation dose). For license renewal, one acceptable method of establishing the 60-year normal radiation dose is to multiply the 40-year normal radiation dose by 1.5 (i.e., 60 years/40 years). The result is added to the accident radiation dose to obtain the total integrated dose for the component. For cyclical aging, a similar approach may be used. Other models may be justified on a case-by-case basis.
Data collection and reduction methods	Reducing excess conservatisms in the component service conditions (for example, temperature, radiation, cycles) used in the prior aging evaluation is the chief method used for a reanalysis. Temperature data used in an aging evaluation should be conservative and based on plant design temperatures or on actual plant temperature data. When used, plant temperature data can be obtained in several ways, including monitors used for technical specification compliance, other installed monitors, measurements made by plant operators during rounds, and temperature sensors on large motors (while the motor is not running). A representative number of temperature measurements are conservatively evaluated to establish the temperatures used in an aging evaluation. Plant temperature data may be used in an aging evaluation in different ways, such as (a) directly applying the plant temperature data in the evaluation or (b) using the plant temperature data to demonstrate conservatism when using plant design temperatures for an evaluation. Any changes to material activation energy values as part of a reanalysis should be justified. Similar methods of reducing excess conservatisms in the component service conditions used in prior aging evaluations can be used for radiation and cyclical aging.
Underlying assumptions	Environmental qualification component aging evaluations contain sufficient conservatisms to account for most environmental changes occurring due to plant modifications and events. When unexpected adverse conditions are identified during operational or maintenance activities that affect the environment of a qualified component, the affected environmental qualification component is evaluated and appropriate corrective actions are taken, which may include changes to the qualification bases and conclusions.
Acceptance criteria and corrective actions	The reanalysis of an aging evaluation should extend the qualification of the component. If the qualification cannot be extended by reanalysis, the component must be refurbished, replaced, or requalified prior to exceeding the current qualified life. A reanalysis should be performed in a timely manner (such that sufficient time is available to refurbish, replace, or requalify the component if the reanalysis is unsuccessful).

Table 4.4-2 Examples of FSAR Supplement for Environmental Qualification of Electric Equipment TLAA Evaluation

10 CFR 54.21(c)(1)(i) Example

TLAA	Description of Evaluation	Implementation Schedule*
Environmental qualification of electric equipment	The original environmental qualification qualified life has been shown to remain valid for the period of extended operation.	Completed

10 CFR 54.21(c)(1)(ii) Example

TLAA	Description of Evaluation	Implementation Schedule*
Environmental qualification of electric equipment	The environmental qualification has been projected to the end of the period of extended operation. Reanalysis addresses attributes of analytical methods, data collection and reduction methods, underlying assumptions, acceptance criteria, and corrective actions.	Completed

10 CFR 54.21(c)(1)(iii) Example

TLAA	Description of Evaluation	Implementation Schedule*
Environmental qualification of electric equipment	The existing environmental qualification process, in accordance with 10 CFR 50.49, will adequately manage aging of environmental qualification equipment for the period of extended operation because equipment will be replaced prior to reaching the end of its qualified life. Reanalysis addresses attributes of analytical methods, data collection and reduction methods, underlying assumptions, acceptance criteria, corrective actions if acceptance criteria are not met, and the period of time prior to the end of qualified life when the reanalysis will be completed.	Existing program

* An applicant need not incorporate the implementation schedule into its FSAR. However, the reviewer should verify that the applicant has identified and committed in the license renewal application to any future aging management activities to be completed before the period of extended operation. The staff expects to impose a license condition on any renewed license to ensure that the applicant will complete these activities no later than the committed date.

4.5   CONCRETE CONTAINMENT TENDON PRESTRESS ANALYSIS

Review Responsibilities

Primary - Branch responsible for TLAA issues

Secondary - Branch responsible for structural engineering

4.5.1   Areas of Review

The prestressing tendons in prestressed concrete containments lose their prestressing forces with time due to creep and shrinkage of concrete and relaxation of the prestressing steel. During the design phase, engineers estimate these losses to arrive at the end of operating life (Refs. 1 and 2), normally 40 years. The operating experiences with the trend of prestressing forces indicate that the prestressing tendons lose their prestressing forces at a rate higher than predicted due to sustained high temperature (Ref. 3). Thus, it is necessary to ensure that the applicant addresses existing TLAAs for the extended period of operation.

The adequacy of the prestressing forces in prestressed concrete containments is reviewed for the period of extended operation.

4.5.2   Acceptance Criteria

The acceptance criteria for the area of review described in Subsection 4.5.1 delineate acceptable methods for meeting the requirements of the NRC's regulations in 10 CFR 54.21(c)(1).

4.5.2.1   Time-Limited Aging Analysis

Pursuant to 10 CFR 54.21(c)(1)(i) - (iii), an applicant must demonstrate one of the following:

   (i) The analyses remain valid for the period of extended operation;

   (ii) The analyses have been projected to the end of the extended period of operation; or

   (iii) The effects of aging on the intended function(s) will be adequately managed for the period of extended operation.

Accordingly, the specific options for satisfying the acceptance criterion are:

4.5.2.1.1   10 CFR 54.21(c)(1)(i)

The existing prestressing force evaluation remains valid because (a) losses of the prestressing force are less than the predicted losses, as evidenced from the trend lines constructed from the recent inspection, (b) the period of evaluation covers the period of extended operation, and (c) the trend lines of the measured prestressing forces remain above the minimum required prestress force specified at anchorages for each group of tendons for the period of extended operation.

4.5.2.1.2   10 CFR 54.21(c)(1)(ii)

The trend line of prestressing forces for each group of tendons developed for 40 years of operation should be extended to 60 years. The applicant should demonstrate that the trend lines of the measured prestressing forces will stay above the design Minimum Required Value (MRV) in the CLB for each group of tendons during the period of extended operation (Ref. 4). If this cannot be done, the applicant should develop a systematic plan for retensioning selected tendons so that the trend lines will remain above the minimum required prestress force specified at anchorages for each group of tendons during the period of extended operation, or perform a reanalysis of containment to demonstrate design adequacy.

4.5.2.1.3   10 CFR 54.21(c)(1)(iii)

In Chapter X of the GALL Report (Ref. 4), the staff evaluated a program that assesses the concrete containment tendon prestressing forces, and has determined that it is an acceptable aging management program to address concrete containment tendon prestress according to 10 CFR 54.21(c)(1)(iii), except for operating experience. The GALL Report recommends further evaluation of the applicant's operating experience related to the containment prestress force. However, the GALL report contains one acceptable way and not the only way to manage aging for license renewal.

The GALL report may be referenced in a license renewal application, and is treated in the same manner as an approved topical report. However, the GALL report contains one acceptable way, but not the only way, to manage aging for license renewal.

In referencing the GALL report, an applicant indicates that the material referenced is applicable to the specific plant involved and should provide the information necessary to adopt the finding of program acceptability as described and evaluated in the report. An applicant also verifies that the approvals set forth in the GALL report for the generic program apply to the applicant's program.

4.5.2.2   FSAR Supplement

The specific criterion for meeting 10 CFR 54.21(d) is:

The summary description of the evaluation of TLAAs for the period of extended operation in the FSAR supplement is appropriate such that later changes can be controlled by 10 CFR 50.59. The description must contain information associated with the TLAAs regarding the basis for determining that the applicant has made the demonstration required by 10 CFR 54.21(c)(1).

4.5.3   Review Procedures

For each area of review described in Subsection 4.5.1, the following review procedures should be followed:

4.5.3.1   Time-Limited Aging Analysis

For a concrete containment prestressing tendon system, the review procedures, depending on the applicant's choice of 10 CFR 54.21(c)(1)(i), (ii), or (iii), are:

4.5.3.1.1   10 CFR 54.21(c)(1)(i)

The results of a recent inspection to measure the amount of prestress loss are reviewed to ensure that the reduction of prestressing force is less than the predicted loss in the existing analysis. The reviewer verifies that the trend line of the measured prestressing force, when plotted on the predicted prestressing force curve, shows that the existing analysis will cover the period of extended operation.

4.5.3.1.2   10 CFR 54.21(c)(1)(ii)

The reviewer reviews the trend lines of the measured prestressing forces to ensure that individual tendon lift-off forces (rather than average lift-off forces of the tendon group) are considered in the regression analysis, as discussed in IN 99-10 (Ref. 3). Either the reviewer verifies that the trend lines will stay above the minimum required prestressing forces for each group of tendons during the period of extended operation or, if the trend lines fall below the minimum required prestressing forces during this period, the reviewer verifies that the applicant has a systematic plan for retensioning the tendons to ensure that the trend lines will return to being, and remain, above the minimum required prestressing forces for each group of tendons during the period of extended operation. If the applicant chooses to reanalyze the containment, the reviewer verifies that the design adequacy is maintained in the period of extended operation.

4.5.3.1.3   10 CFR 54.21(c)(1)(iii)

An applicant may reference the GALL Report in its license renewal application, as appropriate. The reviewer verifies that the applicant has stated that the report is applicable to its plant with respect to its program that assesses the concrete containment tendon prestressing forces. The reviewer verifies that the applicant has identified the appropriate program (i.e., GALL Chapter X.S1) as described and evaluated in the GALL Report. The reviewer also ensures that the applicant has stated that its program contains the same program elements that the staff evaluated and relied upon in approving the corresponding generic program in the GALL Report.

The GALL Report recommends further evaluation of the applicant's operating experience related to the containment prestress force. The applicant's program should incorporate the relevant operating experience that occurred at the applicant's plant as well as at other plants. The applicant considers applicable portions of the experience with prestressing systems described in Information Notice 99-10 (Ref. 3). Tendon operating experience could vary among plants with prestressed concrete containments. The difference could be due to the prestressing system design (for example, button-heads, wedge or swaged anchorages), environment, or type of reactor (PWR or BWR). The reviewer reviews the applicant's program to verify that the applicant has adequately considered plant-specific operating experience.

If the applicant does not reference the GALL Report in its renewal application, additional staff evaluation is necessary to determine whether the applicant's program is acceptable for this area of review. The reviewer uses the guidance provided in Branch Technical Position RLSB-1 of this SRP-LR to ensure that loss of prestress in the concrete containment prestressing tendons is adequately managed for the period of extended operation.

4.5.3.2   FSAR Supplement

The reviewer verifies that the applicant has provided information, to be included in the FSAR supplement, that includes a summary description of the evaluation of tendon prestress TLAA.

Table 4.5-1 contains examples of acceptable FSAR supplement information for this TLAA. The reviewer verifies that the applicant has provided a FSAR supplement with information equivalent to that in Table 4.5-1.

The staff expects to impose a license condition on any renewed license to require the applicant to update its FSAR to include this FSAR supplement at the next update required pursuant to 10 CFR 50.71(e)(4). As part of the license condition, until the FSAR update is complete, the applicant may make changes to the programs described in its FSAR supplement without prior NRC approval, provided that the applicant evaluates each such change pursuant to the criteria set forth in 10 CFR 50.59. If the applicant updates the FSAR to include the final FSAR supplement before the license is renewed, no condition will be necessary.

As noted in Table 4.5-1, an applicant need not incorporate the implementation schedule into its FSAR. However, the reviewer should verify that the applicant has identified and committed in the license renewal application to any future aging management activities, including enhancements and commitments to be completed before the period of extended operation. The staff expects to impose a license condition on any renewed license to ensure that the applicant will complete these activities no later than the committed date.

4.5.4   Evaluation Findings

The reviewer determines whether the applicant has provided sufficient information to satisfy the provisions of Section 4.5 and whether the staff's evaluation supports conclusions of the following type, depending on the applicant's choice of 10 CFR 54.21(c)(1)(i), (ii), or (iii), to be included in the staff's safety evaluation report:

> On the basis of its review, as discussed above, the staff concludes that the applicant has provided an acceptable demonstration, pursuant to 10 CFR 54.21(c)(1), that, for the concrete containment tendon prestress TLAA, [choose which is appropriate] (i) the analyses remain valid for the period of extended operation, (ii) the analyses have been projected to the end of the period of extended operation, or (iii) the effects of aging on the intended function(s) will be adequately managed for the period of extended operation. The staff also concludes that the FSAR supplement contains an appropriate description of the concrete containment tendon prestress TLAA evaluation for the period of extended operation as reflected in the license condition.

4.5.5   Implementation

Except in those cases in which the applicant proposes an acceptable alternative method, the method described herein will be used by the staff in its evaluation of conformance with NRC regulations.

4.5.6   References

1. Regulatory Guide 1.35, Rev. 3, "Inspection of Ungrouted Tendons in Prestressed Concrete Containments," July 1990.

2. Regulatory Guide 1.35.1, "Determining Prestressing Forces for Inspection of Prestressed Concrete Containments," July 1990.

3. NRC Information Notice 99-10, "Degradation of Prestressing Tendon Systems in Prestressed Concrete Containments," April 1999.

4. NUREG-1801, "Generic Aging Lessons Learned (GALL) Report," U.S. Nuclear Regulatory Commission, Revision 2, 2010.

Table 4.5-1 Examples of FSAR Supplement for Concrete Containment Tendon Prestress TLAA Evaluation

10 CFR 54.21(c)(1)(i) Example

TLAA	Description of Evaluation	Implementation Schedule*
Concrete containment tendon prestress	The prestressing tendons are used to impart compressive forces in the prestressed concrete containments to resist the internal pressure inside the containment that would be generated in the event of a LOCA. The prestressing forces generated by the tendons diminish over time due to losses in prestressing forces in the tendons and in the surrounding concrete. The prestressing force evaluation has been determined to remain valid to the end of the period of extended operation, and the trend lines of the measured prestressing forces will stay above the minimum required prestressing forces for each group of tendons to the end of this period.	Completed

10 CFR 54.21(c)(1)(iii) Example

TLAA	Description of Evaluation	Implementation Schedule*
Concrete containment tendon prestress	The prestressing tendons are used to impart compressive forces in the prestressed concrete containments to resist the internal pressure inside the containment that would be generated in the event of a LOCA. The prestressing forces generated by the tendons diminish over time due to losses of prestressing forces in the tendons and in the surrounding concrete. The aging management program developed to monitor the prestressing forces should ensure that, during each inspection, the trend lines of the measured prestressing forces show that they meet the requirements of 10 CFR 50.55a(b)(2)(viii)(B). If the trend lines cross the PLLs, corrective actions will be taken. The program will also incorporate any plant-specific and industry operating experience.	Program should be implemented before the period of extended operation.

* An applicant need not incorporate the implementation schedule into its FSAR. However, the reviewer should verify that the applicant has identified and committed in the license renewal application to any future aging management activities to be completed before the period of extended operation. The staff expects to impose a license condition on any renewed license to ensure that the applicant will complete these activities no later than the committed date.

4.6 CONTAINMENT LINER PLATE, METAL CONTAINMENTS, AND PENETRATIONS FATIGUE ANALYSIS

Review Responsibilities

Primary - Branch responsible for mechanical engineering

Secondary - Branch responsible for structural engineering

4.6.1 Areas of Review

The interior surface of a concrete containment structure is lined with thin metallic plates to provide a leak-tight barrier against the uncontrolled release of radioactivity to the environment, as required by 10 CFR Part 50. The thickness of the liner plates is generally between 1/4 inch (6.2 millimeter) and 3/8-inch (9.5 millimeter). The liner plates are attached to the concrete containment wall by stud anchors or structural rolled shapes or both. The design process assumes that the liner plates do not carry loads. However, normal loads, such as from concrete shrinkage, creep, and thermal changes, imposed on the concrete containment structure, are transferred to the liner plates through the anchorage system. Internal pressure and temperature loads are directly applied to the liner plates. Thus, under design-basis conditions, the liner plates could experience significant strains. Some plants may have metal containments instead of concrete containments with liner plates. The metal containments are designed to carry gravity and seismic loads in addition to the internal pressure and temperature loads. Additionally, the BWR containment suppression pool chamber and the vent system are designed or evaluated for hydrodynamic loads associated with actuation of safety relief valves and the discharge into the suppression pool chamber.

Fatigue of the liner plates or metal containments may be considered in the design based on an assumed number of loading cycles for the current operating term. The cyclic loads include reactor building interior temperature variation during the heatup and cooldown of the reactor coolant system, a LOCA, annual outdoor temperature variations, thermal loads due to the high energy containment penetration piping lines (such as steam and feedwater lines), seismic loads, and pressurization due to periodic Type A integrated leak rate tests. The BWR containment suppression pool chamber and the vent system are designed or evaluated for the hydrodynamic cyclic loads as described in Section 6.2.1.1.C, "Pressure-Suppression Type BWR Containments," of NUREG-0800, "Standard Review Plan" (Ref. 1).

High energy piping penetrations and the fuel transfer tubes in some plants are equipped with stainless steel bellow assemblies. These are designed to accommodate relative movements between the containment wall (including the liner) and the adjoining structures. The penetrations have sleeves (up to 10 feet in length, with a 2- to 3-inch annulus around the piping) to penetrate the concrete containment wall and allow movement of the piping system. Dissimilar metal welds connect the piping penetrations to the bellows or stainless steel plates to provide essentially leak-tight penetrations.

The containment liner plates, metal containments, BWR containment suppression chamber and the vent system, penetration sleeves (including dissimilar metal welds), and penetration bellows may be designed in accordance with requirements of Section III of the ASME Boiler and Pressure Vessel Code. If a plant's code of record requires a fatigue analysis, then this analysis may be a TLAA and must be evaluated in accordance with 10 CFR 54.21(c)(1) to ensure that the effects of aging on the intended functions are adequately managed for the period of extended operation.

The adequacy of the fatigue analyses of the containment liner plates (including welded joints), metal containments, BWR containment suppression chamber and the vent system, penetration sleeves, dissimilar metal welds, and penetration bellows is reviewed in this section for the period of extended operation. The fatigue analyses of the pressure boundary of process piping are reviewed separately following the guidance in SRP-LR Section 4.3, "Metal Fatigue."

4.6.1.1 Time-Limited Aging Analysis

The containment liner plates (including welded joints), metal containments, BWR containment suppression chamber and the vent system, penetration sleeves, dissimilar metal welds, and penetration bellows may be designed and/or analyzed in accordance with ASME code requirements. The ASME code contains explicit metal fatigue or cyclic considerations based on TLAAs. Specific requirements are contained in the design code of reference for each plant.

4.6.1.1.1 ASME Section III, MC or Class 1

ASME Section III, Division 2, "Code for Concrete Reactor Vessel and Containments," Subsection CC, "Concrete Containment," and Division 1, Subsection NE, "Class MC Components," (Ref. 2) require a fatigue analysis for liner plates, metal containments, and penetrations that considers all cyclic loads based on the anticipated number of cycles. Containment components also may be designed to ASME Section III Class 1 requirements. A Section III, MC or Class 1 fatigue analysis requires the calculation of the cumulative usage factor (CUF) based on the fatigue properties of the materials and the expected fatigue service of the component. The ASME code limits the CUF to a value less than or equal to one for acceptable fatigue design. The fatigue resistance of the liner plates or metal containments, and penetrations during the period of extended operation is an area of review.

4.6.1.1.2 Other Evaluations Based on CUF

Other evaluations also contain metal fatigue analysis requirements based on a CUF calculation, such as metal bellows designed to ASME NC-3649.4(e)(3) or NE-3366.2(e)(3). For these cases, the discussion relating to ASME Section III, MC or Class 1, in Subsection 4.6.1.1.1 applies.

4.6.1.2 FSAR Supplement

Detailed information on the evaluation of TLAAs is contained in the renewal application. A summary description of the evaluation of TLAAs for the period of extended operation is contained in the applicant's FSAR supplement. The FSAR supplement is an area of review.

4.6.2 Acceptance Criteria

The acceptance criteria for the areas of review described in Subsection 4.6.1 delineate acceptable methods for meeting the requirements of the NRC's regulations in 10 CFR 54.21(c)(1).

4.6.2.1 Time-Limited Aging Analysis

Pursuant to 10 CFR 54.21(c)(1), an applicant must demonstrate one of the following:

(i) The analyses remain valid for the period of extended operation;

(ii) The analyses have been projected to the end of the extended period of operation; or

(iii) The effects of aging on the intended function(s) will be adequately managed for the period of extended operation.

Specific acceptance criteria for fatigue of containment liner plates, metal containments, liner plate weld joints, dissimilar metal welds, penetration sleeves, and penetration bellows are:

4.6.2.1.1   ASME Section III, MC or Class 1

For containment liner plates, metal containments, BWR containment suppression chamber and the vent system, and penetrations designed or analyzed to ASME MC or Class 1 requirements, the acceptance criteria, depending on the applicant's choice of 10 CFR 54.21(c)(1)(i), (ii), or (iii), are:

4.6.2.1.1.1   10 CFR 54.21(c)(1)(i)

The existing CUF calculations remain valid because the number of assumed cyclic loads will not be exceeded during the period of extended operation.

4.6.2.1.1.2   10 CFR 54.21(c)(1)(ii)

CLB fatigue analysis, per ASME Code Section III, was conducted for a 40-year life. The CUF calculations are re-evaluated based on an increased number of assumed cyclic loads to cover the period of extended operation. All cyclic loads considered in the original fatigue analyses (including Type A and Type B leak rate tests) are re-evaluated and revised, as necessary. The revised analysis shows that the CUF does not exceed one, as required by the ASME code, during the period of extended operation.

4.6.2.1.1.3   10 CFR 54.21(c)(1)(iii)

An aging management program provided by the applicant shall demonstrate that the effects of aging on the component's intended function(s) will be adequately managed during the period of extended operation. If the proposed aging management program relies on mitigation or inspection, it shall be evaluated against the 10 elements described in Branch Technical Position RLSB-1 (Appendix A.1 of this standard review plan). However, if the component is replaced, the CUF for the replacement must be less than or equal to one during the period of extended operation.

4.6.2.1.2   Other Evaluations Based on CUF

The acceptance criteria in Subsection 4.6.2 apply.

4.6.2.2   FSAR Supplement

The specific criterion for meeting 10 CFR 54.21(d) is:

> The summary description of the evaluation of TLAAs for the period of extended operation in the FSAR supplement is appropriate such that later changes can be controlled by 10 CFR 50.59. The description should contain information associated with

the TLAAs regarding the basis for determining that the applicant has made the demonstration required by 10 CFR 54.21(c)(1).

4.6.3   Review Procedures

For each area of review described in Subsection 4.6.1, the following review procedures is followed:

4.6.3.1   Time-Limited Aging Analysis

4.6.3.1.1   ASME Section III, MC or Class 1

For containment liner plates, metal containments, BWR containment suppression chamber and the vent system, and penetrations designed or analyzed to ASME MC or Class 1 requirements, the review procedures, depending on the applicant's choice of 10 CFR 54.21(c)(1)(i), (ii), or (iii), are:

4.6.3.1.1.1   10 CFR 54.21(c)(1)(i)

The number of assumed transients used in the existing CUF calculations for the current operating term is compared to the extrapolation to 60 years of operation of the number of operating transients experienced to date. The comparison confirms that the number of transients in the existing analyses will not be exceeded during the period of extended operation.

4.6.3.1.1.2   10 CFR 54.21(c)(1)(ii)

Operating transient experience and a list of the increased number of assumed cyclic loads projected to the end of the period of extended operation are reviewed to ensure that the cyclic load projection is adequate. The revised CUF calculations based on the projected number of assumed cyclic loads are reviewed to ensure that the CUF remains less than one at the end of the period of extended operation.

The code of record is used for the reevaluation, or the applicant may update to a later code edition pursuant to 10 CFR 50.55a. In the latter case, the reviewer verifies that the requirements in 10 CFR 50.55a are met.

4.6.3.1.1.3   10 CFR 54.21(c)(1)(iii)

The applicant's proposed aging management program to ensure that the effects of aging on the intended function(s) are adequately managed for the period of extended operation is reviewed. If the program relies on mitigation or inspection, it shall be reviewed against the 10 elements described in Branch Technical Position RLSB-1 (Appendix A.1 of this standard review plan). If the applicant proposes a component replacement before its CUF exceeds one, the reviewer verifies that the CUF for the replacement will remain less than or equal to one during the period of extended operation.

Applicant-proposed programs are reviewed on a case-by-case basis.

4.6.3.1.2   Other Evaluations Based on CUF

The review procedures in Subsection 4.6.3.1 apply.

4.6.3.2 FSAR Supplement

The reviewer verifies that the applicant has provided information, to be included in the FSAR supplement that includes a summary description of the evaluation of containment liner plate, metal containments, BWR containment suppression chamber and the vent system, and penetrations fatigue TLAA. Table 4.6-1 contains examples of acceptable FSAR supplement information for this TLAA. The reviewer verifies that the applicant has provided an FSAR supplement with information equivalent to that in Table 4.6-1.

The staff expects to impose a license condition on any renewed license to require the applicant to update its FSAR to include this FSAR supplement at the next update required pursuant to 10 CFR 50.71(e)(4). As part of the license condition, until the FSAR update is complete, the applicant may make changes to the programs described in its FSAR supplement without prior NRC approval, provided that the applicant evaluates each such change pursuant to the criteria set forth in 10 CFR 50.59. If the applicant updates the FSAR to include the final FSAR supplement before the license is renewed, no condition will be necessary.

As noted in Table 4.6-1, the applicant need not incorporate the implementation schedule into its FSAR. However, the review should verify that the applicant has identified and committed in the license renewal application to any future aging management activities, including enhancements and commitments to be completed before the period of extended operation. The staff expects to impose a license condition on any renewed license to ensure that the applicant will complete these activities no later than the committed date.

4.6.4 Evaluation Findings

The reviewer determines whether the applicant has provided sufficient information to satisfy the provisions of this section and to support conclusions of the following type, depending on the applicant's choice of 10 CFR 54.21(c)(1)(i), (ii), or (iii), to be included in the staff's safety evaluation report:

> On the basis of its review, as discussed above, the staff concludes that the applicant has provided an acceptable demonstration, pursuant to 10 CFR 54.21(c)(1), that the containment liner plate or metal containment, BWR containment suppression chamber and the vent system, and penetrations fatigue TLAA, [choose which is appropriate] (i) the analyses remain valid for the period of extended operation, (ii) the analyses have been projected to the end of the period of extended operation, or (iii) the effects of aging on the intended function(s) will be adequately managed for the period of extended operation. The staff also concludes that the FSAR supplement contains an appropriate summary description of the containment liner plate or metal containment, BWR containment suppression chamber and the vent system, and penetrations fatigue TLAA evaluation for the period of extended operation as reflected in the license condition.

4.6.5 Implementation

Except in those cases in which the applicant proposes an acceptable alternative method, the method described herein will be used by the staff in its evaluation of conformance with NRC regulations.

4.6.6   References

1. NUREG-0800, "Standard Review Plan for the Review of Safety Analysis Reports for Nuclear Power Plants," U.S. Nuclear Regulatory Commission, March 2007.

2. ASME Section III, Division 2, "Code for Concrete Reactor Vessels and Containments," Subsection CC, "Concrete Containment," and Division 1, Subsection NE, "MC Components," The ASME Boiler and Pressure Vessel Code, 2004 edition as approved in 10 CFR 50.55a, The American Society of Mechanical Engineers, New York, NY.

Table 4.6-1 Examples of FSAR Supplement for Containment Liner Plates, Metal Containments, and Penetrations Fatigue TLAA Evaluation

10 CFR 54.21(c)(1)(i) Example

TLAA	Description of Evaluation	Implementation Schedule*
Containment liner plates (or metal containment) and penetrations fatigue	The containment liner plates (or metal containment), BWR containment suppression chamber and the vent system, liner weld joints, penetration sleeves, dissimilar metal welds, and penetration bellows that provide an essentially leak-tight barrier. A Section III, MC or Class 1 fatigue analysis limits the CUF to a value less than or equal to one for acceptable fatigue design. The existing CUF evaluation has been determined to remain valid because the number of assumed cyclic loads would not be exceeded during the period of extended operation.	Completed

10 CFR 54.21(c)(1)(ii) Example

TLAA	Description of Evaluation	Implementation Schedule*
Containment liner plates (or metal containment) and penetrations fatigue	The containment liner plates (or metal containment), BWR containment suppression chamber and the vent system, liner weld joints, penetration sleeves, dissimilar metal welds, and penetration bellows that provide an essentially leak-tight barrier. A Section III, MC or Class 1 fatigue analysis limits the CUF to a value less than or equal to one for acceptable fatigue design. The CUF calculations have been reevaluated based on an increased number of assumed cyclic loads to cover the period of extended operation. The revised CUF will not exceed one during the period of extended operation.	Completed

10 CFR 54.21(c)(1)(iii) Example

TLAA	Description of Evaluation	Implementation Schedule*
Containment liner plates (or metal containment) and penetrations fatigue	The containment liner plates (or metal containment), BWR containment suppression chamber and the vent system, liner weld joints, penetration sleeves, dissimilar metal welds, and penetration bellows that provide an essentially leak-tight barrier. A Section III, MC or Class 1 fatigue analysis limits the CUF to a value less than or equal to one for acceptable fatigue design. If the component is replaced, the CUF for the replacement will be shown to be less than one during the period of extended operation.	Program should be implemented before the period of extended operation.

Note: All containment components need not meet the same requirement. It is likely that the liner plate and the bellows may be evaluated per 10CFR54.21(c)(1)(i), while high energy penetrations may be evaluated per 10CFR54.21(c)(1)(ii).

* An applicant need not incorporate the implementation schedule into its FSAR. However, the reviewer should verify that the applicant has identified and committed in the license renewal application to any future aging management activities to be completed before the period of extended operation. The staff expects to impose a license condition on any renewed license to ensure that the applicant will complete these activities no later than the committed date.

4.7   OTHER PLANT-SPECIFIC TIME-LIMITED AGING ANALYSES

Review Responsibilities

Primary - NRR branch responsible for the TLAA issues

Secondary - Other branches responsible for systems, as appropriate

4.7.1   Areas of Review

There are certain plant-specific safety analyses that may have been based on an explicitly assumed 40-year plant life (for example, aspects of the reactor vessel design) and may, therefore, be TLAAs. Pursuant to 10 CFR 54.21(c), a license renewal applicant is required to evaluate TLAAs. The definition of TLAAs is provided in 10 CFR 54.3 and in Section 4.1 of this SRP-LR.

Plant-specific TLAAs may have evolved since issuance of a plant's operating license. As indicated in 10 CFR 54.30, the adequacy of the plant's CLB, which includes TLAAs, is not an area within the scope of the license renewal review. Any questions regarding the adequacy of the CLB must be addressed under the backfit rule (10 CFR 50.109) and are separate from the license renewal process.

License renewal reviews focus on the period of extended operation. Pursuant to 10 CFR 54.30, if the reviews required by 10 CFR 54.21(a) or (c) show that there is not reasonable assurance during the current license term that licensed activities will be conducted in accordance with the CLB, the licensee is required to take measures under its current license to ensure that the intended functions of those systems, structures, or components are maintained in accordance with the CLB throughout the term of the current license. The adequacy of the measures for the term of the current license is not within the scope of the license renewal review.

Pursuant to 10 CFR 54.21(c), an applicant must provide a listing of TLAAs and plant-specific exemptions that are based on TLAAs. The staff reviews the applicant's identification of TLAAs and exemptions separately, following the guidance in Section 4.1 of this SRP-LR.

Based on lessons learned in the review of the initial license renewal applications, the staff has developed review procedures for the evaluation of certain TLAAs. If an applicant identifies these TLAAs as applicable to its plant, the staff reviews them separately, following the guidance in Sections 4.2 through 4.6. The reviewer reviews other TLAAs that are identified by the applicant following the generic guidance in this section. For particular systems, the reviewers from branches responsible for those systems may be requested to assist in the review, as appropriate.

The following areas relating to a TLAA are reviewed:

4.7.1.1   Time-Limited Aging Analysis

The evaluation of the TLAA for the period of extended operation is reviewed.

4.7.1.2   FSAR Supplement

The FSAR supplement summarizing the evaluation of the TLAA for the period of extended operation in accordance with 10 CFR 54.21(d) is reviewed.

4.7.2   Acceptance Criteria

The acceptance criteria for the areas of review described in Subsection 4.7.1 of this section delineate acceptable methods for meeting the requirements of the NRC's regulations in 10 CFR 54.21(c)(1).

4.7.2.1   Time-Limited Aging Analysis

Pursuant to 10 CFR 54.21(c)(1)(i) - (iii), an applicant must demonstrate one of the following for the TLAAs:

   (i)   The analyses remain valid for the period of extended operation;

   (ii) The analyses have been projected to the end of the extended period of operation; or

   (iii) The effects of aging on the intended function(s) will be adequately managed for the period of extended operation.

4.7.2.2   FSAR Supplement

The specific criterion for meeting 10 CFR 54.21(d) is:

The summary description of the evaluation of TLAAs for the period of extended operation in the FSAR supplement is appropriate such that later changes can be controlled by 10 CFR 50.59. The description contains information associated with the TLAAs regarding the basis for determining that the applicant has made the demonstration required by 10 CFR 54.21(c)(1).

4.7.3   Review Procedures

For certain applicants, plant-specific analyses may meet the definition of a TLAA as given in 10 CFR 54.3. The concern for license renewal is that these analyses may not have properly considered the length of the extended period of operation, which may change conclusions with regard to safety and the capability of SSCs within the scope of the Rule to perform or one or more safety functions. The review of these TLAAs provides the assurance that the aging effect is properly addressed through the period of extended operation.

For each area of review described in Subsection 4.7.1, the following review procedures are followed:

4.7.3.1   Time-Limited Aging Analysis

For each TLAA identified, the review procedures depend on the applicant's choice of methods of compliance from those identified in 10 CFR 54.21(c)(1)(i), (ii), or (iii), as follows:

4.7.3.1.1   10 CFR 54.21(c)(1)(i)

Justification provided by the applicant is reviewed to verify that the existing analyses are valid for the period of extended operation. The existing analyses should be shown to be bounding even during the period of extended operation.

The applicant describes the TLAA with respect to the objectives of the analysis, assumptions used in the analysis, conditions, acceptance criteria, relevant aging effects, and intended function(s). The applicant shows that (a) conditions and assumptions used in the analysis already address the relevant aging effects for the period of extended operation, and (b) acceptance criteria are maintained to provide reasonable assurance that the intended function(s) is maintained for renewal. Thus, no reanalysis is necessary for renewal.

In some instances, the applicant may identify activities to be performed to verify the assumption basis of the calculation, such as cycle counting. An evaluation of that activity is provided by the applicant. The reviewer assures that the applicant's activity is sufficient to confirm the calculation assumptions for the 60-year period.

If the TLAA must be modified or recalculated to extend the period of evaluation to consider the period of extended operation, the reevaluation should be addressed under 10 CFR 54.21(c)(1)(ii).

4.7.3.1.2   10 CFR 54.21(c)(1)(ii)

The documented results of the revised analyses are reviewed to verify that their period of evaluation is extended, such that they are valid for the period of extended operation (e.g., 60 years). The applicable analysis technique can be the one that is in effect in the plant's CLB at the time of filing of the renewal application.

The applicant may recalculate the TLAA using a 60-year period to show that the TLAA acceptance criteria continue to be satisfied for the period of extended operation. The applicant also may revise the TLAA by recognizing and reevaluating any overly conservative conditions and assumptions. Examples include relaxing overly conservative assumptions in the original analysis, using new or refined analytical techniques, and performing the analysis using a 60-year period. The applicant shall provide a sufficient description of the analysis and documents the results of the reanalysis to show that it is satisfactory for the 60-year period.

As applicable, the plant's code of record is used for the reevaluation, or the applicant may update to a later code edition pursuant to 10 CFR 50.55a. In the latter case, the reviewer verifies that the requirements in 10 CFR 50.55a are met.

In some cases, the applicant identifies activities to be performed to verify the assumption basis of the calculation, such as cycle counting. An evaluation of that activity is provided by the applicant. The reviewer confirms that the applicant's activity is sufficient to confirm the calculation assumptions for the 60-year period.

4.7.3.1.3   10 CFR 54.21(c)(1)(iii)

Under this option, the applicant proposes to manage the aging effects associated with the TLAA by an aging management program in the same manner as described in the IPA in 10 CFR 54.21(a)(3). The reviewer reviews the applicant's aging management program to verify that the effects of aging on the intended function(s) are adequately managed consistent with the CLB for the period of extended operation.

The applicant identifies the structures and components associated with the TLAA. The TLAA is described with respect to the objectives of the analysis, conditions, assumptions used, acceptance criteria, relevant aging effects, and intended function(s). In cases where a mitigation or inspection program is proposed, the reviewer uses the guidance provided in Branch Technical Position RLSB-1 of this standard review plan to ensure that the effects of aging on the structure- and component-intended function(s) are adequately managed for the period of extended operation.

4.7.3.2   FSAR Supplement

The reviewer verifies that the applicant has provided information to be included in the FSAR supplement that includes a summary description of the evaluation of each TLAA. Each such summary description is reviewed to verify that it is appropriate, such that later changes can be controlled by 10 CFR 50.59. The description should contain information that the TLAAs have been dispositioned for the period of extended operation. Sections 4.2 through 4.6 of this standard review plan contain examples of acceptable FSAR supplement information for TLAA evaluation.

The staff expects to impose a license condition on any renewed license to require the applicant to update its FSAR to include this FSAR supplement at the next update required pursuant to 10 CFR 50.71(e)(4). As part of the license condition, until the FSAR update is complete, the applicant may make changes to the programs described in its FSAR supplement without prior NRC approval, provided that the applicant evaluates each such change pursuant to the criteria set forth in 10 CFR 50.59. If the applicant updates the FSAR to include the final FSAR supplement before the license is renewed, no condition is necessary.

As noted in Sections 4.2 through 4.6, an applicant need not incorporate the implementation schedule into its FSAR. However, the review should verify that the applicant has identified and committed in the license renewal application to any future aging management activities, including enhancements and commitments to be completed before the period of extended operation. The staff expects to impose a license condition on any renewed license to ensure that the applicant completes these activities no later than the committed date.

4.7.4   Evaluation Findings

The reviewer determines whether the applicant has provided sufficient information to satisfy the provisions of Section 4.7 and whether the staff's evaluation supports conclusions of the following type, depending on the applicant's choice of 10 CFR 54.21(c)(1)(i), (ii), or (iii), to be included in the staff's safety evaluation report:

> On the basis of its review, as discussed above, the staff concludes that the applicant has provided an acceptable demonstration, pursuant to 10 CFR 54.21(c)(1), that, for the (name of specific) TLAA, [choose which is appropriate] (i) the analyses remain valid for the period of extended operation, (ii) the analyses have been projected to the end of the period of extended operation, or (iii) the effects of aging on the intended function(s) will be adequately managed for the period of extended operation. The staff also concludes that the FSAR

supplement contains an appropriate summary description of this TLAA evaluation for the period of extended operation as reflected in the license condition.

4.7.5   Implementation

Except in those cases in which the applicant proposes an acceptable alternative method, the method described herein is used by the staff in its evaluation of conformance with NRC regulations.

4.7.6   References

None

APPENDIX A

BRANCH TECHNICAL POSITIONS

A.1 AGING MANAGEMENT REVIEW - GENERIC (BRANCH TECHNICAL POSITION RLSB-1)

A.1.1 Background

Pursuant to 10 CFR 54.21(a)(3), a license renewal applicant is required to demonstrate that the effects of aging on structures and components subject to an Aging Management Review (AMR) are adequately managed so their intended functions will be maintained consistent with the CLB for the period of extended operation. The purpose of this Branch Technical Position (RLSB-1) is to address the aging management demonstration that has not been addressed specifically in Chapters 3 and 4 of this Standard Review Plan.

The license renewal process is not intended to demonstrate absolute assurance that structures and components will not fail, but rather that there is reasonable assurance that they will perform such that the intended functions are maintained consistent with the current licensing basis (CLB) during the period of extended operation.

There are generally four types of aging management programs (AMPS): prevention, mitigation, condition monitoring, and performance monitoring. *Prevention programs* preclude the effects of aging. For example, coating programs prevent external corrosion of a tank. *Mitigation programs* attempt to slow the effects of aging. For example, water chemistry programs mitigate internal corrosion of piping. *Condition monitoring programs* inspect for the presence and extent of aging effects. Examples are the visual examination of concrete structures for cracking and the ultrasonic examination of pipe wall for flow-accelerated corrosion (FAC)-induced wall thinning. *Performance monitoring programs* test the ability of a structure or component to perform its intended function(s). For example, the ability of the tubes on heat exchangers to transfer heat is tested. More than one type of AMP may be implemented to ensure that aging effects are managed. For example, in managing internal corrosion of piping, a mitigation program (water chemistry) may be used to minimize susceptibility to corrosion. However, it may also be necessary to have a condition monitoring program (ultrasonic inspection) to verify that corrosion is indeed insignificant.

A.1.2 Branch Technical Position

A.1.2.1 Applicable Aging Effects

1. The determination of applicable aging effects is based on degradation mechanisms that have occurred and those that potentially could cause structure and component degradation. The materials, environment, stresses, service conditions, operating experience, and other relevant information should be considered in identifying applicable aging effects. The effects of aging on the intended function(s) of structures and components also should be considered.

2. Relevant aging information may be contained in, but is not limited to, the following documents: plant-specific maintenance and inspection records; plant-specific site deviation or issue reports; plant-specific NRC and Institute of Nuclear Power Operations (INPO) inspection reports; plant-specific licensee self-assessment reports; plant-specific and other licensee event reports (LERs); NRC, INPO, and vendor generic communications; GSIs/unresolved safety issues (USIs); NUREG reports; and Electric Power Research Institute (EPRI) reports.

3. If operating experience or other information indicates that a certain aging effect may be applicable and an applicant determines that it is not applicable to its specific plant, the reviewer may question the absence of this aging effect unless the applicant has provided the basis for this determination in its license renewal application. However, in questioning the absence of the aging effect, a reference and/or basis which aided the applicant in addressing the question should be provided. For example, the question could cite a previous application review, NRC generic communications, engineering judgment, relevant research information, or other industry experience as the basis for the question. Simply citing that the aging effect is listed in the GALL Report is not a sufficient basis. For example, the aging effect is applicable to a PWR component, but the applicant's plant is a BWR and does not have such a component. In this example, using the GALL Report merely as a checklist is not relevant.

4. An aging effect may not have been identified in the GALL Report, if it arises out of industry experience after the issuance of the GALL Report. The reviewer should ensure that the applicant has evaluated the latest industry experience to identify all applicable aging effects.

5. An aging effect should be identified as applicable for license renewal even if there is a prevention or mitigation program associated with that aging effect. For example, water chemistry, a coating, or use of cathodic protection could prevent or mitigate corrosion, but corrosion should be identified as applicable for license renewal, and the AMR should consider the adequacy of the AMP referencing water chemistry, coating, or cathodic protection.

6. Specific identification of aging mechanisms is not a requirement; however, it is an option to identify specific aging mechanisms and the associated aging effects in the integrated plant assessment (IPA).

7. The applicable aging effects to be considered for license renewal include those that could result from normal plant operation, including plant/system operating transients and plant shutdown. Specific aging effects from abnormal events need not be postulated for license renewal. However, if an abnormal event has occurred at a particular plant, its contribution to the aging effects on structures and components for license renewal should be considered for that plant. For example, if a resin intrusion has occurred in the reactor coolant system at a particular plant, the contribution of this resin intrusion event to aging should be considered for that plant.

Design basis events (DBEs) are abnormal events; they include design basis pipe break, loss of coolant accident (LOCA), and safe shutdown earthquake (SSE). Potential aging effects resulting from DBEs are addressed, as appropriate, as part of the plant's CLB. There are other abnormal events which should be considered on a case-by-case basis. For example, abuse due to human activity is an abnormal event; aging effects from such abuse need not be postulated for license renewal. When a safety-significant piece of equipment is accidentally damaged by a licensee, the licensee is required to take immediate corrective action under existing procedures (see 10 CFR Part 50 Appendix B) to ensure functionality of the equipment. The equipment degradation is not due to aging; corrective action is not necessary solely for the period of extended operation. However, leakage from bolted connections should not be considered as abnormal events. Although bolted connections are not supposed to leak, experience shows that leaks do occur, and the leakage could cause corrosion. Thus, the aging effects from leakage of bolted connections should be evaluated for license renewal.

An aging effect due to an abnormal event does not preclude that aging effect from occurring during normal operation for the period of extended operation. For example, a certain PWR licensee observed clad cracking in its pressurizer, and attributed that to an abnormal dry out of the pressurizer. Although dry out of a pressurizer is an abnormal event, the potential for clad cracking in the pressurizer during normal operation should be evaluated for license renewal. This is because the pressurizer is subject to extensive thermal fluctuations and water level changes during plant operation, which may result in clad cracking given sufficient operating time. The abnormal dry out of the pressurizer at that certain plant may have merely accelerated the rate of the aging effect.

A.1.2.2   Aging Management Program for License Renewal

1. An acceptable AMP should consist of the 10 elements described in Table A.1-1, as appropriate (Ref. 1). These program elements/attributes are discussed further in Position A.1.2.3 below.

2. All programs and activities that are credited for managing a certain aging effect for a specific structure or component should be described. These AMPs/activities may be evaluated together for the 10 elements described in Table A.1-1, as appropriate.

3. The risk significance of a structure or component could be considered in evaluating the robustness of an AMP. Probabilistic arguments may be used to develop an approach for aging management adequacy. However, use of probabilistic arguments alone is not an acceptable basis for concluding that, for those structures and components subject to an AMR, the effects of aging will be adequately managed in the period of extended operation. Thus, risk significance may be considered in developing the details of an AMP for the structure or component for license renewal, but may not be used to conclude that no AMP is necessary for license renewal.

A.1.2.3   Aging Management Program Elements

A.1.2.3.1   Scope of Program

The specific program necessary for license renewal should be identified. The scope of the program should include the specific structures and components, the aging of which the program manages.

A.1.2.3.2   Preventive Actions

1. The activities for prevention and mitigation programs should be described. These actions should mitigate or prevent aging degradation.

2. Some condition or performance monitoring programs do not rely on preventive actions and thus, this information need not be provided.

3. In some cases, condition or performance monitoring programs may also rely on preventive actions. The specific prevention activities should be specified.

A.1.2.3.3 Parameters Monitored or Inspected

1. This program element should identify the aging effects that the program manages and should provide a link between the parameter or parameters that will be monitored and how the monitoring of these parameters will ensure adequate aging management.

2. For a condition monitoring program, the parameter monitored or inspected should be capable of detecting the presence and extent of aging effects. Some examples are measurements of wall thickness and detection and sizing of cracks.

3. For a performance monitoring program, a link should be established between the degradation of the particular structure or component-intended function(s) and the parameter(s) being monitored. An example of linking the degradation of a passive component-intended function with the performance being monitored is linking the fouling of heat exchanger tubes with the heat transfer-intended function. This could be monitored by periodic heat balances. Since this example deals only with one intended function of the tubes (heat transfer), additional programs may be necessary to manage other intended function(s) of the tubes, such as pressure boundary. Thus, a performance monitoring program must ensure that the structure and components are capable of performing their intended functions by using a combination of performance monitoring and evaluation (if outside acceptable limits of acceptance criteria) that demonstrate that a change in performance characteristic is a result of an age-related degradation mechanism.

4. For prevention or mitigation programs, the parameters monitored should be the specific parameters being controlled to achieve prevention or mitigation of aging effects. An example is the coolant oxygen level that is being controlled in a water chemistry program to mitigate pipe cracking.

A.1.2.3.4 Detection of Aging Effects

1. Detection of aging effects should occur before there is a loss of the structure- and component-intended function(s). The parameters to be monitored or inspected should be appropriate to ensure that the structure- and component-intended function(s) will be adequately maintained for license renewal under all CLB design conditions. Thus, the discussion for the "detection of aging effects" program element should address (a) how the program element would be capable of detecting or identifying the occurrence of age-related degradation or an aging effect prior to a loss of structure and component (SC)-intended function or (b) for preventive/mitigative programs, how the program would be capable of preventing or mitigating their occurrence prior to a loss of a SC-intended function. The discussion should provide information that links the parameters to be monitored or inspected to the aging effects being managed.

2. Nuclear power plants are licensed based on redundancy, diversity, and defense-in-depth principles. A degraded or failed component reduces the reliability of the system, challenges safety systems, and contributes to plant risk. Thus, the effects of aging on a structure or component should be managed to ensure its availability to perform its intended function(s) as designed when called upon. In this way, all system level-intended function(s), including redundancy, diversity, and defense-in-depth consistent with the plant's CLB, would be maintained for license renewal. A program based solely

on detecting structure and component failure should not be considered as an effective AMP for license renewal.

3. This program element describes "when," "where," and "how" program data are collected (i.e., all aspects of activities to collect data as part of the program).

4. For condition monitoring programs, the method or technique (such as visual, volumetric, or surface inspection), frequency, and timing of new, one-time inspections may be linked to plant-specific or industrywide operating experience. The discussion should provide justification, including codes and standards referenced, that the technique and frequency are adequate to detect the aging effects before a loss of SC-intended function. A program based solely on detecting SC failures is not considered an effective AMP.

   For a condition monitoring program, when sampling is used to represent a larger population of SCs, applicants should provide the basis for the inspection population and sample size. The inspection population should be based on such aspects of the SCs as a similarity of materials of construction, fabrication, procurement, design, installation, operating environment, or aging effects. The sample size should be based on such aspects of the SCs as the specific aging effect, location, existing technical information, system and structure design, materials of construction, service environment, or previous failure history. The samples should be biased toward locations most susceptible to the specific aging effect of concern in the period of extended operation. Provisions on expanding the sample size when degradation is detected in the initial sample should also be included.

5. For a performance monitoring program, the "detection of aging effects" program element should discuss and establish the monitoring methods that will be used for performance monitoring. In addition, the "detection of aging effects" program element should also establish and justify the frequency that will be used to implement these performance monitoring activities.

6. For a prevention or mitigation program, the "detection of aging effects" program element should discuss and establish the monitoring methods that the program will use to monitor for the preventive or mitigative parameters that the program controls and should justify the frequency of performing these monitoring activities.

A.1.2.3.5   Monitoring and Trending

1. Monitoring and trending activities should be described, and they should provide a prediction of the extent of degradation and thus effect timely corrective or mitigative actions. Plant-specific and/or industrywide operating experience may be considered in evaluating the appropriateness of the technique and frequency.

2. This program element describes "how" the data collected are evaluated and may also include trending for a forward look. This includes an evaluation of the results against the acceptance criteria and a prediction regarding the rate of degradation in order to confirm that timing of the next scheduled inspection will occur before a loss of SC-intended function. Although aging indicators may be quantitative or qualitative, aging indicators should be quantified, to the extent possible, to allow trending. The parameter or indicator trended should be described. The methodology for analyzing the inspection or test results against the acceptance criteria should be described. Trending is a comparison of

the current monitoring results with previous monitoring results in order to make predictions for the future.

A.1.2.3.6   Acceptance Criteria

1. The quantitative or qualitative acceptance criteria of the program and its basis should be described. The acceptance criteria, against which the need for corrective actions are evaluated, should ensure that the structure- and component-intended function(s) are maintained consistent with all CLB design conditions during the period of extended operation. The program should include a methodology for analyzing the results against applicable acceptance criteria.

    For example, carbon steel pipe wall thinning may occur under certain conditions due to FAC. An AMP for FAC may consist of periodically measuring the pipe wall thickness and comparing that to a specific minimum wall acceptance criterion. Corrective action is taken, such as piping replacement, before deadweight, seismic, and other loads, and this acceptance criterion must be appropriate to ensure that the thinned piping would be able to carry these CLB design loads. This acceptance criterion should provide for timely corrective action before loss of intended function under these CLB design loads.

2. Acceptance criteria could be specific numerical values, or could consist of a discussion of the process for calculating specific numerical values of conditional acceptance criteria to ensure that the structure- and component-intended function(s) will be maintained under all CLB design conditions. Information from available references may be cited.

3. It is not necessary to justify any acceptance criteria taken directly from the design basis information that is included in either the final safety analysis report (FSAR), plant Technical Specifications, or other codes and standards incorporated by reference into NRC regulations; they are a part of the CLB. Nor is it necessary to justify the acceptance criteria that have been established in either NRC-accepted or NRC-endorsed methodology, such as those that may be given in NRC-approved or NRC-endorsed topical reports or NRC-endorsed codes and standards; the acceptance criteria referenced in these types of documents have been subject to an NRC review process and have been approved or endorsed for their application to an NRC-approved or NRC-endorsed evaluation methodology. Also, it is not necessary to discuss CLB design loads if the acceptance criteria do not permit degradation because a structure and component without degradation should continue to function as originally designed. Acceptance criteria, which do permit degradation, are based on maintaining the intended function under all CLB design loads.

A.1.2.3.7   Corrective Actions

1. Actions to be taken when the acceptance criteria are not met should be described in appropriate detail or referenced to source documents. Corrective actions, including root cause determination and prevention of recurrence, should be timely.

2. If corrective actions permit analysis without repair or replacement, the analysis should ensure that the structure- and component-intended function(s) are maintained consistent with the CLB.

3. For safety-related components, an applicant's 10 CFR Part 50, Appendix B, Quality Assurance Program, is an acceptable means to confirm that the corrective actions are done in a manner consistent with the condition monitoring program, preventive program, mitigative program, or performance monitoring program that is credited for aging management. For example, for a plant-specific condition monitoring program that is based on ASME Section XI requirements, the implementation of the Appendix B program should ensure that any corrective actions are performed in accordance with applicable Code requirements or NRC-approved Code cases.

A.1.2.3.8 Confirmation Process

1. The confirmation process should be described. The process ensures that preventive actions are adequate and that appropriate corrective actions have been completed and are effective.

2. The effectiveness of prevention and mitigation programs should be verified periodically. For example, in managing internal corrosion of piping, a mitigation program (water chemistry) may be used to minimize susceptibility to corrosion. However, it also may be necessary to have a condition monitoring program (ultrasonic inspection) to verify that corrosion is indeed insignificant.

3. When corrective actions are necessary, there should be follow-up activities to confirm that the corrective actions have been completed, a root cause determination was performed, and recurrence will be prevented.

A.1.2.3.9 Administrative Controls

1. The administrative controls of the program should be described. Administrative controls provide a formal review and approval process.

2. Any AMPs to be relied on for license renewal should have regulatory and administrative controls. That is the basis for 10 CFR 54.21(d) to require that the FSAR supplement include a summary description of the programs and activities for managing the effects of aging for license renewal. Thus, any informal programs relied on to manage aging for license renewal must be administratively controlled and included in the FSAR supplement.

A.1.2.3.10 Operating Experience

1. Consideration of future plant-specific and industry operating experience relating to aging management programs should be discussed. Reviews of operating experience by the applicant in the future may identify areas where aging management programs should be enhanced or new programs developed. An applicant should commit to a future review of plant-specific and industry operating experience to confirm the effectiveness of its aging management programs or indicate a need to develop new aging management programs. This information should provide objective evidence to support the conclusion that the effects of aging will be managed adequately so that the structure and component intended function(s) will be maintained during the period of extended operation.

2. Operating experience with existing programs should be discussed. The operating experience of AMPs that are existing programs, including past corrective actions

resulting in program enhancements or additional programs, should be considered. A past failure would not necessarily invalidate an AMP because the feedback from operating experience should have resulted in appropriate program enhancements or new programs. This information can show where an existing program has succeeded and where it has failed (if at all) in intercepting aging degradation in a timely manner. This information should provide objective evidence to support the conclusion that the effects of aging will be managed adequately so that the structure- and component-intended function(s) will be maintained during the period of extended operation.

3. For new AMPs that have yet to be implemented at an applicant's facility, the programs have not yet generated any operating experience (OE). However, there may be other relevant plant-specific OE at the plant or generic OE in the industry that is relevant to the AMP's program elements even though the OE was not identified as a result of the implementation of the new program. Thus, for new programs, an applicant may need to consider the impact of relevant OE that results from the past implementation of its existing AMPs that are existing programs and the impact of relevant generic OE on developing the program elements. Therefore, operating experience applicable to new programs should be discussed. Additionally, an applicant should commit to a review of future plant-specific and industry operating experience for new programs to confirm their effectiveness.

A.1.3   References

1. NEI 95-10, "Industry Guideline for Implementing the Requirements of 10 CFR Part 54 – The License Renewal Rule," Nuclear Energy Institute, Revision 6.

Table A.1-1 Elements of an Aging Management Program for License Renewal	
Element	**Description**
1. Scope of Program	Scope of program includes the specific structures and components subject to an AMR for license renewal.
2. Preventive Actions	Preventive actions should prevent or mitigate aging degradation.
3. Parameters Monitored or Inspected	Parameters monitored or inspected should be linked to the degradation of the particular structure or component-intended function(s).
4. Detection of Aging Effects	Detection of aging effects should occur before there is a loss of structure or component-intended function(s). This includes aspects such as method or technique (i.e., visual, volumetric, surface inspection), frequency, sample size, data collection, and timing of new/one-time inspections to ensure timely detection of aging effects.
5. Monitoring and Trending	Monitoring and trending should provide predictability of the extent of degradation, and timely corrective or mitigative actions.
6. Acceptance Criteria	Acceptance criteria, against which the need for corrective action will be evaluated, should ensure that the structure or component-intended function(s) are maintained under all CLB design conditions during the period of extended operation.
7. Corrective Actions	Corrective actions, including root cause determination and prevention of recurrence, should be timely.
8. Confirmation Process	Confirmation process should ensure that preventive actions are adequate and that appropriate corrective actions have been completed and are effective.
9. Administrative Controls	Administrative controls should provide a formal review and approval process.
10. Operating Experience	If the AMP is an existing program, operating experience of the AMP, including past corrective actions resulting in program enhancements or additional programs, should provide objective evidence to support the conclusion that the effects of aging will be managed adequately so that the structure- and component-intended function(s) will be maintained during the period of extended operation.

A.2 QUALITY ASSURANCE FOR AGING MANAGEMENT PROGRAMS (BRANCH TECHNICAL POSITION IQMB-1)

A.2.1 Background

The license renewal applicant is required to demonstrate that the effects of aging on structures and components subject to an Aging Management Review (AMR) will be managed adequately to ensure that their intended functions are maintained consistent with the current licensing basis (CLB) of the facility for the period of extended operation. Therefore, those aspects of the AMR process that affect quality of safety-related structures, systems, and components are subject to the quality assurance (QA) requirements of 10 CFR Part 50 Appendix B. For nonsafety-related structures and components (SCs) subject to an AMR, the existing 10 CFR Part 50 Appendix B QA program may be used by the applicant to address the elements of corrective actions, the confirmation process, and administrative controls, as described in Branch Technical Position RLSB-1 (Appendix A.1 of this standard review plan for license renewal [SRP-LR]). The confirmation process ensures that preventive actions are adequate and that appropriate corrective actions have been completed and are effective. Administrative controls should provide for a formal review and approval process. Reference 1 describes how a license renewal applicant can rely on the existing requirements in 10 CFR Part 50 Appendix B, "Quality Assurance Criteria for Nuclear Power Plants and Fuel Reprocessing Plants," to satisfy these program elements/attributes. The purpose of this branch technical position (IQMB-1) is to describe an acceptable process for implementing the corrective actions, the confirmation process, and administrative controls of aging management programs for license renewal.

A.2.2 Branch Technical Position

1. Safety-related SCs are subject to 10 CFR Part 50 Appendix B requirements, which are adequate to address all quality-related aspects of an aging management program consistent with the CLB of the facility for the period of extended operation.

2. For nonsafety-related SCs that are subject to an AMR for license renewal, an applicant has the option to expand the scope of its 10 CFR Part 50 Appendix B program to include these SCs and to address corrective actions, the confirmation process, and administrative controls for aging management during the period of extended operation. The reviewer verifies that the applicant has documented such a commitment in the Final Safety Analysis Report supplement in accordance with 10 CFR 54.21(d).

3. If an applicant chooses an alternative means to address corrective actions, the confirmation process, and administrative controls for managing aging of nonsafety-related SCs that are subject to an AMR for license renewal, the applicant's proposal is reviewed on a case-by-case basis following the guidance in Branch Technical Position RLSB-1 (Appendix A.1 of this SRP-LR).

A.2.3 References

1. NUREG-1801, "Generic Aging Lessons Learned (GALL) Report," U.S. Nuclear Regulatory Commission, Revision 2, 2010.

A.3 GENERIC SAFETY ISSUES RELATED TO AGING (BRANCH TECHNICAL POSITION RLSB-2)

A.3.1 Background

Unresolved Safety Issues (USIs) and Generic Safety Issues (GSIs) are identified and tracked in the NRC's formal resolution process set forth in NUREG-0933, "Resolution of Generic Safety Issues," which is updated periodically (Ref. 1). Appendix B to NUREG-0933 contains a listing of those issues that are applicable to operating and future plant. NUREG-0933 is a source of information on generic concerns identified by the NRC. Some of these concerns may be related to the effects of aging or Time-Limited Aging Analyses (TLAAs) for systems, structures, or components within the scope of license renewal review. The purpose of this branch technical position (RLSB-2) is to address the license renewal treatment of an aging effect or a TLAA which is a subject of an USI or a GSI (60 FR 22484).

Table A.3-1 provides examples to help determine whether a USI or GSI should or should not be specifically addressed for license renewal, based on lessons learned from the staff review of the initial license renewal applications. However, two of these examples (GSI-23 and -190) have been resolved by the staff. They are included in the examples for illustrative purposes.

A.3.2 Branch Technical Position

A.3.2.1 Treatment of GSIs

1. The license renewal rule requires that aging effects be managed to ensure that the structure- and component-intended function(s) are maintained and that TLAAs are evaluated for license renewal. Thus, all applicable aging effects of structures and components subject to an AMR and all TLAAs must be evaluated, regardless of whether they are associated with GSIs or USIs.

2. USIs and HIGH- and MEDIUM-priority issues described in NUREG-0933 Appendix B (Ref. 1) that involve aging effects for structures and components subject to an AMR or TLAAs are specifically addressed. The version of NUREG-0933 that is current on the date 6 months before the date of the license renewal application is used to identify such issues. Prior to Safety Evaluation Report (SER) completion, any new issues contained in later versions of NUREG-0933 should be reviewed and resolved if applicable to the applicant's plant. New issues are addressed by using one of the approaches described in Position A.3.2.2 below.

3. New generic safety issues, designated as USI, HIGH-, or MEDIUM- priority after the application has been submitted, that involve aging effects for structures and components subject to an aging management review or TLAA should be submitted in the annual update of the application.

4. During the preparation and review of a license renewal application, an applicant or the NRC may become aware of an aging management or TLAA issue that is generically applicable to other nuclear plants. If issues have generic applicability (but are not yet part of the formal GSIs resolution process as identified in NUREG-0933), an applicant should still address the issue to demonstrate that the effects of aging are or will be managed adequately or that TLAAs are evaluated for the period of extended operation.

A.3.2.2   Approaches for Addressing GSIs (60 FR 22484)

One of the following approaches may be used:

1. If resolution has been achieved before issuance of a renewed license, implementation of that resolution is incorporated within the license renewal application. The plant-specific implementation information is provided.

2. A technical rationale is provided that demonstrates that the CLB will be maintained until some later time in the period of extended operation, at which point one or more reasonable options (for example, replacement, analytical evaluation, or a surveillance/maintenance program) become available to adequately manage the effects of aging. An applicant describes the basis for concluding that the CLB is maintained during the period of extended operation and briefly describes options that are technically feasible during the period of extended operation to manage the effects of aging, but does not have to preselect which option to use.

3. An aging management program is developed that, for that plant, incorporates a resolution to the aging effects issue.

4. An amendment of the CLB (as a separate action outside the license renewal application) is proposed that, if approved, removes the intended function(s) from the CLB. The proposed CLB amendment is reviewed under 10 CFR Part 50 and is not a review area for license renewal.

A.3.3   References

1. NUREG-0933, "Resolution of Generic Safety Issues," Supplement 32, U.S. Nuclear Regulatory Commission, August 2008.

2. NRC Regulatory Issue Summary 2000-02, "Closure of Generic Safety Issue 23, Reactor Coolant Pump Seal Failure," February 15, 2000.

3. Letter from Ashok C. Thadani of the Office of Nuclear Regulatory Research, NRC, to William D. Travers, Executive Director of Operations, NRC, dated December 26, 1999.

4. SECY 94-225, "Issuance of Proposed Rulemaking Package on GSI-23, Reactor Coolant Pump Seal Failure," August 26, 1994.

5. Information Notice 93-61, "Excessive Reactor Coolant Leakage Following a Seal Failure in a Reactor Coolant Pump or Reactor Recirculation Pump," August 9, 1993.

6. Deleted.

7. NRC Regulatory Issue Summary 2003-09, "Environmental Qualification of Low-Voltage Instrumentation and Control Cables" dated May 2, 2003.

Table A.3-1   Examples of Generic Safety Issues that Should/Should Not Be Specifically Addressed for License Renewal and Basis for Disposition

Example	Disposition
GSI-23, "Reactor Coolant Pump Seal Failures"	This issue relates to reactor coolant pump seal failures, which challenge the makeup capacity of the emergency core cooling system in PWRs. Although GSI-23 originally addressed seal performance both during normal operation and during loss of seal cooling conditions, it has been modified to address only seal performance during loss of seal cooling conditions (Refs. 4 and 5). Loss of all seal cooling may cause the reactor coolant pump seals to fail or leak excessively. Because the reactor coolant pump seal performance during loss of seal cooling conditions is not an issue that involves AMR or TLAA, GSI-23 need not be specifically addressed for license renewal (Ref. 2).
GSI-168, "Environmental Qualification of Electrical Equipment"	This issue relates to aging of electrical equipment that is subject to environmental qualification requirements. Environmental qualification is a TLAA for license renewal. Regulatory Issue Summary (RIS) 2003-09 was issued on May 2, 2003, to inform addressees of the results of the technical assessment of GSI-168, "Environmental Qualification of Electrical Equipment." This RIS requires no action on the part of the addressees (Ref. 7).
GSI-173.A, "Spent Fuel Storage Pool: Operating Experience"	This issue relates to the potential for a sustained loss of spent fuel pool cooling capacity and the potential for a substantial loss of spent fuel pool coolant inventory. The staff evaluated the issue and concluded that no actions will be taken for operating plants. As indicated in NUREG-0933, the staff is pursuing regulatory improvement changes to RG 1.13, "Spent Fuel Storage Facility Design Basis," and NUREG-0800, "Standard Review Plan for the Review of Safety Analysis Reports for Nuclear Power Plants." Thus, GSI-173.A need not be specifically addressed for license renewal.
GSI-190, "Fatigue Evaluation of Metal Components for 60-Year Plant Life"	This issue relates to environmental effects on fatigue of reactor coolant system components for 60 years. Fatigue is also a TLAA for license renewal. Thus, GSI-190 was specifically addressed for license renewal by the initial license renewal applicants. This GSI has now been resolved (Ref. 3).

NRC FORM 335 (9-2004) NRCMD 3.7	U.S. NUCLEAR REGULATORY COMMISSION	1. REPORT NUMBER (Assigned by NRC, Add Vol., Supp., Rev., and Addendum Numbers, if any.)
BIBLIOGRAPHIC DATA SHEET *(See instructions on the reverse)*		NUREG 1800, Revision 2

2. TITLE AND SUBTITLE	3. DATE REPORT PUBLISHED	
Standard Review Plan for Review of License Renewal Applications for Nuclear Power Plants	MONTH December	YEAR 2010
	4. FIN OR GRANT NUMBER	

5. AUTHOR(S)	6. TYPE OF REPORT
U.S. Nuclear Regulatory Commission	
	7. PERIOD COVERED *(Inclusive Dates)*

8. PERFORMING ORGANIZATION - NAME AND ADDRESS *(If NRC, provide Division, Office or Region, U.S. Nuclear Regulatory Commission, and mailing address, if contractor, provide name and mailing address.)*

Division of License Renewal

Office of Nuclear Reactor Regulation

U.S. Nuclear Regulatory Commission

Washington, DC 20555-0001

9. SPONSORING ORGANIZATION - NAME AND ADDRESS *(If NRC, type "Same as above"; if contractor, provide NRC Division, Office or Region, U.S. Nuclear Regulatory Commission, and mailing address.)*

Same as item 8, above.

10. SUPPLEMENTARY NOTES

11. ABSTRACT *(200 words or less)*

The Standard Review Plan of Review of License Renewal Applications for Nuclear Power Plants (SRP-LR) provides guidance to Nuclear Regulatory Commission staff reviewers in the Office of Nuclear Reactor Regulation. These reviewers perform safety reviews of applications to renew nuclear power plant licenses in accordance with Title 10 of the Code of Federal Regulations Part 54. The principal purposes of the SRP-LR are to ensure the quality and uniformity of staff reviews and to present a well-defined base from which to evaluate applicant programs and activities for the period of extended operation. The SRP-LR is also intended to make information about regulatory matters widely available, to enhance communication with interested members of the public and nuclear power industry, and to improve public and industry understanding of the staff review process. The safety review is based primarily on the information provided by the applicant in a license renewal application. Each of the individual SRP-LR sections address (1) the matters that are reviewed, (2) the bases for the review, (3) the way the review is accomplished, and (4) the conclusions that are sought.

12. KEY WORDS/DESCRIPTORS *(List words or phrases that will assist researchers in locating the report.)*	13. AVAILABILITY STATEMENT
License Renewal Aging Nuclear Safety Aging Mechanisms Aging Effects	unlimited
	14. SECURITY CLASSIFICATION
	(This Page) unclassified
	(This Report) unclassified
	15. NUMBER OF PAGES
	16. PRICE